T3-BOW-437

MAGNETIC RESONANCE IN BIOLOGY

VOLUME TWO

Magnetic Resonance in Biology

VOLUME TWO

edited by

JACK S. COHEN
National Institutes of Health
Bethesda, Maryland

A WILEY-INTERSCIENCE PUBLICATION

JOHN WILEY & SONS
New York • Chichester • Brisbane • Toronto • Singapore

Library of Congress Cataloging in Publication Data:

(Revised for vol. 2)
Main entry under title:

Magnetic resonance in biology.

"A Wiley-Interscience publication."
Includes bibliographies and index.
1. Nuclear magnetic resonance. 2. Electron paramagnetic resonance. 3. Biology—Technique. I. Cohen. Jack S. [DNLM: 1. Biology—Periodicals. 2. Electron spin resonance—Methods—Periodicals. 3. Molecular biology—Periodicals. 4. Nuclear magnetic resonance—Methods—Periodicals. W1 MA341]
QH324.9.N8M33 574.19′285 80-13070

ISBN 0-471-05176-4 (V.1)
ISBN 0-471-05175-6 (V.2)

Printed in the United States of America

10 9 8 7 6 5 4 3 2 1

Contributors

JACK S. COHEN	Developmental Pharmacology Branch National Institute of Child Health and Human Development National Institutes of Health Bethesda, MD 20205
CHARLES W. DeBROSSE	Department of Chemistry Pennsylvania State University University Park, PA 16802
LOU J. HUGHES	Department of Chemistry The American University Washington, D.C. 20016
LAWRENCE C. KUO	Department of Biophysics and Theoretical Biology Cummings Life Science Center University of Chicago Chicago, IL 60637
L. J. LYNCH	CSIRO Physical Technology Unit Ryde NSW 2112 Australia
MARVIN W. MAKINEN	Department of Biophysics and Theoretical Biology Cummings Life Science Center University of Chicago Chicago, IL 60637
HEISABURO SHINDO	Tokyo College of Pharmacy Horinouchi, Hachioji Tokyo 192-03 Japan

JOSEPH J. VILLAFRANCA Department of Chemistry
Pennsylvania State University
University Park, PA 16802

JAN B. WOOTEN Phillip Morris Research Center
Richmond, VA 23261

Magnetic Resonance in Biology

The aim of this series is to provide a forum where experts in magnetic resonance methods can describe the results of applications to biological systems in a format that is comprehensible to the nonexpert.

The need for such a forum is two-fold: first, there are an ever-increasing number and range of magnetic resonance applications to biology appearing in the scientific literature; second, there is an apparent reluctance by biologists and biochemists to assimilate these results into the general corpus of their specific fields. Such results are often not referenced except by other spectroscopists. One reason for this attitude may be the relative ease with which some spectra can be run and the superficial nature of many of the earlier analyses in the area. Also, the mathematical-physical basis of the method and the specialized technology involved are unfamiliar to most biologists. It is my intention that researchers in a given field will be able to find an accessible account of what information magnetic resonance methods have contributed to their subject in the articles published in this series, *Magnetic Resonance in Biology*. The articles will be written by those actively pursuing research and publishing articles in the scientific literature.

The methods in question, nuclear magnetic resonance (NMR) and electron spin resonance (ESR), have provided valuable, and often unique, information on a wide range of systems, from small molecules (drugs and hormones), through macromolecules (proteins and nucleic acids), to membranes, cells, and whole organisms. The nature of the information obtained is dynamic as well as structural and spatial. It is intended to cover the whole range of biological applications currently to be found in the literature. This includes studies of cell metabolism (with NMR observation of ^{31}P, ^{13}C, etc.), whole organism imaging by NMR (so-called zeugmatography), combinations of NMR and ESR approaches (including use of spin labels and paramagnetic metal ions), as well as the more familiar molecular studies. That is why the title of the series was chosen to include the term "biology," since the methods in question are indeed adding to our knowledge of biology in general.

March 1983

JACK S. COHEN

Preface

Volume 2 continues the format established in Volume 1 in this series by including in-depth articles on topics of significant biological interest, but seen from an NMR perspective.

Chapters 1 and 2 cover magnetic resonance methods applied to enzyme mechanisms; the first, by Debrosse and Villafranca, utilizes isotope effects on phosphorus chemical shifts, and the second, by Makinen and Kuo, utilizes ESR studies of spin labels. Both of these techniques give detailed quantitative information regarding the mechanisms of action of enzymes in solution.

The two main types of biological macromolecules, DNA and proteins, are treated in Chapters 3 and 4, respectively. The review by Shindo considers details of DNA conformation and dynamics, a subject of great current research activity, in a way distinct from the general review of nucleic acid structure by Schweizer in Volume 1. Similarly, whereas the review of London in Volume 1 focused on the topic of NMR relaxation measurements and mobility of groups in proteins, Chapter 4 by Cohen *et al.* in this volume covers the more general aspects of protein structure. However, the organization of this review is unusual in that it is centered around each type of amino acid side chain, a format that it is hoped will be useful for the non–NMR expert. This is particularly important when NMR studies of proteins, both ^{1}H and ^{13}C, have become essentially routine, and examples occur in practically every issue of major biochemical journals. This review also contains a detailed analysis of what will presumably be the final situation with regard to the status of the histidine residues in the active site of ribonuclease, a subject of some controversy. This matter is now essentially resolved in light of the recent reinvestigation by X-ray diffraction and the neutron diffraction results on RNase (see Chapter 4).

The subject of Chapter 5 "Water Relaxation in Heterogeneous and Biological Systems," by Lynch, may seem somewhat technical and less biologically relevant to some. Yet it covers a subject of major importance in relation to the clinical diagnostic use of NMR in imaging (described by Hoult in Chapter 2 of Volume 1). In relation to the beautiful pictures of the insides of people's heads and bodies that are now being widely displayed, it is as well to remember that they are representations of water distribution and relaxation observed by ^{1}H NMR. If we are truly to understand these images and the reasons for the superior contrast between certain types of tissues observed by NMR imaging compared with that observed by CAT scanning, then we should certainly not lose sight of the fundamental processes actually occurring in the tissues under study.

There is no chapter in this volume devoted to the very active field of ^{31}P and other NMR studies of metabolic activity in cells, intact tissues, and whole organisms. This is largely because of the extensive publishing activity already in that area, including the volume edited by myself (Cohen 1982) and the recent book by Gadian (1982).

Cohen, J. S. (1982), *Non-invasive Probes of Tissue Metabolism*, Wiley, New York.

Gadian, D. (1982), *Nuclear Magnetic Resonance Studies of Living Biological Systems*, Clarendon Press, Oxford.

JACK S. COHEN

Bethesda, Maryland
April 1983

Contents

ONE

Isotope Effects on Phosphorus Chemical Shifts: Applications to Enzyme Mechanisms 1

CHARLES W. DEBROSSE AND JOSEPH J. VILLAFRANCA

1. Introduction, 2
2. Theoretical Considerations, 2
3. Experimental Requirements, 8
4. Chiral Phosphoryl Transfer, 9
5. Orientation States in P–O Bond-Forming and Cleavage Reactions, 26
6. Exchange Studies, 29
7. Kinetic Considerations, 30
8. Kinetic Studies Reported, 33
9. Kinetic Studies Proposed, 40

References, 50

TWO

Spin-Label Probes of Enzyme Action 53

MARVIN W. MAKINEN AND LAWRENCE C. KUO

1. Introduction, 54
2. Molecular Structure of Nitroxide Spin-Labels, 55
3. Studies of Spin-Labeled Proteins and Enzymes in Crystals, 61
4. Application of Cryoenzymology for Structural Characterization, 80

References, 91

THREE

DNA Backbone Conformation and Dynamics 95

HEISABURO SHINDO

1. Introduction, 95
2. NMR Methods, 96
3. Phosphodiester-Backbone Conformation of DNA, 99
4. DNA Dynamics in Solution, 113
5. Concluding Remarks, 126

References, 127

FOUR

Observations of Amino Acid Side Chains in Proteins by NMR Methods 130

JACK S. COHEN, LOU J. HUGHES, AND JAN B. WOOTEN

1. Introduction, 131
2. NMR Methods Applied to Proteins, 133
3. Effects on NMR Parameters of the Incorporation of Amino Acid Side Chains into Proteins, 131
4. NMR Titration Curves, 167
5. Studies of Specific Types of Amino Acid Side Chains in Proteins, 173
6. Conclusions and Prognosis, 235

References, 237

FIVE

Water Relaxation in Heterogeneous and Biological Systems 248

L. J. LYNCH

1. Introduction, 248
2. Relaxation Theory, 251
3. Applications of NMR to Studies of Water in Heterogeneous Systems, 255
4. Concluding Remarks, 294

List of Symbols, 296

References, 300

Index 305

One

Isotope Effects on Phosphorus Chemical Shifts: Applications to Enzyme Mechanisms

Charles W. DeBrosse and Joseph J. Villafranca

Department of Chemistry
Pennsylvania State University
University Park, PA 16802

1. **Introduction** 2
2. **Theoretical Considerations** 2
3. **Experimental Requirements** 8
4. **Chiral Phosphoryl Transfer** 9
   4.1. 5′ Nucleotidase, 15
   4.2. ATPases, 15
   4.3. Phosphodiesterases and Kinases, 19
   4.4. Additional Group-Transfer Reactions, 22
   4.5. Adenylate Cyclase, 24
5. **Orientation Studies in P—O Bond-Forming and Cleavage Reactions** 26
   5.1. Phosphatases, 26
   5.2. Phosphorylases and Phosphohydrolases, 27
   5.3. Acetyl CoA Synthetase, 28
6. **Exchange Studies** 29
   6.1. Pyruvate Kinase, 30
   6.2. Creatine Kinase, 30
   6.3. ATPase, 30
7. **Kinetic Considerations** 30
8. **Kinetic Studies Reported** 33
   8.1. Carbamyl Phosphate Synthetase, 33
   8.2. Glutamine Synthetase, 37

9. Kinetic Studies Proposed 40

9.1. Guanosine 5′-Phosphate Synthetase, 40
9.2. Pyruvate-Phosphate Dikinase, 42
9.3. Phosphoenolpyruvate Carboxylase, 44
9.4. Phosphoenolpyruvate Carboxykinase, 47
9.5. Adenylosuccinate Synthetase, 47
9.6. S-Adenosylmethionine Synthetase, 48

Acknowledgments 49

References 50

1. INTRODUCTION

In recent years, a number of complex biochemical problems have been solved through application of new techniques utilizing ^{31}P nuclear magnetic resonance (NMR) spectroscopy. This article deals specifically with two techniques that involve high-resolution ^{31}P NMR: (1) the analysis of chirality at phosphorus in the mechanisms of phosphoryl-transfer enzymes, and (2) the study of oxygen isotope-exchange phenomena to analyze individual mechanistic events in enzyme reactions. Both methods entail the use of specific substitution of the rare oxygen isotopes (^{17}O, ^{18}O) for ^{16}O and observations of perturbations in the ^{31}P NMR spectra of the organophosphate species. Both methods have greatly enhanced our knowledge of the subtleties encountered in the wide range of biochemical reactions involving the phosphate bond. We deal with each of these topics in depth and include most of the literature through early 1982. In the last section we speculate on how the mechanisms of several heretofore unstudied enzymes might be probed with these methods.

2. THEORETICAL CONSIDERATIONS

There are two major phenomena underlying the utility of isotope substitution in ^{31}P NMR spectroscopy. These are (1) isotope-mediated effects on the ^{31}P nuclear shielding (chemical shift effects) and (2) quadrupolar-broadening effects. The two rare isotopes of oxygen, ^{18}O and ^{17}O, when substituted for ^{16}O at phosphorus can give rise to effects 1 and 2, respectively. At this point, we will discuss the physical origins of both effects.

The chemical shift of any phosphorus resonance, relative to those of other phosphorus atoms, is determined by minute magnetic fields due to circulation of electrons in its immediate environment. These local fields are due in large part to the proximal electron distribution, which makes the contribution from atoms directly attached to the nucleus under observation quite important. Thus it is apparent that substitution of different isotopes, for example ^{18}O for ^{16}O, at a given phosphorus could perturb its electronic environment and induce a measurable change in the chemical shift.

The physical basis for this effect has been understood for some time, and it has been widely observed in proton and ^{13}C NMR spectra on substitution of ^{2}H for ^{1}H. The effect was predicted by Ramsey (1952) as resulting from perturbations of zero-point vibrational levels by substitution of one isotope for another. Gutowsky (1959) holds that the effect of changes in the vibrational levels manifests itself at the probe nucleus in an electrostatic sense, and that the change in shielding should be proportional to the inverse fourth power of the difference between average distances of the two isotopes from the observed nucleus. He points out that this electrostatic effect would have far more impact on a chemical shift than nuclear–electron interaction differences between two different isotopes could cause.

Bernheim and Batiz-Hernandez (1967) provided a qualitative explanation for the isotope-induced shift, based on the changes in vibrational levels of bonds between the observed nucleus and the isotope. The absolute shielding at the observed nucleus is diminished by the electrical fields emanating from the bonded atoms. These fields arise from the oscillations of the bonded atoms in the applied magnetic field. When a heavier isotope is substituted for a lighter one, its average distance to the observed nucleus is slightly shorter, and its vibratory amplitude is smaller. It then must give rise to smaller electric fields, which, when felt at the probe nucleus, do not deshield it as strongly as would those from the lighter isotope.

This accounts for the almost universally observed trend of shifts to higher field on substitution of a heavier for a lighter isotope at the probe nucleus. Theory predicts that the magnitude of such shifts should be on the order of 0.01 ppm (Gutowsky, 1959) and should reflect the number of such substitutions.

The isotopic shift has been observed only recently in ^{31}P NMR spectroscopy. The earliest reports dealt with phosphate ion into which ^{18}O had been randomly incorporated. The resonances of all the various P $^{18}O_n$ $^{16}O_{(4-n)}$ species were resolved at 36.4 MHz (Lutz *et al.*, 1978) and 145.7 MHz (Cohn and Hu, 1978), and a shift of 0.0203 ppm (0.019 ppm) to higher field was observed for each successive incorporation of ^{18}O. The magnitude of the effect is, as expected, well correlated with the bond order between oxygen and phosphorus, as illustrated by Lowe (Lowe *et al.*, 1979), who observed the indicated isotopic shifts for the methyl phosphate esters shown in Figure 1.

This correlation of isotopic shifts with P—O bond order has been used to good advantage in studies on nucleotide phosphate esters, since the effect allows incorporation of an ^{18}O on a ^{31}P to be distinguished as bridging to another phosphorus or nonbridging, in addition to indicating the number of such incorporations. Cohn and Hu (1980) synthesized ATP and ADP with ^{18}O specifically located at various sites in the polyphosphate chain and

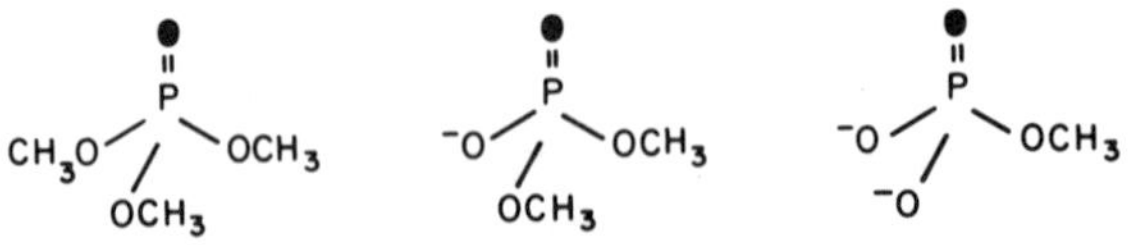

Δδ 0.036 0.029 0.023

Figure 1. ^{18}O isotope shifts observed in the ^{31}P NMR spectra of the indicated methyl phosphate esters (Δδ, ppm from the ^{16}O species; ^{18}O is represented by ●, ^{17}O is represented by ϴ). The charges and multiple bonds on phosphates will be omitted for clarity throughout the rest of this chapter.

reported the isotopic shifts of the ^{31}P resonances. The isotope shift for the bridging (P—O—P) substitution of ^{18}O averages nearly 0.017 ppm, whereas the nonbridging substitution increases the upfield shift to ~0.028 ppm. These results are tabulated in Table I. The chemical syntheses for these molecules are described in the report of Cohn and Hu (1980). We will discuss below specific instances of the use of the bridge/nonbridge isotope-shift differences in problems associated with mechanistic pathways.

The $^{18}O/^{16}O$ isotopic shift for ^{31}P resonances has been reported for several thiophosphoryl nucleotides. Although the precise effect of sulfur substitution on the vibratory modes of geminal P—O bonds is imperfectly understood, the incorporation of sulfur would seem to impose greater double-bond character on the P—O bonds (Corbridge, 1980; Milolajczyk *et al.*, 1976). The ^{18}O predictably induces a slightly larger shift in these compounds than in the respective oxo analogs (Tsai and Chang, 1980; Tsai, 1980). For example, the isotopic shift of the β ^{31}P (^{18}O, nonbridging) in A is 0.035 ppm, whereas in B the value is 0.028 ppm. The larger shift difference makes the observation

AMP −O−P(●)(S)−O−P(O)(O)−O

A

AMP −O−P(●)(O)−O−P(O)(O)−O

B

of the effect easier, in addition to the other advantages that accrue from the use of thiophosphoryl nucleotides (see below). The second rare isotope of oxygen, ^{17}O, can also be selectively incorporated into P—O linkages of interest, producing useful effects in the ^{31}P NMR spectrum. The isotope ^{17}O has a magnetically active nucleus, whose spin ($I = \frac{5}{2}$) confers on it a nonspherical nuclear charge distribution. This property is known as quadrupole moment and usually provides an extremely efficient mechanism for the spin-lattice relaxation of the nucleus possessing it. It is this relaxation effect that leads to quadrupolar broadening of directly bonded nuclei.

The quadrupolar nucleus, here ^{17}O, possessing various spin states, is spin-coupled to any directly attached ^{31}P nucleus. This would give rise to a six-line ($2I + 1$) multiplet in the ^{31}P spectrum in the absence of any quadrupole relaxation effects. The quadrupole relaxation of ^{17}O can cause rapid loss of identity for the ^{17}O spin states, in effect decoupling it from the ^{31}P spin. This decoupling can vary in effectiveness from broadening of the multiplet to its complete collapse. The effectiveness depends on the magnitude of (1) the quadrupole relaxation time and (2) the scalar coupling constant between ^{31}P and ^{17}O. When the decoupling is *intermediate* in effectiveness, the ^{31}P resonance is extremely broad and is more or less obscured in the baseline. This *apparent* loss of signal intensity can be quantitatively related to the degree of ^{17}O incorporation on that phosphorus, and it is this phenomenon that has made ^{17}O substitution useful in NMR studies of phosphate-containing compounds. Tsai and co-workers (Tsai *et al.*, 1980) have pointed out the advantages of this approach and have shown the conditions under which it may be applied.

Tsai notes that in the case of intermediate decoupling, the following condition governs the ^{31}P linewidth:

$$\Delta^{31}P = \frac{1}{\pi T_{2(^{31}P)}} = a\ \pi T_q J^2 \tag{1}$$

where $\Delta^{31}P$ is the width of the resonance at half-height, T_q is the relaxation time of the ^{17}O nucleus, J is the one-bond P—O coupling constant, and $a = \frac{4}{3}I(I + 1)$, where $I = \frac{5}{2}$, the spin number of ^{17}O. As a rule, $\Delta^{31}P$ must be larger than 20 Hz before the loss in integrated intensity can be quantitated.

Generally, J is not easily measured but should be around 160 Hz (Gray and Albright, 1977). The value of T_q depends on both electric fields around ^{17}O and the diffusional properties of the molecule, and is given by

$$\frac{1}{T_q} = \frac{3}{40}\frac{2I + 3}{I^2(2I - 1)}\left(1 + \frac{\eta^2}{3}\right)\left(\frac{e^2qQ}{\hbar}\right)^2\tau_r \tag{2}$$

where τ_r is the isotropic rotational correlation time, Q is the quadrupole moment of ^{17}O, η is the asymmetry parameter associated with the electric field (value generally unknown, but ranges from zero to unity), and $(e^2qQ/\hbar)$ is the quadrupole coupling constant. This last term is a measure of the interaction of the nucleus with the electric field and increases as the symmetry of the environment decreases. To favor large $\Delta^{31}P$ then, one wishes T_q to be long, so a short τ_r (small molecule; rapid tumbling) and a relatively small quadrupole coupling constant (high symmetry in the P—O bond, for example, a nonbridge P—O) are desirable.

Tsai has shown that in various nucleotides and thio nucleotides (both with and without added Mg^{2+}), the $(\pi T_q{}^1)^{-1}$ values range upward from 400 Hz, and this results in predictable diminution of any ^{31}P resonance in these molecules, proportionate to the relative incorporation of even a single ^{17}O. This method for detecting oxygen incorporation has found several applications, which are discussed in the following sections.

Table I. ^{18}O isotope shifts at the indicated ^{31}P for various ^{18}O-labeled nucleotides[a]

Nucleotide Species[b]		α	β	γ
I. $[\beta\text{-}^{18}O]ADP$	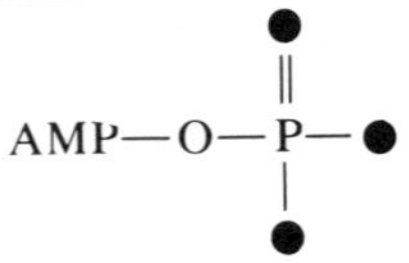 AMP—O—P(=●)(—●)—●		0.0220	
II. $[\gamma\text{-}^{18}O_3]ATP$	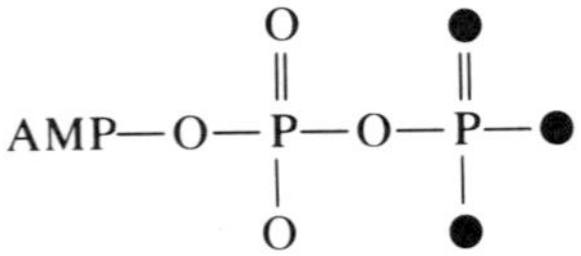 AMP—O—P(=O)(—O)—O—P(=●)(—●)—●			0.0220
III. $[\beta\text{-}^{18}O_4]ADP$	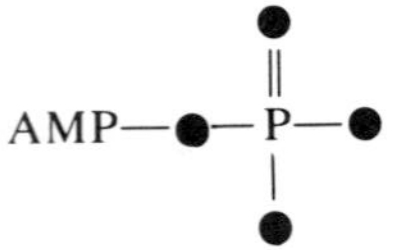 AMP—●—P(=●)(—●)—●	0.0166(α,βB)[c]	0.0215	

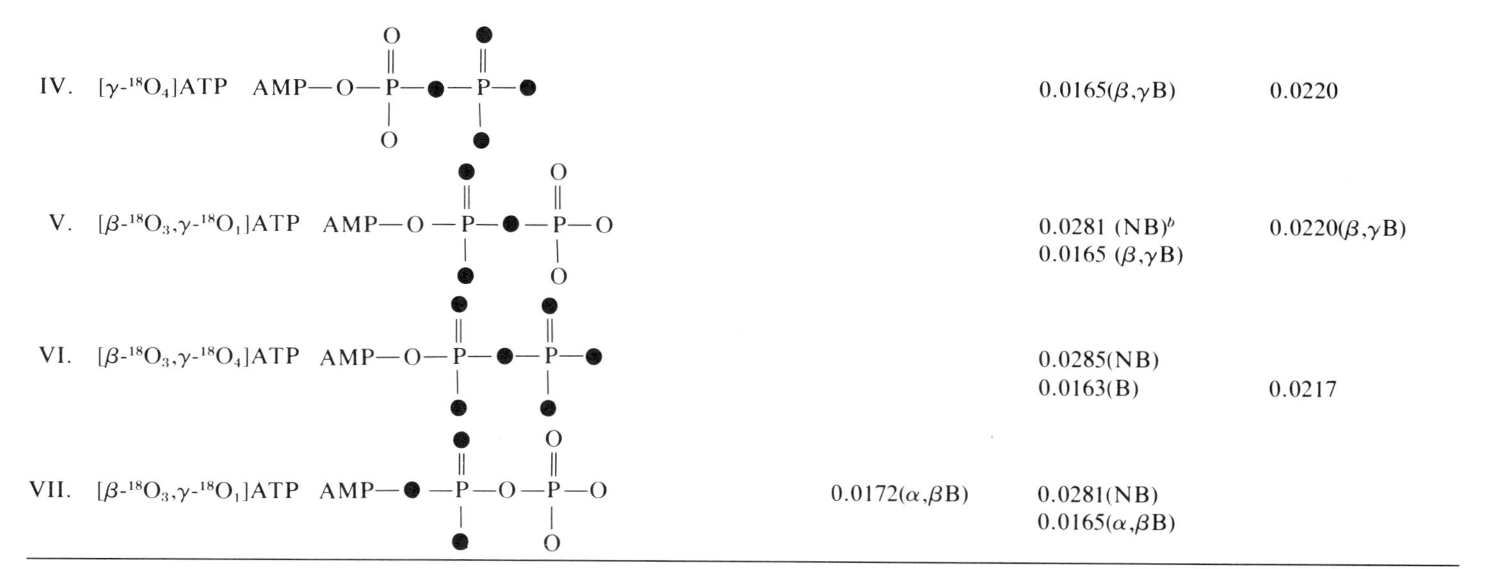

IV.	[γ-$^{18}O_4$]ATP		0.0165(β,γB)	0.0220
V.	[β-$^{18}O_3$,γ-$^{18}O_1$]ATP		0.0281 (NB)[b] 0.0165 (β,γB)	0.0220(β,γB)
VI.	[β-$^{18}O_3$,γ-$^{18}O_4$]ATP		0.0285(NB) 0.0163(B)	0.0217
VII.	[β-$^{18}O_3$,γ-$^{18}O_1$]ATP	0.0172(α,βB)	0.0281(NB) 0.0165(α,βB)	

[a]Reproduced with permission from Cohn, M., and A. Hu (1980), *J. Am. Chem. Soc.* **102,** 913, copyright © 1980, The American Chemical Society.

[b]O = ^{16}O, ● = ^{18}O.

[c]B = bridging; NB = nonbridging.

3. EXPERIMENTAL REQUIREMENTS

The small magnitude of chemical shift changes expected at ^{31}P on substitution of ^{18}O for ^{16}O made quantitative observation of differentially substituted species extremely difficult until the advent of high-field superconducting magnets in NMR spectrometer systems. At field strengths of 8.5 T (for which the resonance frequency of ^{31}P is 145.7 MHz), the frequency separation of signals differing by 0.02 ppm is 2.9 Hz. This quality of resolution is routinely attainable for aqueous samples of organophosphorus compounds. For metal–free ATP solutions, the ^{31}P linewidth is 1.2 Hz at this field strength (Cohn and Hu, 1980), so baseline resolution of the different isotopically substituted phosphorus resonances in ATP is straightforward using commercially available 360-MHz instruments (Brüker WM-360, Nicolet NT-360, among others), and possible, with more care, using 200-MHz systems.

The ^{31}P line broadening due to chemical effects, which can be especially pronounced for nucleotide triphosphates, is extremely deleterious to the observation of an isotopic shift. However, in most cases the effects can be obviated by the careful choice of experimental conditions and care in sample preparation. It is necessary to eliminate broadening due to proton exchange among the phosphate sites by maintaining the pH of the samples between 8.0 and 9.5. Conventional buffering agents have been shown to be satisfactory for this purpose.

Metal ions in these solutions can lead to exchange broadening, which is especially apparent at high field strengths. Cohn (Cohn and Hu, 1980) has noted that Mg^{2+} exchange among the various coordination sites in ATP is sufficiently slow on the NMR time scale that the linewidths of the β and the α ^{31}P are approximately 4 Hz at a field strength of 8.5 T. Ca^{2+} ions, for which the exchange lifetime is nearly 1000-fold less, cause no broadening in these signals. Unfortunately, most ATP-dependent biological systems require the Mg^{2+} form of the nucleotide triphosphate. The addition of EDTA to the NMR samples in amounts sufficient to complex the Mg^{2+} eliminates broadening from this source quite effectively. Concentrations of EDTA ranging from 1 to 20 m*M* have been used to achieve acceptable resolution of isotopically shifted ATP resonances. Usually this procedure necessitates quenching a reaction before one can observe the result, although several cases of continuous reaction monitoring of isotopic shifts have been reported.

Paramagnetic impurities must be scrupulously excluded in these experiments, though traces of these ions should be tightly complexed by the EDTA. Pretreatment of solutions (short Chelex columns, etc.) has been advocated in some cases.

The situation in nonaqueous solutions is considerably eased. For example, cyclic phosphate triesters of sugars give sufficiently resolved ^{31}P spectra in DMSO and DMSO–Ch_3OH mixtures to observe the difference between ^{18}O substitution in the P—O—C or the P═O bond (Jarvest *et al.*, 1980).

To ensure that accurate peak areas are obtained for the isotopically shifted ^{31}P signals, it is necessary to optimize both the field homogeneity and the instrumental resolution of the spectrometer. The former is accomplished by the adjustment of the shim coils to achieve the best possible resolution of the deuterium lock signal. In aqueous samples, a concentration of approximately 20% v/v of D_2O is sufficient for this purpose.

It is crucial that the spectrometer's computer be able to define all the peaks in its memory with sufficient precision to calculate the areas properly. This ability is termed *digital resolution* and is determined by (1) the size of computer memory available for data storage (i.e., 32K data points) and (2) the sampling rate, which is determined by the spectral width selected. To achieve the optimal digital resolution, one must use the smallest spectral window able to accommodate the peaks under observation, and the largest possible data table.

For example, a spectral width of 3012 Hz will encompass all the ^{31}P signals from ATP at 145.7 MHz. If 32,768 data points (32K) are specified, 16,384 points will define the real spectrum. This leads to a digital resolution of 0.183 Hz/point. For peaks of linewidth (at half-height) equal to 1.2 Hz, this means that above its half-height, any signal would be defined by seven data points, which is adequate for accurate computer calculation of peak areas.

The use of small sweep widths and large data tables necessarily requires lengthy acquisition times; therefore, longer periods of data accumulation are needed to obtain the same signal-to-noise ratio that could be expected under more-routine conditions. Concentrations of ~1 mM in nucleotide provide usable spectra in 1000–2000 scans.

4. CHIRAL PHOSPHORYL TRANSFER

One of the most elegant applications of isotope substitution in ^{31}P NMR probes of mechanistic problems involves the synthesis of chiral phosphate-containing compounds and subsequent analysis by ^{31}P NMR. A tetravalent phosphorus possesses tetrahedral geometry, and when four different substituents are present, the phosphorus is chiral. In principle, if one can determine the configuration at the phosphorus, then the effects of a chemical reaction on the configuration at such a center can be assessed. In the case of a nucleotide triphosphate, substitution of a sulfur for an oxygen of either the α or β phosphorus renders the phosphorus chiral. Given the chirality already present in the ribose moiety, the substitution actually results in the phosphorus enantiomers' being diastereomers of one another. The diastereomeric phosphorus atoms are often distinguishable by ^{31}P NMR, and this has allowed this stereochemical course of several enzymatic transformations to be studied. Fortunately the substitution of sulfur for oxygen is tolerated by many biological processes; thiophosphoryl nucleotides are substrates for many enzymes, with sizable preference for one of the diastereomers usually exhibited.

The terminal phosphoryl in a nucleotide or di- or triphosphate does not become chiral on the substitution of sulfur for one oxygen. If, however, one of the remaining ^{16}Os is replaced by an ^{18}O, the terminal thiophosphoryl is rendered chiral. This has provided an avenue for the investigation of several phosphatase and ATPase enzymes. The method relies on both the isotopic shift produced by ^{18}O and the quadrupolar broadening of ^{31}P by ^{17}O. The strategy involves several stereospecific enzymatic reactions and was developed independently by Trentham and Tsai and their co-workers. If one prepares a terminal thiophosphoryl compound of known configuration, for example, Figure 2, the transfer of the phosphoryl group to a nucleophile inverts the configuration at the phosphorus. If the ultimate acceptor of phosphoryl is water (as in a phosphatase) and the reaction is carried out in ^{17}O-labeled water, the resulting thiophosphate is itself chiral and may be analyzed to determine whether its overall configuration has been retained or inverted. Inversion would imply a direct single-step displacement on the phosphoryl (or an odd number of successive displacements), whereas retention would indicate double (or an even number of) displacements. The latter could be taken as evidence for participation by a nucleophile on the enzyme (so-called EP intermediate) in the hydrolysis. However, it was recently pointed out (Benkovic and Gerlt, private discussions) that if an EP intermediate results from E—CO_2^- as the nucleophile, that is, an acylphosphate EP, then hydrolysis could occur by water attack on the *carbon*, and the stereochemical result would appear to be inversion. Only single-turnover experiments would be able to distinguish this case from single displacement. Racemization of the thiophosphate might suggest the intermediacy of a dissociative (metaphosphate) pathway in which the phosphoryl group is accessible from both faces, or more generally, a nonspecific mechanism. Frontside attack by water followed by pseudorotation would also show retention at the phosphorus. The analysis of chirality for the inorganic thiophosphate requires that it be stereospecifically incorporated into the β position of ATP(βS) followed by determination via ^{31}P NMR of whether the ^{18}O label is bridging to another phosphorus or nonbridging. In the procedure

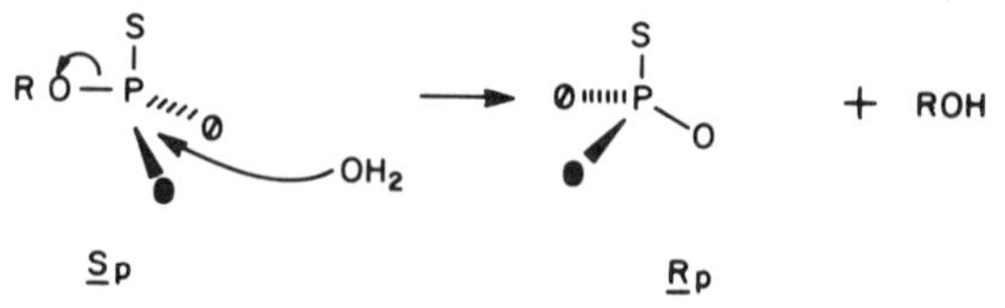

Figure 2. Hydrolysis of a chiral thiophosphate ester, proceeding through backside attack to produce chiral P_{si}, whose configuration has inverted.

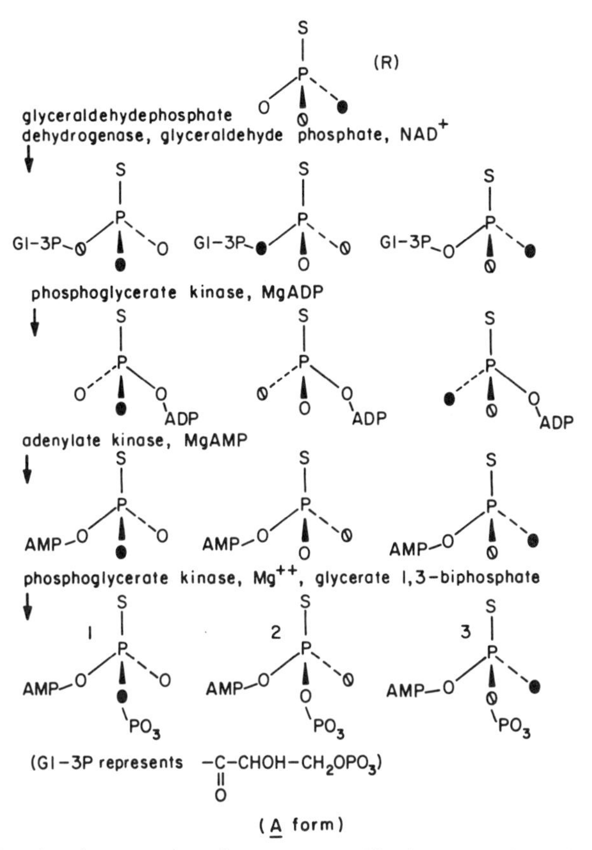

Scheme 1. Trentham's procedure for stereospecific incorporation of chiral P_{si} into [β $^{17}O^{18}O$]ATP(βS).

of Trentham (Webb and Trentham, 1980; Scheme 1) illustrated here for the *R* enantiomer of thiophosphate, the chiral inorganic phosphorothioate (P_{si}) is used with glyceraldehyde phosphate dehydrogenase to phosphorylate glyceraldehyde 3-phosphate. This reaction is known to proceed with integrity of the P—O bond and thus produces three different glycerate 1,3-bisphosphate isomers, each having a different isotope of oxygen bridging the phosphoryl group and the former aldehyde carbon. (These isomers will be conveniently referred to as *isotopomers*.)

In the next step, phosphoglycerate kinase transfers the thiophosphoryl group to ADP, with loss of the oxygen that had bridged to the glycerate. This stereochemistry was first demonstrated by Webb and Trentham (1980) in developing this procedure. In the following step, the ATP(γS) is used as a substrate for adenylate kinase and AMP to form ADP and ADP(βS), which can be separated. This reaction was previously shown (Sheu and Frey, 1977) to occur with inversion at P_γ of ATP.

The ADP(βS) thus formed is treated with phosphoglycerate kinase and glycerate 1,3-bisphosphate to form ATP(βS). This reaction was previously shown by Cohn (Jaffe and Cohn, 1978) to result in phosphorylation of ADP(βS) stereospecifically at one of its prochiral oxygens, producing exclusively the *A* diastereomer, for which the absolute configuration at P_β is as shown in Scheme 1.

In each of the above steps, three isotopomers are produced, each related to the particular isotope of oxygen that formed the original C—O—P bridge. These species are not separable chromatographically, but this is not necessary since the ^{17}O present in two of them causes the $^{31}P_\beta$ resonance to be so broadened as to be unobservable. The remaining isotopomer has lost its ^{17}O in the initial reaction, and its ^{31}P spectrum is observed readily. Under conditions of high resolution, the $^{31}P_\beta$ NMR signal will show an isotope shift due to the presence of ^{18}O. However, the position of this label, that is, bridging to P_γ or nonbridging, is governed by the original configuration of the thiophosphate. The *S* enantiomer gives the ^{18}O in the nonbridge position; the *R* leads to ^{18}O enrichment in the bridge. This distinction is due to the stereoselectivity for one of the prochiral ADP(S) oxygens by phosphoglycerate kinase. The bridge–nonbridge difference in chemical shift for [^{18}O]ATP(βS) is 0.016 ppm. Figure 3 shows the relative enrichments arising from *R*, *S*, and racemic P_{si}. (In actual practice, the ^{18}O impurity in $H_2{}^{17}O$ results in the presence of both bridge and nonbridge substitutions, but the enrichment in the predicted position is obvious.)

Tsai (1980) independently developed a similar strategy with different enzymes to incorporate chiral thiophosphate into the β position in ATP(βS). His route is shown in Scheme 2 and results in the spectra for ATP(βS) derived from *R*, *S*, and racemic P_{si} as given in Figure 4. He diverged from the Trentham scheme after formation of the chiral ADP(βS) isotopomer. This is stereospecifically phosphorylated in the reaction with acetate kinase and acetylphosphate to produce specifically the *B* diastereomer of ATP(βS). This phosphorylation shows (Richard *et al.*, 1978) the reverse of the stereoselectivity of the phosphoglycerate kinase reaction used in the scheme devised by Trentham and leads ultimately to the opposite results for the *R* and *S* enantiomers of P_{si} in the ^{31}P spectrum.

There are noteworthy differences between the ATP(βS) spectra obtained in the Trentham and Tsai strategies. The large peak in the Trentham spectrum shown corresponds to completely unlabeled $^{16}O_2$ material. This washout occurred in his synthesis of chiral P_{si} and is considerably reduced in spectra in subsequent reports employing this strategy. Tsai's initial report using synthetic chiral P_{si} exhibited the same, large $^{16}O_2$ resonance.

The key ratios between the bridged and nonbridged central peaks ($^{18}O_B : {}^{18}O_{NB}$) are slightly greater in the Tsai reports. These ratios are governed by, among other things, (1) the quality of chromatographic separation of the ADP(αS) diastereomers in the P_{si} syntheses (each used a different ion-exchange resin), and (2) the relative degrees of stereospecificity in the

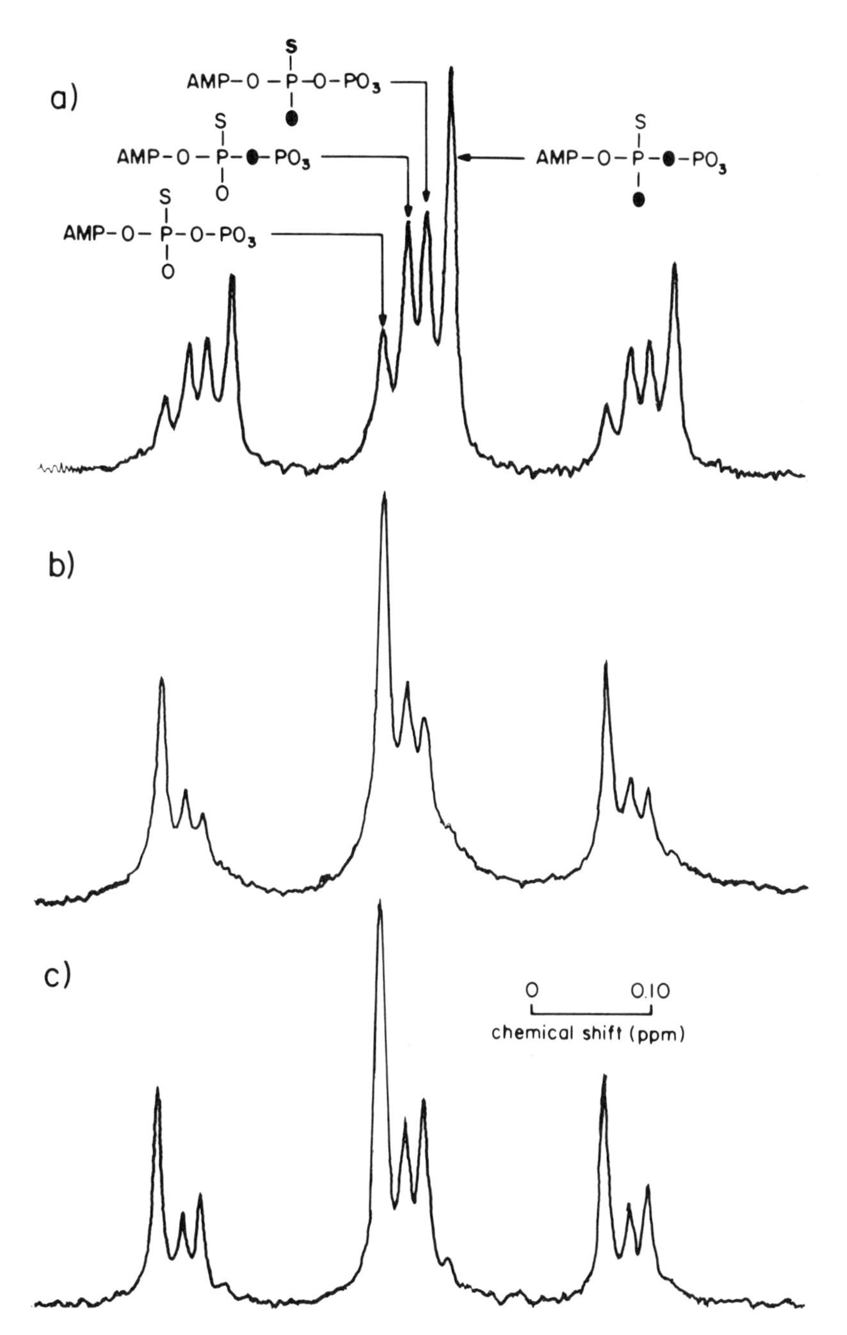

Figure 3. The ^{31}P spectra of P_β in ATP (βS) produced in the Trentham strategy from (*a*) racemic, (*b*) R_p, and (*c*) S_p P_{si}. Reproduced with permission of the author.

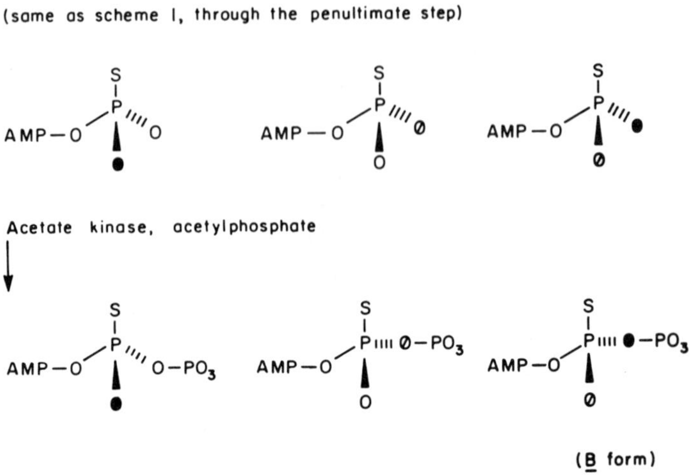

(same as scheme I, through the penultimate step)

AMP—O—P(S)(O)(●) AMP—O—P(S)(Ø)(O) AMP—O—P(S)(●)(Ø)

Acetate kinase, acetylphosphate

↓

AMP—O—P(S)(●)—O—PO_3 AMP—O—P(S)(O)—Ø—PO_3 AMP—O—P(S)(Ø)—●—PO_3

(B form)

Scheme 2. The last step in Tsai's procedure for stereospecific synthesis of [β $^{17}O^{18}O$]ATP(βS) from chiral P_{si}.

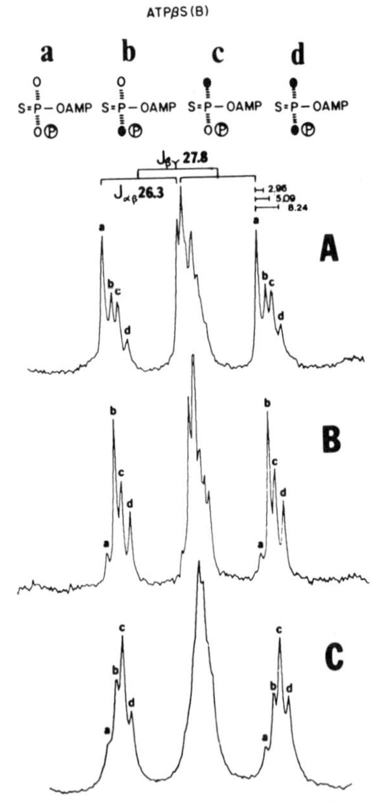

Figure 4. The ^{31}P spectra of P_β in ATP(βS) produced in the Tsai strategy from (*a*) racemic, (*b*) S_p, and (*c*) R_p P_{si}. Reproduced with permission, from Tsai, M. D., (1980), *Biochemistry*. **19,** 5310, copyright (1980) The American Chemical Society.

phosphoglycerate kinase and acetate kinase reactions. Since phosphoglycerate kinase is used twice in Trentham's method, there is great reliance on two stereospecificities for this enzyme. Both procedures outlined above certainly provide decisive and useful avenues for investigating the stereochemistry of thiophosphate hydrolysis, provided that one can prepare chirally pure (at phosphorus) substrates. The reports that have used these strategies are summarized in the following sections.

4.1. 5′ Nucleotidase

Tsai, in the first report using this methodology (Tsai 1980), studied the hydrolysis of AMP(S) to adenosine and P_{si} by 5′ nucleotidase. The [$^{18}O_2$]AMPS is prepared by the reaction of adenosine with $PSCl_3$ followed by hydrolysis in $H_2{}^{18}O$. Nonspecific phosphorylation of the [$^{18}O_2$]AMPS provides both diastereomers of [α-$^{18}O_2$]ADP(αS). Pyruvate kinase–phosphoenolpyruvate acts to phosphorylate stereospecifically the *A* form of ADP(αS) to give *A* ATP(αS) and *B* ADP(αS), which are readily separated. This constitutes *resolution* of the "enantiomers" of the α phosphorus in the starting mixture. Reaction of both the *A* [α-^{18}O]ATP(αS) and *B* [α-^{18}O]ADP(αS) with alkaline phosphatase give respectively the S_p and R_p enantiomers of [^{18}O]AMPS, for which the configuration can be assigned given the established selectivity of pyruvate kinase (Eckstein and Goody, 1976). When each is hydrolyzed by 5′ nucleotidase from snake venom in $H_2{}^{17}O$ and the product thiophosphate derivatized as described above, the ATP(βS) is found to have opposite configuration from its starting [^{18}O]AMP(S); therefore hydrolysis involves a single displacement at phosphorus, and a phosphoenzyme is excluded.

4.2. ATPases

Similarly, Trentham and his co-workers (Webb *et al.*, 1980) showed that hydrolysis of ATP(γS) by beef heart mitochondrial ATPase proceeds without formation of a phosphoenzyme. The same inversion of configuration was demonstrated by this group for the ATPase activity of myosin subfragment I (Webb and Trentham, 1980).

They showed the opposite result (retention) for the ATPase found in the sarcoplasmic reticulum membrane (Webb and Trentham, 1981). In this example, the intermediacy of a phosphoenzyme had previously been demonstrated.

The stereospecific synthesis of terminally chiral nucleotide triphosphates is of critical importance in these strategies. Here we describe the strategy for synthesis of chiral [β,γ-^{18}O;γ-^{16}O,^{18}O]ATP(γS) devised by Frey and used by Trentham. Adenosine is treated with $PSCl_3$ and hydrolyzed in $H_2{}^{18}O$ to provide [$^{18}O_2$]AMPS. This is converted to [α-^{18}O;α,β-^{18}O]ADP(αS) by reaction with adenylate kinase–ATP, and with pyruvate kinase–PEP–NADH–LDH, yielding *A* ATP(αS). This is converted completely to ADP(αS) by

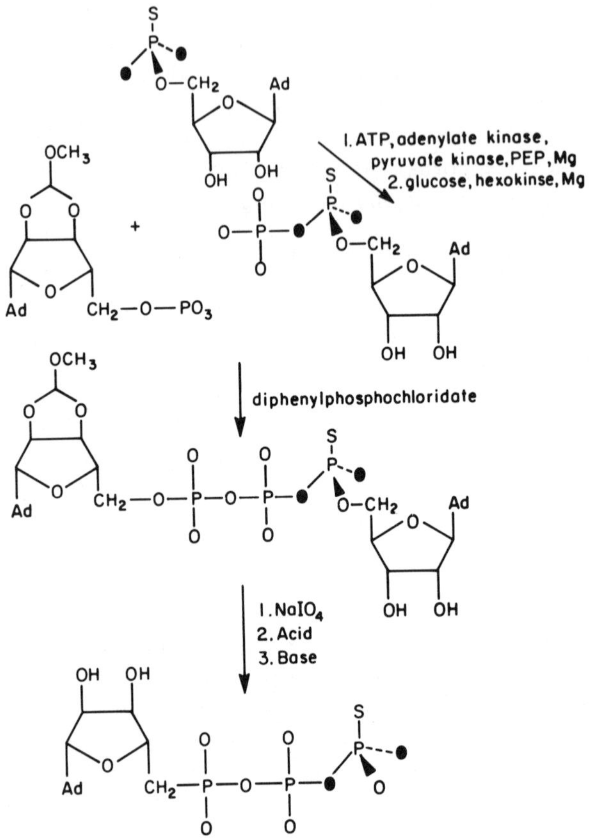

Scheme 3. Strategy of Frey, used by Trentham for the synthesis of [β,γ^{18}O;γ^{18}O]ATP(γS) chiral at P_γ.

reaction with hexokinase and glucose. The isomer formed is the *A* form of [α-^{18}O;α,β^{18}O]ADP(αS) and this stereospecificity is key to the ultimate chirality in the ATP(αS).

The *A* [α-^{18}O;α,β-^{18}O]ADP(αS) is then chemically coupled with AMP (protected as the 2′,3′-methoxymethylidine orthoester) by use of diphenylphosphochloridate. This places the protected adenosine ring distal to P(S) in the A·P·P·P(S)·A intermediate. Periodate cleavage of the nonprotected adenosine sugar results in its facile removal in base, yielding [γ-^{18}O;β,γ-^{18}O]ATP(γS), with the absolute configuration at P_γ being R_p. The overall reaction scheme is outlined in Scheme 3. Webb and Eccleston (1981) similarly synthesized chiral [β,γ-^{17}O;γ-^{17}O,^{18}O]GTP(γS), which was used to investigate the GTPase reaction associated with the ribosome-dependent elongation factor G. The ribosome-factor G complex facilitates the translocation process following the addition of each amino acid in bacterial protein biosynthesis. The chiral P_{si} produced by the GTPase reaction was incorporated into ATP(βS), and analysis of the ^{31}P NMR spectrum showed complete

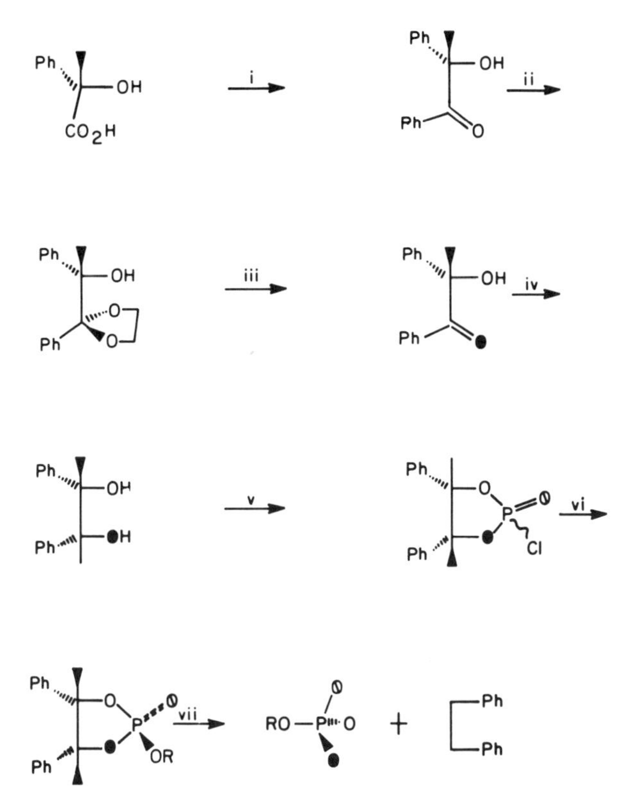

Scheme 4. Synthesis from Lowe *et al.* (1979) of the organic phosphate compounds of known isotopic configuration at phosphorus: *i*, PhLi; *ii*, $LiAlH_4$; *iii*, CH_2OHCH_2OH-*p*-MeC_6H_4—SO_3H; *iv*, $H_2{}^{18}O$-dioxane-*p*-MeC_6H_4—SO_3H; *v*, $P^{17}OCl_3$–pyridine; *vi*, *ROH*–pyridine; *vii*, H_2–Pd. The starting material is *S* mandelic acid.

inversion of configuration, supporting earlier failures to detect (Eckstein *et al.*, 1975) a phosphorylated intermediate in this complex.

Strategies based on similar reasoning have been advanced for the use of ^{31}P NMR to probe the stereochemistry of transformations involving the α phosphorus in nucleotides. Lowe *et al.* have prepared chiral [^{16}O,^{17}O,^{18}O]AMP and [^{16}O,^{18}O]AMP(S) (Jarvest and Lowe, 1979; Cullis and Lowe, 1978; Cullis *et al.*, 1981) of known configuration via reaction of the appropriately protected nucleoside with the phosphorochloridate derived from *meso*-*R*,*S*[^{16}O,^{18}O]hydrobenzoin and $P^{17}OCl_3$ (or $PSCl_3$). This yields, under thermodynamic conditions, the *trans* diastereomeric triester, as shown in Scheme 4.

Subsequent reduction of this product under Birch conditions provides the requisite chiral [^{16}O,^{17}O,^{18}O]AMP, whose absolute configuration has been transferred from the starting hydrobenzoin and hence is known to be S_p. The method appears to be generally applicable for synthesis of chiral organophosphate esters.

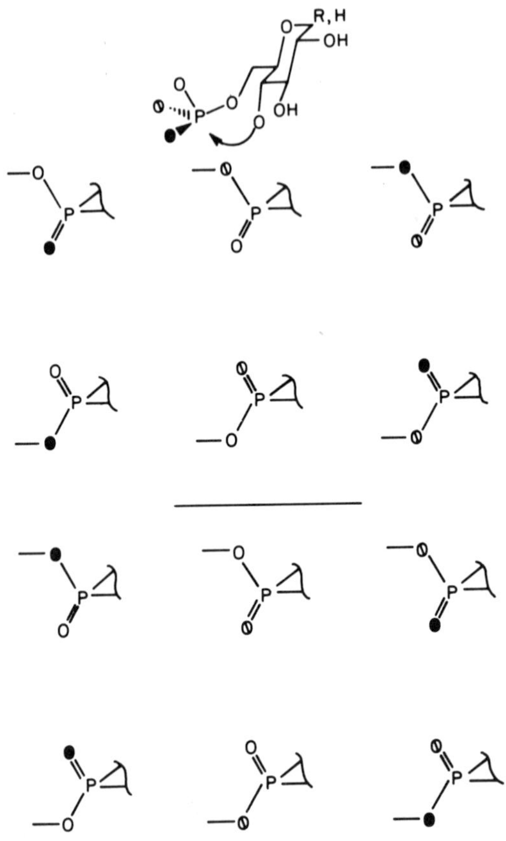

Scheme 5. Mixtures of isotopomers for each of the diastereomeric pairs generated by inversion or retention at phosphorus in the chemical cyclization–methylation procedure of Lowe (Jarvest *et al.*, 1980).

In a displacement at the phosphorus, it is obvious that as before, three isotopomers will be produced, but again, the presence of ^{17}O in two of these so broadens their ^{31}P resonances that the resonances are entirely obliterated. Lowe has investigated the chemical cyclization of chiral AMP and shown that it proceeds with inversion of configuration (Cullis *et al.*, 1981) at phosphorus. In his strategy, the cyclic AMP product is methylated to produce the two epimeric methyl triesters. The axial and equatorial triesters show different ^{31}P chemical shifts (axial $^{31}P—OCH_3$ shifted to ~2 ppm higher field), and this is the key to Lowe's strategy.

For the S_p enantiomer of AMP, it is evident that for the cyclized, methylated product, the axial ^{31}P resonance should show enrichment of ^{18}O in the nonbridge position, providing a single inversion has occurred. The opposite result (bridge enrichment) should obtain if the reaction involves overall retention. These situations are reversed for the equatorially methylated species. Scheme 5 illustrates this effect.

4.3. Phosphodiesterase and Kinases

The stereochemical course of bovine heart cyclic AMP phosphodiesterase was investigated using this method (Cullis *et al.*, 1981; Jarvest *et al.*, 1980) by carrying out the enzymatic hydrolysis of chiral cyclic AMP of known epimeric configuration in ^{17}O-enriched water (see below). The isolated AMP was chemically recyclized and methylated and showed enrichment of ^{18}O in the bridging position of the equatorial triester. This indicated that overall inversion occurs during the phosphodiesterase reaction.

The initial reports employing this method were subsequently found to be erroneous, because of the mistaken assignment in the literature of the stereochemistry in the thermodynamic product of alcohol attack on the phosphorochloridate mentioned above (2-alkoxy-2-oxo-4,5-diphenyl-1,3,2-dioxaphospholane). Lowe was able to show that the *trans* rather than the *cis*

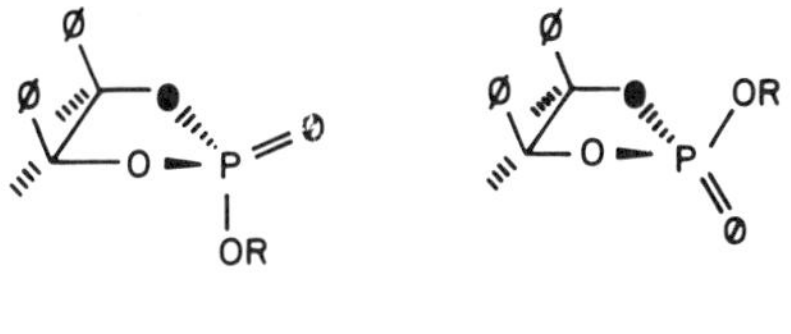

trans cis

triester forms under equilibrating conditions; he did this by a spectral (NMR) comparison of this material with an authentic sample of the *trans* product (synthesized by an alternate route), for which an X-ray diffraction structure was available. He was thus able to revise his earlier findings of retention at phosphorus in the cyclization and phosphodiesterase reactions and confirm that the cyclic phosphodiesterase hydrolysis of both cyclic 3′,5′-AMP and its thiophosphoryl analog proceed via single-displacement mechanisms. Lowe has extended his method for the transfer of chiral phosphoryl groups to the synthesis of [γ-^{16}O,^{17}O,^{18}O]ATP of known S_p configuration at P_γ (Cullis and Lowe, 1981) by reaction of appropriately protected ADP with phosphorochloridate I. This chiral ATP was used to examine the stereochemistry of phosphoryl transfer to sugars, which were cyclized and methylated and the ^{31}P NMR spectrum analyzed as described above.

Lowe and Potter (1981) investigated the 6′ phosphorylation of glucose by hexokinase and [γ-^{16}O,^{17}O,^{18}O]ATP and found the transfer to occur with inversion at phosphorus. In another example, polynucleotide kinase (Lowe, private communication, 1981) was used with P-chiral ATP to phosphorylate the 5′ position of 3′-AMP. The resulting 3′,5′-ADP was dephosphorylated at the 3′ position with nuclease S1. The [^{16}O,^{17}O,^{18}O]5′-AMP when analyzed as described above showed that this enzyme also transfers the phosphoryl with inversion of configuration. Rat liver glucokinase exhibits unusual kinetics compared with other hexokinases. However, Lowe (Pollard-Knight *et al.*, 1982) used his method to show that this enzyme also mediates phos-

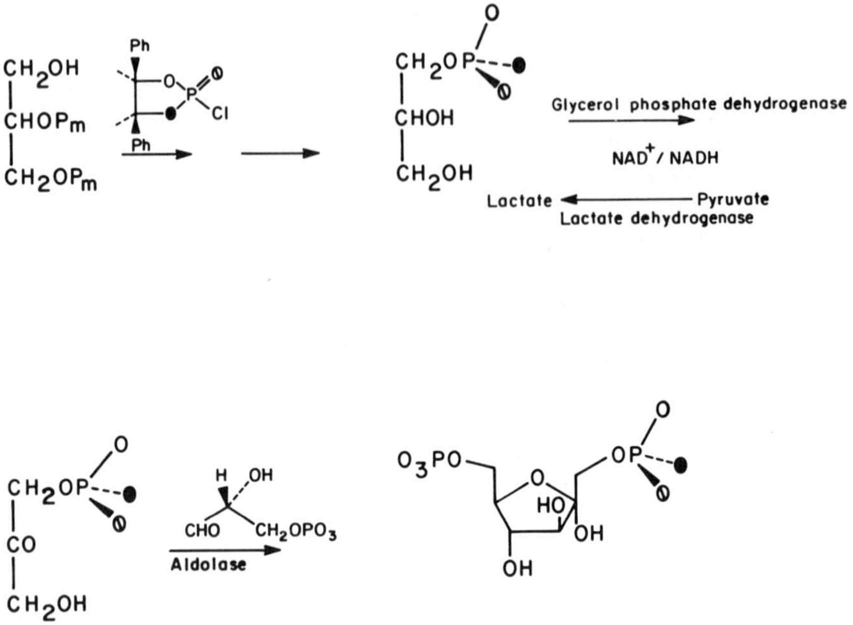

Scheme 6. Stereospecific synthesis of 1,6-bisphosphofructose made chiral at P-1 (Jarvest and Lowe, 1981).

phoryl transfer from ATP with inversion at phosphorus; thus the basic mechanism for these enzymes would appear to be the same, with differences in the rate constants for substrate binding and the chemical step accounting for the diverse kinetics observed.

Jarvest and Lowe (1981) used the hexokinase result to probe the mechanism of phosphofructokinase-catalyzed phosphoryl transfer between fructose 1,6-bisphosphate and ADP. They employed fructose 1,6-bisphosphate with chiral $[^{16}O,^{17}O,^{18}O]$P-1 and analyzed the chirality of the γPO_3 of the product ATP by subsequent hexokinase-mediated transfer to glucose, followed by the usual cyclization, methylation, and ^{31}P NMR analysis. Since the reaction is unfavorable in the direction of ATP synthesis, they drove the reaction by removal of the fructose 6-phosphate as it was formed. This was facilitated by the action of phosphoglucose isomerase, glucose 6-phosphate dehydrogenase and $NADP^+$, which convert fructose 6-phosphate to D-gluconolactone 6-phosphate, which becomes irreversibly hydrolyzed.

The P-1 chiral bisphosphofructose was synthesized using Scheme 6.

The ^{31}P spectrum of cyclic glucose 3,6-phosphate methyl triester ultimately produced by the chiral ATP by phosphofructokinase (PFK) transfer showed that the transfer of phosphoryl from fructose 1,6-bisphosphate to ADP occurs with inversion at P-1; a second inversion occurs in the hexokinase transfer, resulting in net retention for the combined processes.

The cyclic glucose 3,6-phosphate analyzed here exists as an equilibrium mixture of *anomers* at C-1, whose ^{31}P resonances are shifted by an amount comparable to the ^{18}O-induced shift. The judicious selection of solvent mixture for the NMR analysis (Jarvest *et al.*, 1980) (DMSO:MeOH, 1:1) results in sufficient separation of the anomeric resonances, so that analysis of the isotopic shifts is straightforward. The phosphofructokinases from both rabbit muscle and *B. stearothermophilus* proceed with simple inversion at phosphorus during transphosphorylation.

The action of phosphoglucomutase in converting glucose 1′-P to glucose 6′-P was known to involve an equilibrium mixture of two distinguishable phosphoenzyme intermediates. If each intermediate represents phosphorylation of a different enzymatic nucleophile, and if both are obligatory in the transition from glucose 1′-P to glucose 6′-P, then overall inversion at phosphorus should be observed.

$$\text{Glu 1-P} \overset{\text{inv}}{\rightleftharpoons} E_{n1}P \overset{\text{inv}}{\rightleftharpoons} E_{n2}P' \overset{\text{inv}}{\rightleftharpoons} \text{Glu 6-P}$$

(3 inversions = net inversion)

This assumes that adjacent attack/pseudorotatory processes do not pertain. However, if the two states differ only by a conformation change, for example, the overall result should be retention. In his study of this system Lowe (Lowe and Potter, 1981b) employed glucose 1-[^{16}O,^{17}O,^{18}O]P that had been prepared chiral with known configuration at the phosphorus. The spectrum of the final triester shows net retention at the phosphorus, indicating that the two EP intermediates are likely equilibrating conformers, or else only one of them is an obligatory participant in the process.

Recently, these workers have expanded their efforts in probing phosphodiesterase mechanisms. Having established phosphoryl inversion in the Mg^{2+}-requiring bovine cyclophosphodiesterase (Lowe, private communication, 1981) they used chiral cyclic AMPS as a substrate for the Zn^{2+}-dependent yeast cyclophosphodiesterase.

Though both P(S) epimers are accepted by this enzyme, there is a strong preference for the R_p diastereomer. The strategy for this study included (1) enzymatic hydrolysis of each epimer in 50% $H_2{}^{18}O$, (2) stereospecific conversion of the chiral 5′-AMPS product into the *A* form of ATP(αS), (3) ^{31}P NMR analysis of the β phosphorus to determine whether the ^{18}O label occurred in the α,β bridge or not. The results in each case indicated inversion as the stereochemical course of the reaction. Jarvest and Lowe (1981) also examined snake venom phosphodiesterase hydrolysis of ATP made chiral at the α phosphorus from [^{16}O,^{17}O,^{18}O]AMP. Hydrolysis in $H_2{}^{17}O$ produces a mixture of three isotopomers (only one of which is chiral). Chemical cyclization and methylation as before produce the chiral phosphotriesters, which on ^{31}P NMR analysis indicate that retention occurs in the phosphodiesterase reaction. This result is in accord with the previous discoveries of retention with thio analogs using this enzyme and ATP(αS) (Bryant and

Benkovic, 1979) or adenosine 5′-phosphorothioate-*p*-nitrophenyl ester (Burgers *et al.*, 1979).

4.4. Additional Group-Transfer Reactions

The enzyme ATP-sulfate adenylyltransferase catalyzes the production of AMP sulfate (APS) from ATP and sulfate. In this system Bicknell, *et al.* (1982) were able to study the stereochemistry of the reverse displacement of SO_4^{2-} by PP_i by preparing the chiral AMP as before and treating this with $SO_3 \cdot Et_3N$ to form three indiscriminately sulfurated isotopomers. Expulsion of SO_4^{2-} by PP_i produces the three corresponding ATP molecules, of which only one is chiral, that is, that which results from expulsion of $S^{16}O_4$. These are hydrolyzed in $H_2{}^{17}O$ by snake venom phosphodiesterase (retention) to AMP (mixture of isotopomers; one chiral, the other two having two ^{17}O's) then cyclized, methylated, and analyzed as before, to demonstrate that inversion occurs in the original reaction of PP_i with APS.

Knowles had earlier reported a different synthetic approach to chiral organophosphates via chiral cyclic phosphoramidate esters, followed by transfer to S-propane-1,2-diol (Abbott *et al.*, 1978; Scheme 7). When the intermediate is subjected to in-line cyclization, followed by methylation, a

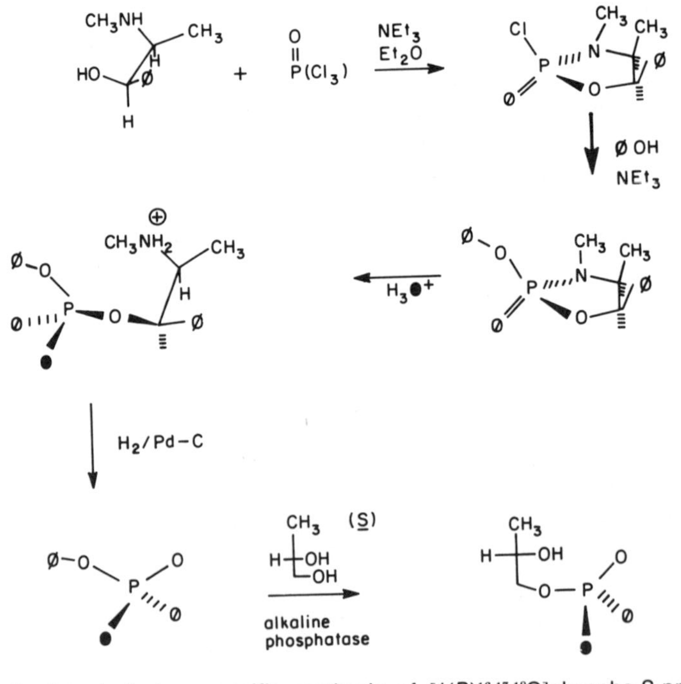

Scheme 7. Knowles' stereospecific synthesis of $[1(R)^{16,17,18}O]$phospho-S-propane-1,2 diol (Hansen and Knowles, 1981).

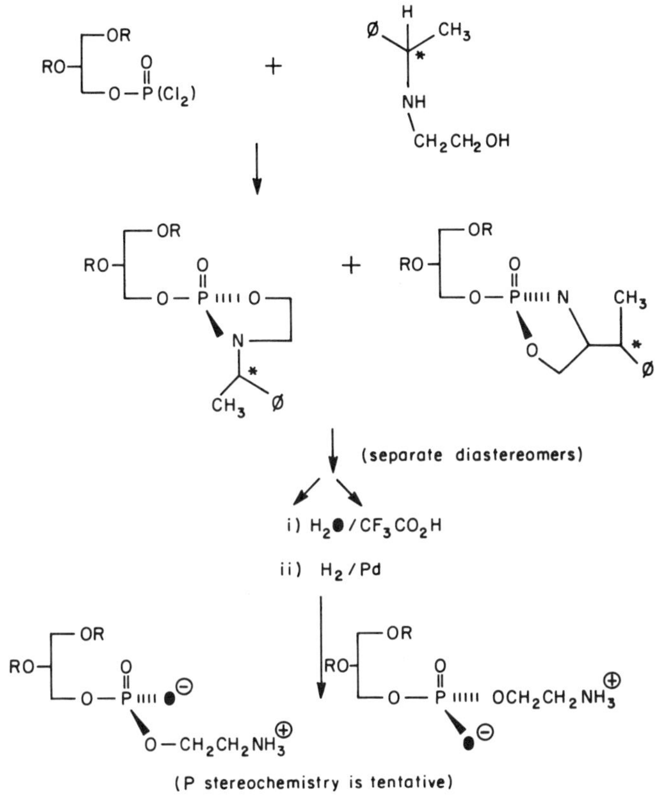

Scheme 8. Synthesis by Tsai and Bruzik (1982) of P-chiral DPPE. The absolute configurations have not been assigned since the stereochemical identities of the phosphoramidates have not been confirmed.

mixture of diastereomeric isotopomers is produced, for which Knowles has developed elegant mass-spectral probes of chirality. He recently showed that the diastereomeric methyl triesters are resolved in the ^{31}P spectrum (101.3 MHz) (Hansen and Knowles, 1981). Knowles used this observation in a manner reminiscent of Lowe's analysis to probe the stereochemistry of creatine kinase–mediated phosphoryl transfer from [^{16}O,^{17}O,^{18}O]ATP to creatine. Following alkaline phosphatase transfer of PO_3^- (net retention; well-known E—P intermediate) to S-propane-1,2-diol, the ^{31}P spectrum of the methyl triesters showed that the creatine kinase transfer proceeds with inversion at P_γ of ATP.

Tsai has extended these studies to reactions involving phospholipids (Tsai and Bruzik, 1982). Normally the ^{31}P signals in aqueous solutions of phospholipids such as dipalmitoylphosphatidylethanolamine (DPPE) would be fairly broad because of micellar-type aggregation. Tsai and co-workers have avoided this problem by silylation of the phosphoryl oxygen, yielding a structure with a nonamphiphilic head group. This results in loss of aggregate structure and narrowed ^{31}P resonances. They studied the transphosphati-

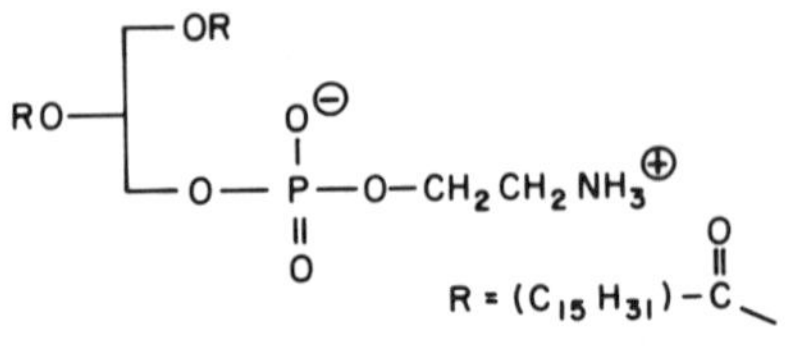

RO—[—OR; —O—P(=O)(O$^{\ominus}$)—O—$CH_2CH_2NH_3^{\oplus}$]

R = $(C_{15}H_{31})$—C(=O)—

DPPE

dylation reaction of phospholipase D using chirally ^{18}O-labeled DPPE (Scheme 8). Treatment of each enantiomer of chiral DPPE with hexamethyldisilazene produces a mixture of diastereomeric monotrimethylsilyl esters, which differ in disposition of the ^{18}O, which is either bridging to silicon or nonbridging. The diastereomers give rise to separate ^{31}P signals in $CDCl_3$, and the ^{18}O induces a shift of 0.018 and 0.038 ppm for bridge and nonbridge substitution, respectively. When one of the chiral DPPE enantiomers is converted to the N,N,N-trimethyl derivative, then allowed to react with ethanolamine and phospholipase D, the *product* DPPE-silyl ester shows the same isotopic distribution as starting material, implying overall retention at phosphorus. From this, Tsai has inferred the participation of an EP intermediate.

4.5. Adenylate Cyclase

Gerlt (Coderre and Gerlt, 1980) and co-workers have described a method for the determination of the oxygen chirality of either the α or β phosphorus in ADP, using the isotope shift in the ^{31}P NMR spectrum. The method is based on the methodology of Cleland (Cornelius *et al.*, 1977; Dunaway-Mariano and Cleland, 1980) for the preparation of the so-called substitution-inert chelates of nucleotide diphosphates with $Co(NH_3)_4^{3+}$. These complexes are epimeric at the α phosphorus, and the diastereomeric mixture is separable chromatographically. Thus the ^{31}P resonances are identifiable and can be assigned to either the Δ or the Λ diastereomer, for which the absolute configurations at the α phosphorus are known to be S_p and R_p, respectively. It is apparent that the introduction of a single ^{18}O label renders that phosphorus chiral, and that the diastereomeric chelates are thus each a mixture of two isotopomers, which differ in the disposition of the ^{18}O label, that is, bridging to the Co or nonbridging. Figure 5 illustrates the structures of the species involved.

Gerlt and Coderre (1980) were able to prepare the chiral dADP (^{18}O) in Figure 5 via the reaction (known to proceed stereospecifically, with inversion of configuration) of chiral cyclic dAMP with pyrophosphate and adenylate cyclase from *B. liquifaciens* coupled to a glycerol kinase–catalyzed dephosphorylation of the product dATP. Both the R_p and S_p enantiomers of

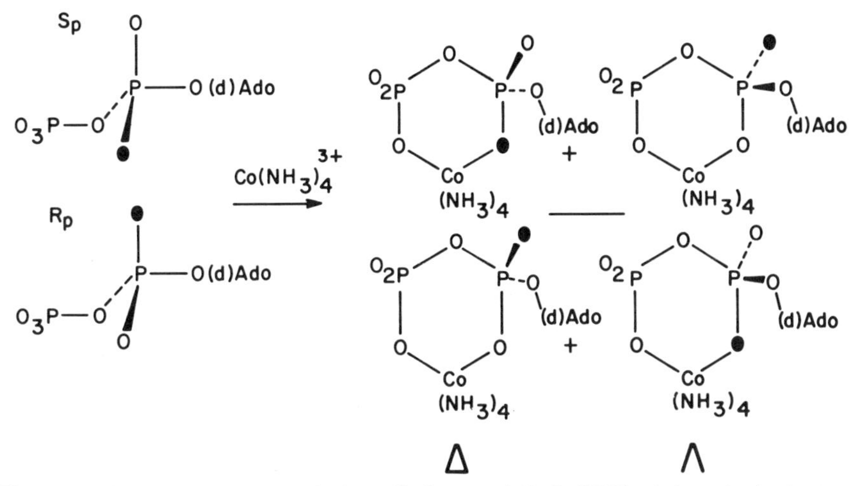

Figure 5. Stereospecific synthesis from Coderre and Gerlt (1980) of the substitution-inert tetramine Co^{3+} complexes with chiral ADP($\alpha^{18}O$).

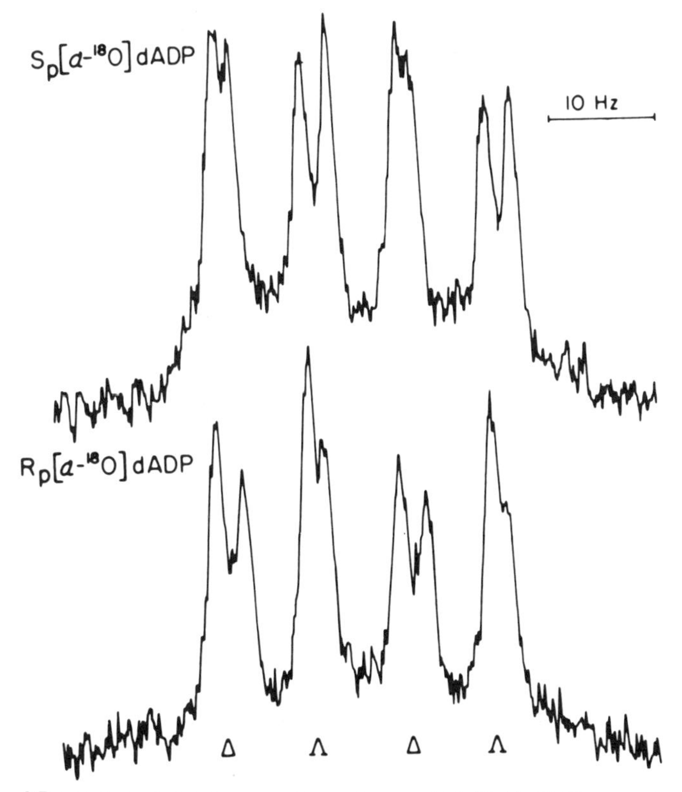

Figure 6. ^{31}P spectra of diastereomeric mixtures of $Co(NH_3)_4(ADP)$ complexes arising from S_p and R_p ADP($\alpha^{18}O$). Reproduced with permission, from Coderre, J. A., and J. A. Gerlt, (1980), *J. Am. Chem. Soc.* **102,** 6597, copyright (1980) The American Chemical Society.

cyclic dAMP were accessible through the recent methodology from Gerlt's laboratory and Stec's laboratory (Baraniak *et al.*, 1980). This chemistry involves base-catalyzed frontside displacement at phosphorus by $C^{18}O_2$ on the appropriate epimeric cyclic dAMP phosphoroanilidate.

In line with the preceding discussion, the larger isotope shift in the $^{31}P_\alpha$ resonance for each pair of isotopomers should be observed for the nonbridging substitution of ^{18}O. In other words, the chelate derived from the S_p enantiomer of dADP (the Λ diastereomer) has ^{18}O nonbridging, and should exhibit a larger isotope shift than the Δ diastereomer. The R_p-derived chelate, conversely, should give a larger shift for the Δ species. This is observed in the spectrum of each isomer (Figure 6). The nonbridge shift for these complexes is 0.031 ppm: the bridging ^{18}O gives a shift of only 0.010 ppm (Figure 6).

5. ORIENTATION STUDIES IN P—O BOND-FORMING AND CLEAVAGE REACTIONS

The detection of isotopic perturbations on ^{31}P NMR signals discussed above, that is, the ^{18}O-induced chemical shift and/or ^{17}O-induced line broadening, has obvious utility for probing the mechanisms of reactions in which P—O bonds are formed or broken. Specifically, one can observe directly whether a given P—O bond has maintained its integrity in a cleavage process, or whether a label has become incorporated into a newly formed bond (these examples require the lack of phosphoryl exchange with solvent water).

The first example of such an application was reported by Cohn and Hu (1978), who investigated the exchange reaction between ADP and P_i, which is catalyzed by polynucleotide phosphorylase (previously demonstrated by ^{32}P exchange). Continuous monitoring of this reaction by ^{31}P NMR revealed the gradual depletion of $P^{18}O_4^{3-}$ in favor of $P^{16}O_4^{3-}$ concomitant with buildup of $[\alpha,\beta\text{-}^{18}O;\beta\text{-}^{18}O_3]$ADP. This proved that the exchange necessarily proceeded with cleavage of the P—O—P bond at P_α.

5.1. Phosphatases

Valyl-tRNA synthetase has been shown to catalyze the reversible adenylylation of valine by ATP. Cohn and Hu (1980) observed that when $[\gamma^{18}O_3]$ATP is the substrate for this reaction, the α ^{31}P signal shows that ^{18}O is slowly incorporated into the α,β bridge, indicating that the free PP_i was incorporated symmetrically on reversal of the reaction to valine and ATP. Moreover, the same reaction using $[\beta,\gamma\text{-}^{18}O;\beta\text{-}^{18}O_2]$ATP shows ^{18}O incorporation into the α,β bridge. This scrambling of the P_β labels is ~20% faster than the PP_i reversal, suggesting the participation of a partially encumbered (nonsymmetrical) PP_i in the reverse reaction.

The exchange reactions between P_i and water as catalyzed by phosphatase enzymes were early candidates for these studies. Cohn and Hu (1978) studied the inorganic pyrophosphatase–catalyzed exchange of $P^{18}O_4^{3-}$ with $H_2{}^{16}O$, using ^{31}P NMR at 24.3 MHz. The data were not resolved enough for quantitation but clearly showed the progressive substitutions of ^{16}O into phosphate during the course of this reaction. Van Etten and Risley (1978) examined the same process under catalysis by human prostatic acid phosphatase, by observing the 40-MHz ^{31}P NMR spectrum. Through use of a curve resolver, they were able to assess the isotopic composition of phosphate as a function of time and thus were able to extract an overall rate constant for the catalysis. This number compares favorably with that from an analysis of the isotopic distribution in P_i by mass spectrometry. When the relative quantity of each isotopomer is plotted as a function of time, the data coincide with a model (Eargle *et al.*, 1977) for a random, noncoupled exchange mechanism. The technique has similarly been applied to the myosin-catalyzed P_i–H_2O exchange by Trentham (Webb *et al.*, 1978).

5.2. Phosphorylases and Phosphohydrolases

The direction of bond cleavage in the phosphorolysis reversibly catalyzed by purine-nucleoside phosphorylase was determined by the ^{18}O-shift method. Jordan *et al.* (1979) reported that when an equilibrium mixture of inosine, hypoxanthine, ribose 1-phosphate and $P^{18}O_4^{3-}$ is incubated with this enzyme, ribose 1-phosphate becomes enriched with the $^{18}O_4$ species at the expense of the original $^{16}O_4$ material. Simultaneously, the $P^{18}O_4^{3-}$ peak diminishes in intensity and the $P^{16}O_4^{3-}$ signal develops. This experiment proved that the phosphorolysis involves C—O bond scission and is likely a simple in-line displacement. These workers verified that this enzyme does not catalyze $P^{18}O_4^{3-}$–H_2O exchange in the absence of other substrates, indicating that a phosphoryl enzyme does not form. The displacement need not be concerted, however, since the enzyme catalyzes exchange of $^{32}PO_4^{3-}$ into ribose 1-phosphate in the absence of nucleotide, implicating a stepwise dissociative mechanism (ionic or ion-pair mechanisms).

Gerlt and Wan (1979) have used the incorporation of ^{18}O from solvent water to examine the stereochemistry of hydrolysis of cyclic 2′,3′-UMPS by a nonspecific phosphohydrolase (Cd^{2+} form) isolated from *Enterobacter aerogenes*. When the cyclic endo UMPS(S) is hydrolyzed eyzymatically in H_2O (20% ^{18}O), the product is almost exclusively 3′-UMP(S) whose ^{31}P resonance shows a second peak ($\Delta\delta$ 0.034 ppm), corresponding to the incursion of a single (20%) ^{18}O. Chemical recyclization by treatment with diethylphosphochloridate in pyridine produces both exo and endo cyclic diesters, whose ^{31}P resonances are easily resolved in the NMR spectrum (obviating the need for a chromatographic separation). Since the chemical cyclization is known from Lowe's work to involve a similar in-line displacement, the endo recyclization product should expel the ^{18}O label provided that hydrolysis involves a single, in-line displacement by water. The endo product spectrum indeed shows no ^{18}O label, whereas the exo has maintained its enrichment, confirming the initial in-line attack. This deduction is reminiscent of the study by Usher *et al.* (1970) of the cUMP(S) hydrolysis by ribonuclease A, for which the same result obtained, although those experiments involved chromatographic separation of the diastereomers followed by chemical degradation, derivatization, and mass spectrometry. The relative ease of observing directly whether the ^{18}O label is incorporated and expelled using ^{31}P NMR certainly heightens the appeal of this method.

The ^{17}O effect, that is, extreme line broadening of ^{31}P signals when ^{17}O is directly bonded, can be applied to these kinds of problems. The incorporation of an ^{17}O label is manifested as a diminution of intensity (peak height) for the affected ^{31}P signals. Several advantages accrue from the ^{17}O method. Since the experiment does not demand high resolution (as required to observe the ^{18}O shift), extremely high magnetic fields are not necessary. Since broader signals can be tolerated, exhaustive field shimming is not necessary, and lower digital resolution can be used. The latter condition permits acquisition of larger sweep widths, permitting simultaneous monitoring of all ^{31}P species. Solution pH and ionic content are for the same reason not restrictive; thus the ^{17}O method may be preferable for continuous analysis of a reaction. This certainly is desirable when the investigator is limited to a small amount of material. The relatively high cost and low percentage enrichment of the ^{17}O water commercially available ($\sim$50% atom ^{17}O; $175.00/0.5 g, 1982 price), at present limits the attractiveness of this method.

5.3. Acetyl CoA Synthetase

Tsai *et al.* (1980) have used the line-broadening effect of ^{17}O in ^{31}P NMR biochemical studies. In an elegant example, Tsai (1979) investigated the mechanism of the intermediate acetate adenylylation by acetyl CoA synthetase, which employs the *B* diastereomer of ATP(αS) (not the *A* form). The *B* form is known to have the R_p configuration at P_α.

He carried out the reaction of *B* ATP(S) with ^{17}O-labeled acetate and coenzyme A and analyzed the product AMP(S) to determine whether the oxygen derived from acetate occupies the pro-*R* or pro-*S* position. As illustrated in Scheme 9, a single inversion of configuration at P_α of *B* ATP(S) would leave the ^{17}O at the pro-*R* position of AMP(S).

Taking advantage of the existing methodology for converting AMP(S) to the *A* form of ATP(αS), which involves necessarily the phosphorylation of the pro-*R* oxygen, he reasoned that if the ^{17}O resided at the pro-*R* site, it would be incorporated into the α,β bridge of the final *A* ATP(S). This would result in equally diminished intensities for both the α- and β-phosphorus resonances. The result of retention (i.e., intermediacy of an adenylylenzyme) would place the ^{17}O at the pro-*S* site of AMP(S), and the final ATP(αS) would show decreased intensity only for P_α.

He found that the integrated intensities of *both* the P_α and P_β resonances

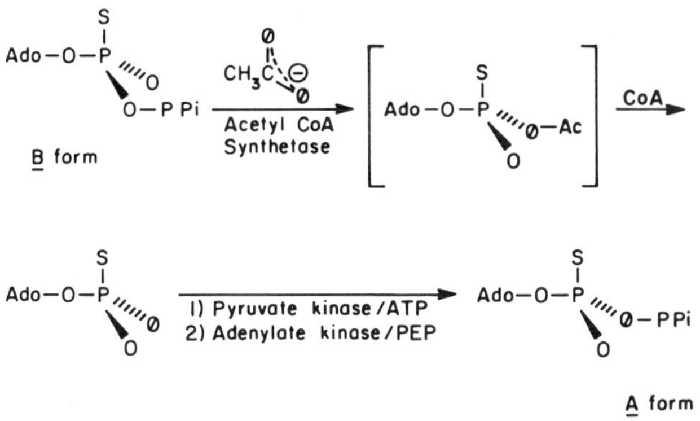

Scheme 9. Stereochemical course of the intermediate adenylylation of acetate by acetyl CoA synthetase, as determined by Tsai (1979).

decrease by 20% in the product ATP(αS). This amount corresponds to the original enrichment of ^{17}O in the acetate, indicating that inversion of P_α is the stereochemical fate of ATP in this adenylylation.

6. EXCHANGE STUDIES

A particularly attractive application of the ^{18}O isotopic-shift phenomenon is the study of reversible phosphoryl transfer between ATP and various substrates. When a substrate or an enzymatic nucleophile attacks the terminal phosphoryl group, one recognizes that the β phosphorus of the putative ADP is potentially torsionally symmetrical, that is, if rotation about the $P_\alpha—O—P_\beta$ bond is not restricted, the reverse displacement to reform ATP is equally likely to involve any of the three equivalent P_β oxygens. Selectively ^{18}O-labeled ATP, for example, C, would incorporate an ^{18}O into the β,γ bridge. This would be detectable in the ATP released from the enzyme through either NMR or mass-spectral analysis. Similarly D would show a gradual depletion of ^{18}O from P_γ to P_β. The process has been termed *positional isotope exchange* or PIX by Rose (1978). The experimental protocols for mass-spectral (Midelfort and Rose, 1976) and NMR observation of the

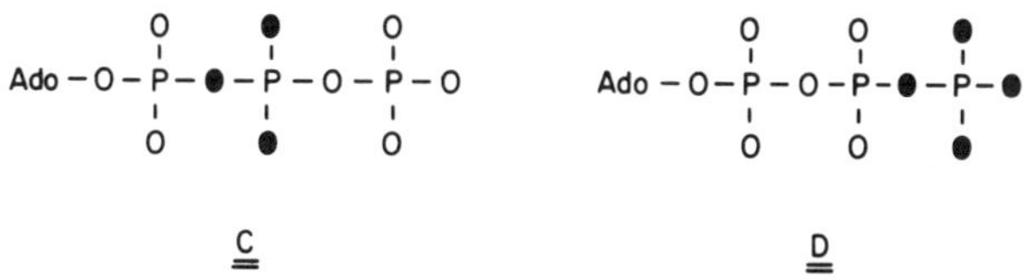

effect have been reported. To observe the PIX described, release of ATP must occur at a rate faster than the overall release of phosphorylated product. Both the detection of PIX and measurement of its time course (or rate of approach toward equilibrium scrambling) can provide useful information on the mechanistic role of ATP.

6.1. Pyruvate Kinase

Lowe and his co-workers have contributed substantially to the development and use of the PIX–NMR method. Using [α,β-^{18}O;β-$^{18}O_2$]ATP, they studied the interaction of ATP with rabbit muscle pyruvate kinase (Lowe and Sproat, 1978a,b). ^{18}O scrambling into the β,γ bridge was observed when ATP and the enzyme were incubated with pyruvate, oxalate (a competitive inhibitor), and no second substrate. This last would suggest either that a phosphoenzyme is formed or that ATP binds, then dissociates to the elusive metaphosphate and ADP. Since two lines of evidence, (1) lack of ^{32}P labeling by [$\gamma^{32}P$]ATP and (2) absence of back reaction with phosphoenolpyruvate, tend to discount an E—P intermediate, these workers argue for the $S_N1(P)$ dissociative (metaphosphate) pathway.

6.2. Creatine Kinase

A similar investigation of creatine phosphorylation by rabbit muscle creatine kinase (Lowe and Sproat, 1980) showed isotope scrambling in ATP only in the presence of the actual substrate, creatine. This suggests that the phosphorylation is reversible but only when all the substrates are bound, and that it involves a direct associative [$S_N2(P)$] displacement by creatine.

6.3. ATPase

Boyer and co-workers used the ^{31}P isotope shift to study the chloroplast- and light-promoted synthesis of ATP from ADP and $P^{18}O_4^{3-}$ (Hackney *et al.*, 1978). They coupled ATP production to the hexokinase phosphorylation of glucose and examined the ^{31}P spectrum of P_i and the product glucose 6-phosphate. At low ADP levels, extensive exchange of ^{16}O from water into the glucose 6-phosphate (i.e., into P_γ of ATP) was observed. At high levels of ADP, the glucose 6-phosphate showed nearly complete ^{18}O labeling. In neither case was loss of ^{18}O from P_i detected. Since ADP is known to control the rate of product (ATP) release in the oxidative phosphorylation, they interpreted their result to mean that slow release of ATP permits exchange with solvent to occur. Rapid ATP release in the presence of high ADP levels would preclude this exchange.

7. KINETIC CONSIDERATIONS

We have discussed several types of reactions in which enzyme-bound phosphorylated intermediates may be involved in the reaction mechanism. The

PIX experimental strategy is ideally suited to test for formation of such intermediates. However ^{31}P NMR detection of PIX alone does not allow one to establish whether an important kinetic criteria is met, that is, whether the rate of formation and breakdown of these intermediates is fast enough to be considered in the normal reaction pathway. To appreciate this, we will develop the mathematical expressions for PIX and then include an additional experimental method, rapid-quench kinetics, to demonstrate that a combination of these techniques permits one to evaluate all rate constants for formation of an enzyme-bound intermediate. The combination of kinetic and chemical (PIX) methods constitutes proof for the involvement of one or more intermediates in the normal reaction mechanism.

Consider the following simplified kinetic scheme

$$\mathrm{E} \underset{k_2}{\overset{k_1\mathrm{S}}{\rightleftharpoons}} \mathrm{ES} \underset{k_4}{\overset{k_3}{\rightleftharpoons}} \mathrm{EP} \overset{k_5}{\rightleftharpoons} \mathrm{E} + \mathrm{P} \tag{3}$$

where E is free enzyme, S is the substrate that can undergo reaction to produce PIX, and P is the product. In the steady state, under initial velocity conditions, there is insufficient P present, so no reversal from products need be considered.

A general reaction involving ATP is depicted in Equation (4). Since

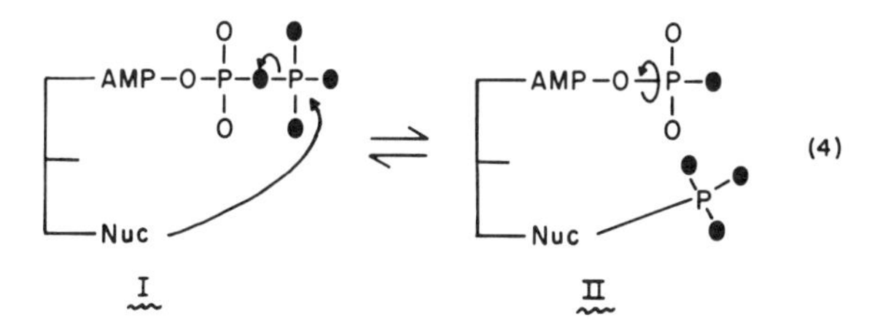

observation of PIX in this reaction involves detection of an ^{18}O shift in the ^{31}P NMR spectrum of ATP in solution, the chemical reaction involves conversion of **I** to **II** followed by reversal and release of ATP from **I.** The PIX is dependent upon rotation of the P_β on the enzyme in **II,** and thus the maximum statistical amount of ^{18}O in the P_β nonbridge oxygens in **I** is $\frac{2}{3}$ of that originally in the β,γ bridge position. In reality **I** consists of molecules that may have, but not necessarily must have, undergone PIX. It follows that the "rate" of PIX is determined by the net rate constant k_4' for reversal of the reaction from EP, and this rate expression is given in Equation (5). The evaluation of net rate constants follows the method of Cleland (1975).

$$V_{\mathrm{PIX}} = k_4'[\mathrm{EP}] = \frac{k_2 k_4[\mathrm{EP}]}{k_2 + k_3} \tag{5}$$

In Equation (4), *Nuc* can be an amino acid residue involved in intermediate phosphorylation of the enzyme, it can be another substrate, or it can be H_2O. Specific examples will be given later for many of these possibilities.

Using the steady-state assumption for the reaction in Equation (3), one can show that

$$\frac{d[\text{EP}]}{dt} = k_3(\text{ES}) - (k_4 + k_5)[\text{EP}] = 0 \tag{6}$$

and

$$k_3[\text{ES}] = (k_4 + k_5)[\text{EP}] \tag{7}$$

$$E_T = \text{ES} + \text{EP} \tag{8}$$

when ATP and Nuc are at saturating concentrations and E_T is the total enzyme concentration. Rearrangement and combination of equations leads to

$$[\text{EP}] = \frac{k_3(\text{E}_T)}{k_3 + k_4 + k_5} \tag{9}$$

and therefore

$$\frac{V_{\text{PIX}}}{\text{E}_T} = k_x = \frac{k_2 k_3 k_4}{(k_2 + k_3)\,(k_3 + k_4 + k_5)} \tag{10}$$

The above derivation was used to obtain the equations given in Raushel and Villafranca (1979) and Meek *et al.* (1982). The limiting condition given by Rose (1979) is for the case where $k_5 \ll k_3, k_4$, and rearrangement of Equation (10) gives his expression

$$\frac{\text{E}_T}{\text{V}_{\text{PIX}}} = \frac{1}{k_2} + \frac{1}{k_3} + \frac{(k_2 + k_3)}{k_2 k_4} \tag{11}$$

With PIX data alone, only ratios of rate constants for partitioning of **I** and **II** can be determined, but all the rate constants can be evaluated if these data are combined with rapid-quench experiments.

For the reaction given in Equation (3), rapid-quench experiments would consist of following radioactive ^{32}P label from [^{32}P$_\gamma$]ATP into product. A scheme such as that given in Equation (3) predicts a ''burst'' of ^{32}P due to formation of the EP complex, with the amplitude of the burst dependent on the magnitude of the individual rate constants. In practice, at various times the reaction is stopped by acid quench, or another method, followed by separation of unreacted [^{32}P$_\gamma$]ATP and the ^{32}P$_\gamma$ that has been transferred. An exponential rise in formation of EP is expected followed by the steady-state rate of formation of P from EP. The appropriate equations are

$$\frac{V}{\text{E}_T} = \frac{k_3 k_5}{k_3 + k_4 + k_5} \tag{12}$$

$$\text{Transient burst amplitude} = \frac{\beta}{\text{E}_T} = \frac{k_3(k_3 + k_4)}{(k_3 + k_4 + k_5)^2} \tag{13}$$

$$\text{Transient burst rate} = \lambda = k_3 + k_4 + k_5 \tag{14}$$

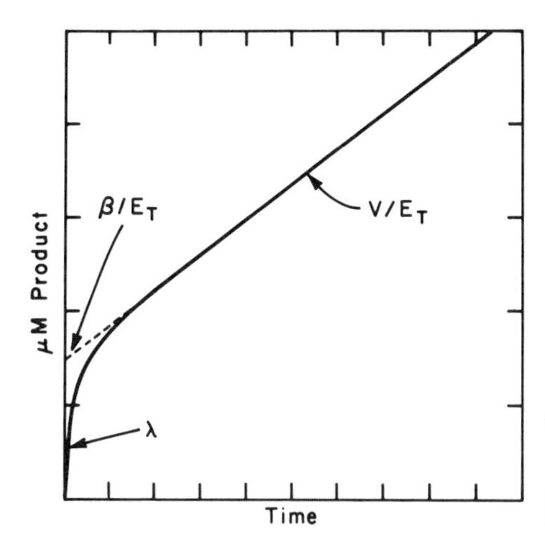

Figure 7. Representation of "burst" kinetics showing the transient burst rate λ, transient burst amplitude β/E_T, and steady-state rate V/E_T.

Figure 7 shows these various phases of the reaction. Since the three equations [(12)–(14)] have three unknowns, all three rate constants can be evaluated. With values for k_3, k_4, and k_5 and a value for V_{PIX}/E_T from ^{31}P NMR experiments, k_2 can be evaluated from Equation (10). Thus the combination of these two methods provides values for all the rate constants of interest. Under certain conditions k_1 can also be evaluated, provided the dissociation constant for ATP is known under the conditions employed for the PIX and kinetic experiments.

8. KINETIC STUDIES REPORTED

Specific examples of the application of PIX and rapid-quench experiments follow.

8.1. Carbamyl Phosphate Synthetase

The reaction that *Escherichia coli* carbamylphosphate synthetase catalyzes is

$$2ATP + HCO_3^- + \text{L-Gln} \rightarrow 2ADP + \text{carbamyl-P} + P_i + \text{L-Glu} \quad (15)$$

Our laboratory has been investigating the formation of intermediates in this reaction for several years (Raushel and Villafranca, 1979,1980). Early work by Anderson and Meister (1965,1966) led to the suggestion of the following sequence of reactions catalyzed by this enzyme:

$$ATP + HCO_3^- \rightleftharpoons ADP + {}^-O_2COPO_3^{2-} \quad (16)$$

$$O_2COPO_3^{2-} + NH_3 \longrightarrow NH_2CO_2^- + P_i \quad (17)$$

$$NH_2CO_2^- + ATP \rightleftharpoons ADP + NH_2COPO_3^{2-} \quad (18)$$

Since there are two distinct reactions that utilize ATP, PIX experiments are ideally suited to test for the formation of intermediates.

The first set of experiments that was conducted involved establishing the order of substrate addition and product release. Equation (19) shows the

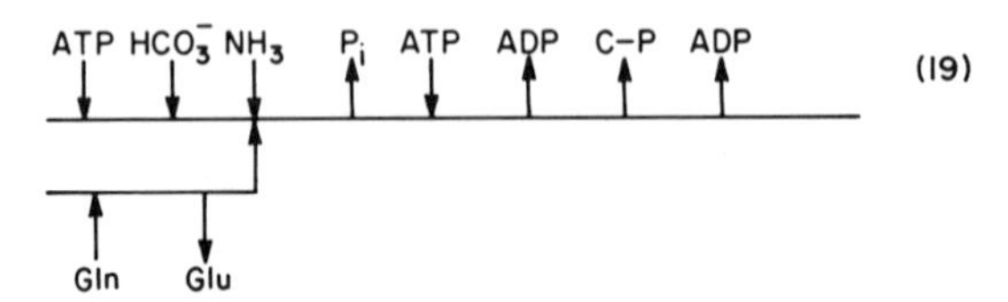

result of our steady-state kinetic experiments (Raushel *et al.*, 1978). There is an ordered addition of ATP, HCO_3^-, and NH_3 (or the NH_3 derived from Gln) followed by P_i release. Up to this point we only know that a bond has to be broken prior to P_i release, but we do not know if the reaction in Equation (16) takes place before or after NH_3 addition. The next set of experiments involved rapid-quench and PIX experiments to test for formation of carboxyphosphate ($^-O_2COPO_3^{2-}$) and if it is formed, to establish whether it is formed fast enough to be along the normal catalytic reaction pathway.

Figure 8 shows the results of rapid-quench experiments for the ATPase reaction catalyzed by carbamyl phosphate synthetase in the absence of NH_3 [Equation (16)]. The transient rate constant λ is 4.5 s^{-1}, and the steady-state rate is 0.2 s^{-1}. Since the reaction is characterized by a "burst" in the formation of P_i, the data are consistent with formation of an intermediate such as carboxyphosphate followed by rate-determining release of P_i or $^-O_2COPO_3^{2-}$. The rate constants evaluated from this experiment are k_3 = 4.2 s^{-1}, k_4 = 0.10 s^{-1}, and k_5 = 0.21 s^{-1}.

The rapid-quench experiments strongly suggest that PIX should be ob-

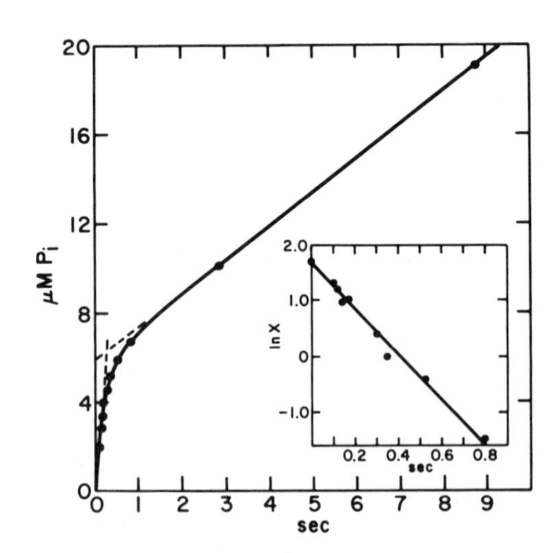

Figure 8. Burst kinetics for the ATPase reaction catalyzed by carbamyl-P synthetase (Raushel and Villafranca, 1979).

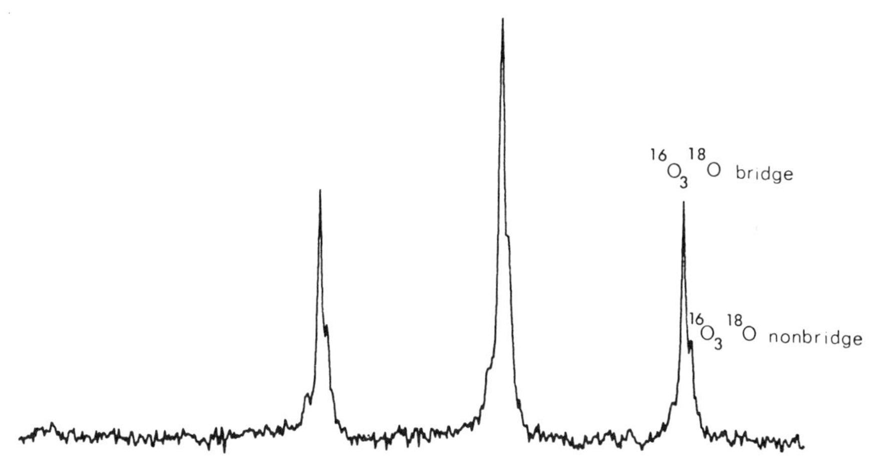

Figure 9. The ^{31}P NMR spectrum of the P_β of ATP after a PIX experiment with carbamyl-P synthetase (Raushel and Villafranca, 1980).

served in ATP. Experimentally this is verified as shown in Figure 9. Since the ATP utilized for this experiment had >90% ^{18}O in all four oxygens of the P_γ of ATP, monitoring the P_γ of ATP is the easiest way to follow exchange of ^{16}O for ^{18}O in the β,γ bridge position. Figure 9 shows that the same exchange can be followed by observing the P_β resonance, but the small chemical-shift difference between $^{16}O_3^{18}O$ bridge (III) and $^{16}O_3^{18}O$ nonbridge (IV) species is

```
          O   O   ●                  O   O   ●
          |   |   |                  |   |   |
Ado-O-P-O-P-●-P-●         Ado-O-P-O-P-O-P-●
          |   |   |                  |   |   |
          O   O   ●                  O   ●   ●

         III                              IV

               ● = ¹⁸O,  O = ¹⁶O
```

insufficient for quantitative experiments at a magnetic field strength of 4.7 T. These PIX experiments for the ATP reaction resulted in evaluation of k_x. Utilizing the results from rapid-quench experiments and combining Equations (10) and (12), one arrives at Equation (20), from which evaluation of $k_2 = 120\ s^{-1}$ was made:

$$\frac{k_x}{k_{cat}} = \frac{k_2 k_4}{k_5(k_2 + k_3)} = 0.46 \tag{20}$$

The numerical value of k_2 compared with k_3 suggests that at short reaction times ($\leqslant$ 50 ms), the ATP dissociates rapidly and cannot be trapped. But at longer reaction times ($\geqslant$ 1.5 s), essentially all the ATP can be trapped as P_i since at least one turnover has occurred. This was also experimentally verified in pulse-chase experiments using [γ-^{32}P]ATP (Raushel and Villafranca, 1979).

Thus from the combination of PIX and rapid-quench experiments, the formation of carboxyphosphate ($k_3 = 4.2\ s^{-1}$) is at least partially rate deter-

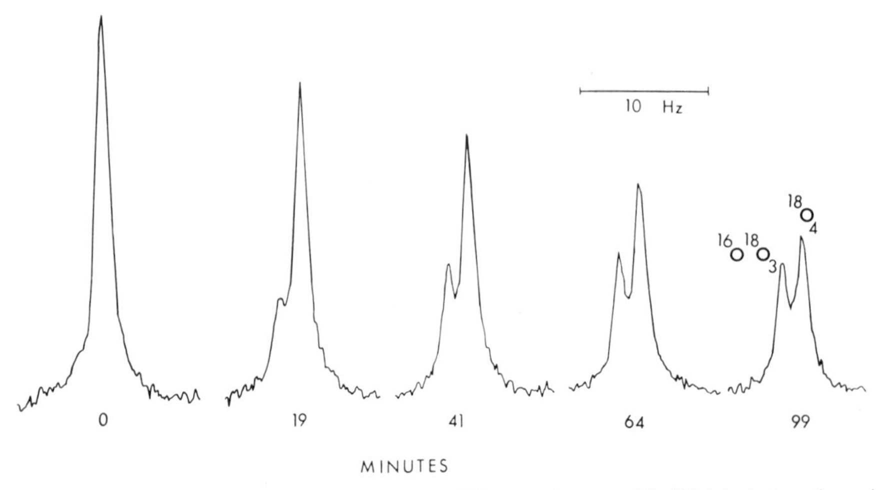

Figure 10. The ^{31}P NMR spectra during a PIX experiment with ^{18}O-labeled carbamyl phosphate and ADP [equation (18)] and carbamyl-P synthetase (Raushel and Villafranca, 1980).

mining in the overall reaction (plus Gln) for which $k_{cat} = 3.1\ s^{-1}$. These data therefore show that carboxyphosphate is an intermediate in the catalytic reaction, since its rate of formation meets the minimum criterion of kinetic competence.

The overall reaction given in Equation (19) also predicts that if carboxyphosphate is formed prior to NH_3 addition (from NH_3 or Gln), the addition of Gln should suppress the PIX rate. This arises because observation of PIX requires complete reversal to E·ATP, from which ATP is released into solution and monitored by ^{31}P NMR. The prediction of Equation (19) is borne out, since k_x is reduced by a factor of ~8 in the presence of Gln. Thus the PIX, rapid-quench, pulse-chase, and steady-state kinetic data are all internally consistent.

Another intermediate, proposed by Anderson and Meister (1965,1966) [Equations (17) and (18)], is carbamate, $NH_2CO_2^-$. We designed a PIX experiment to test for its formation based on ^{18}O–^{16}O positional isotope exchange in carbamyl-P (Raushel and Villafranca, 1980). Structures V and VI show how the ^{18}O–^{16}O shift in the ^{31}P NMR spectrum of carbamyl-P can be used to monitor formation of carbamate.

$$\begin{array}{lcl}
\text{ADP} + NH_2\overset{O}{\overset{\|}{C}}-\bullet-\overset{\bullet}{\overset{|}{\underset{\underset{\bullet}{|}}{P}}}-\bullet & \rightleftharpoons & \text{ATP} + NH_2\overset{O}{\overset{\|}{C}}-\bullet^- \\
\text{V} & & \qquad\downarrow\uparrow \\
\text{ADP} + NH_2\overset{\bullet}{\overset{\|}{C}}-O-\overset{\bullet}{\overset{|}{\underset{\underset{\bullet}{|}}{P}}}-\bullet & \rightleftharpoons & \text{ATP} + NH_2\overset{\bullet}{\overset{\|}{C}}-O^- \\
\text{VI} & &
\end{array} \qquad (21)$$

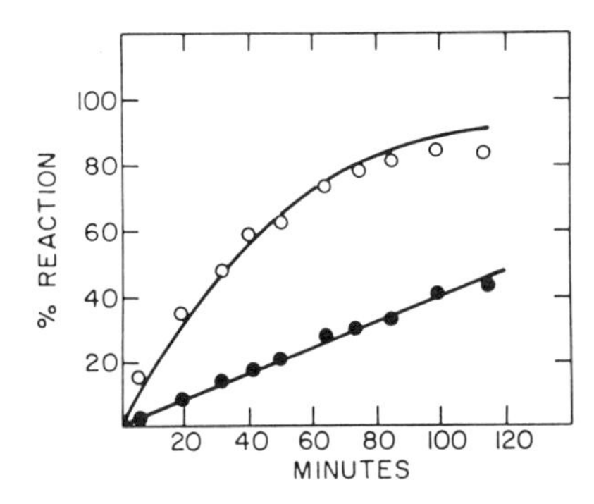

Figure 11. Time course for the experiment shown in Figure 10. Open circles represent the ^{31}P NMR data for PIX, and the closed circles show the time course for synthesis of ATP in the reverse reaction given in equation (18).

In this scheme rotation about the N—C bond in enzyme-bound carbamate is required to observe a maximum of 50% isotope exchange in the C—O—P bond of carbamyl-P. Figures 10 and 11 present data showing that the prediction is verified by experiment. Figure 11 also shows the steady-state rate of formation of ATP in the reaction depicted in Equation (18). The PIX rate is approximately four times faster than the rate of ATP formation, and equilibration of ^{18}O and ^{16}O in carbamyl-P almost reaches the theoretical maximum of 50% (Figures 10 and 11). Thus carbamate is also an intermediate in the overall reaction catalyzed by *E. coli* carbamyl-P synthetase.

8.2. Glutamine Synthetase

The enzyme from *E. coli* carries out the reaction in Equation (22):

$$\text{ATP} + \text{L-Glu} + \text{NH}_3 \longrightarrow \text{ADP} + \text{P}_i + \text{L-Gln} \tag{22}$$

In addition to this biosynthetic reaction, the enzyme also catalyzes a glutamate-dependent ATPase reaction in the absence of NH_3 [Equation (23)]:

$$\text{ATP} + \text{L-Glu} \longrightarrow \text{ADP} + \text{P}_i + \text{pyrrolidonecarboxylate} \tag{23}$$

Data from Meister's laboratory and others (Krishnaswamy *et al.*, 1962; Gass and Meister, 1970; Tsuda *et al.*, 1971; Todhunter and Purich, 1975) led to a mechanism involving the intermediate formation of γ-glutamyl phosphate [Equation (24)]:

$$\text{ATP} + \text{Glu} \longrightarrow \text{ADP} + \gamma\text{-Glu-P} \tag{24}$$

This reaction serves to activate the γ carboxyl of glutamate to facilitate NH_3 attack to complete the reaction. The first description of the PIX method (Midelfort and Rose, 1976) tested for formation of γ-Glu-P with this enzyme, and the results (in the absence of NH_3) were consistent with the reaction in Equation (24). However, as mentioned before, PIX gives ratios of rate constants, and the question of whether γ-Glu-P is formed fast enough to be considered an integral part of the overall mechanism awaited studies from our laboratory.

We began our studies (Meek and Villafranca, 1980) with a determination of the steady-state kinetic mechanism, which showed an ordered addition of substrates and of product release [Equation (25)]. These kinetic data do not

(25)

provide evidence for formation of γ-Glu-P, nor do they tell us whether this intermediate is formed or whether it occurs before or after NH_3 binds. Our work addressed these questions using rapid-quench kinetics and PIX experiments (Meek *et al.*, 1982).

Figure 12 shows the results of rapid-quench studies (at 25°C) for the ATPase reaction in the absence of NH_3. There is a burst in the formation of P_i with the transient burst rate $\lambda = 10.3\ s^{-1}$ and steady-state rate equal to $0.017\ s^{-1}$. Using equations (12)–(14) to analyze these data, $k_5 = 6.9\ s^{-1}$, $k_6 = 3.4\ s^{-1}$, and $k_7 = 0.017\ s^{-1}$. From PIX data shown in Figure 13, a value of $k_x = 1.8\ s^{-1}$ was calculated. Comparison of k_5 with k_{cat} ($14\ s^{-1}$) demonstrates that this step is *not* equal to or faster than the turnover number determined in the presence of NH_3. Therefore, evidence that γ-Glu-P is formed in the ATPase reaction cannot be used to conclude that this intermediate is involved in the overall reaction shown in Equation (25).

In order to demonstrate whether γ-Glu-P is an intermediate in the biosynthetic reaction [Equation (22)], rapid-quench and PIX experiments with NH_3 present were required. The reaction is more conveniently followed at 10°C in the presence of NH_3, and Figure 14 shows the results of these experiments. Again, a burst is observed ($\lambda = 88\ s^{-1}$) followed by a steady-state rate of $4\ s^{-1}$. From these data and Equations (12)–(14), $k_{11} = 54\ s^{-1}$, $k_{12} = 28\ s^{-1}$, and $k_{13} = 9\ s^{-1}$. Thus k_{11}, which is the rate of formation of γ-Glu-P, is faster than k_{cat} ($4\ s^{-1}$). These data demonstrate (1) that γ-Glu-P is formed in the biosynthetic reaction and (2) that it is formed after NH_3 binds in the

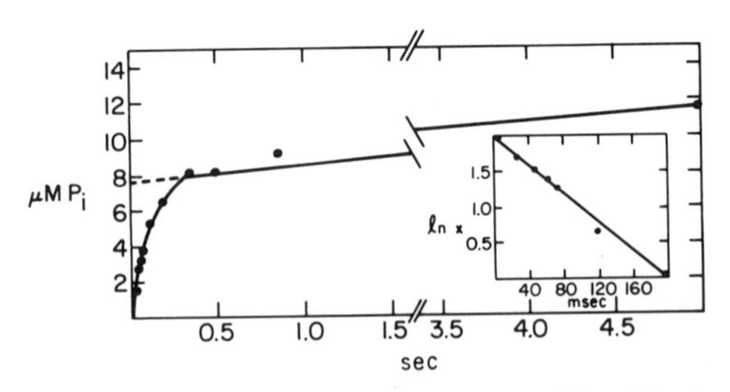

Figure 12. Burst kinetics for the ATPase reaction catalyzed by glutamine synthetase at 25°C (Meek *et al.*, 1982).

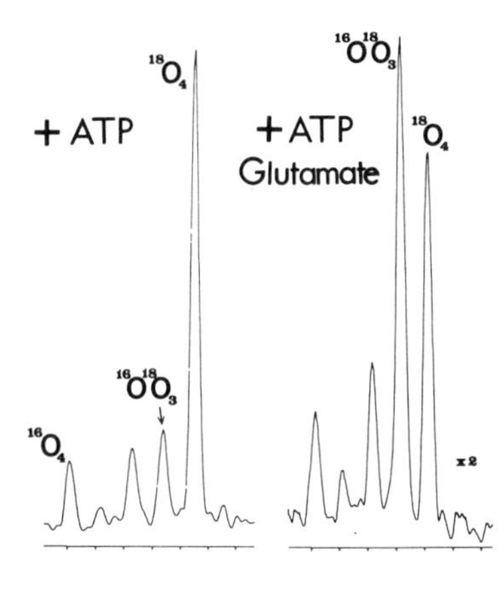

Figure 13. The ^{31}P NMR spectra of one peak of the P_γ doublet of ATP in a PIX experiment with glutamine synthetase, with and without glutamate present.

quaternary E·ATP·Glu·NH_3 complex in a stepwise manner, that is, γ-Glu-P formation follows NH_3 attack to form glutamine. However the reaction could also proceed by the concerted pathway shown in Equation (26) (Wed-

$$\text{ADP}{-}\text{O}{-}\overset{\text{O}}{\underset{\text{O}}{\text{P}}}{-}\text{O} \quad \text{R}{-}\text{C}(\text{=O}){-}\text{O} \leftarrow \text{NH}_3 \rightleftharpoons \text{ADP} \quad \text{O}{-}\overset{\text{O}}{\underset{\text{O}}{\text{P}}}{-}\text{O} \quad \text{R}{-}\text{C}(\text{=O}){-}\text{NH}_2 \qquad (26)$$

$$\text{R} = {-}\text{CH}_2\text{CH}_2\text{CH}(\text{NH}_2)\text{CO}_2\text{H}$$

ler and Boyer, 1972), and PIX would be observed. However, this has been ruled out by the data of Midelfort and Rose (1976) and our work (Meek *et al.*, 1982). The ^{18}O–^{16}O isotopic shift in the ^{31}P NMR spectrum of ATP was

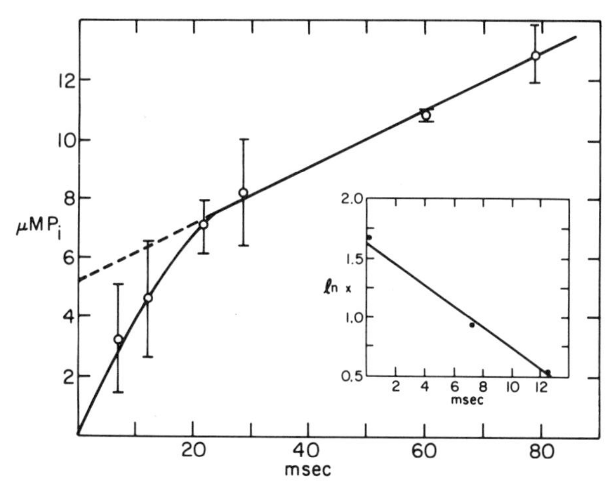

Figure 14. Burst kinetics for the biosynthetic reaction catalyzed by glutamine synthetase (Meek *et al.*, 1982).

used to monitor the consequences of the concerted mechanism [Equation (26)]. Equation (27) predicts that not only is PIX expected from a concerted

(27)

mechanism, but depletion ("washout") of ^{18}O from the P_γ of the reisolated ATP is also expected, since the ^{18}O in P_i would be randomly lost to the γ carboxyl of glutamate. Positional isotope exchange is observed, but the "washout" of ^{18}O in the P_γ of ATP is not found in the reaction catalyzed by glutamine synthetase. Other ^{31}P NMR data from our laboratory demonstrated that the enzyme-bound P_i is free to rotate on the enzyme surface and loses its oxygens indiscriminately to glutamate (Balakrishnan *et al.*, 1978).

Thus for *E. coli* glutamine synthetase, experiments utilizing the ^{18}O–^{16}O isotopic shift on ^{31}P NMR not only were used to suggest the intermediate formation of γ-Glu-P but also to rule out a concerted reaction pathway.

9. KINETIC STUDIES PROPOSED

Several enzyme systems will be considered whose reaction mechanisms can be examined by application of the ^{18}O–^{16}O isotopic shift in the ^{31}P NMR spectrum of phosphorus-containing compounds. Each of these will be examined in turn, and predictions based on the proposed mechanism will be considered for the ^{31}P NMR spectra.

9.1. Guanosine 5′-Phosphate Synthetase

Several groups have conducted preliminary investigations on the mechanism of this reaction (Hartman and Prusiner, 1973; Abrams and Bentley, 1959; Fukuyama, 1966), shown in Equation (28).

$$\text{ATP} + \text{XMP} + \text{NH}_3 \rightleftharpoons \text{AMP} + \text{PP}_i + \text{GMP} \quad (28)$$

Transfer of ^{18}O from XMP to AMP was detected as well as evidence for formation of an adenyl-XMP intermediate. In addition, no ATP–PP_i exchange in the absence of NH_3 could be detected. A simplified mechanism based on these data is given in Equation (29).

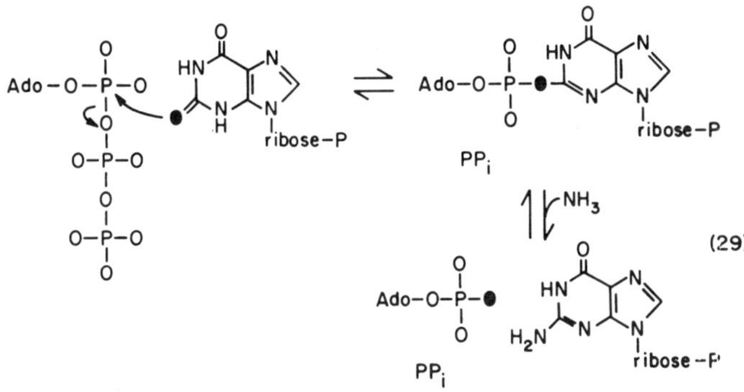

(29)

The fact that suggestive evidence exists for formation of an adenyl-XMP intermediate leads to two types of PIX experiments. With ^{18}O in the α,β bridge position of ATP, PIX might be observed in the absence of NH_3, as shown in Equation (30). The reisolated ATP would show depletion of ^{18}O

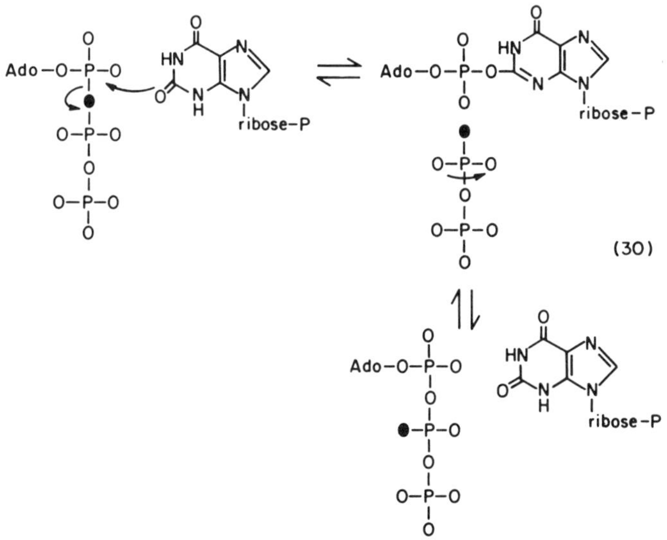

(30)

from the P_α resonance, and a PIX rate could be measured if the mechanism proceeds according to Equation (30). Addition of NH_3 should result in a diminution of the PIX rate.

If ^{18}O were utilized in the XMP, then one additional step beyond that shown in Equation (30) would be required to observe ^{18}O–^{16}O exchange. A possible mechanism depicting this prediction is given in Equation (31). The

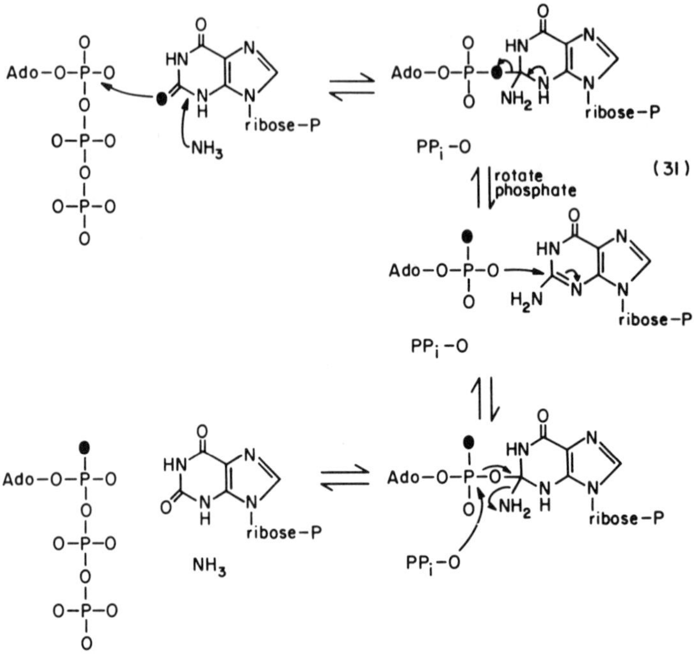

(31)

same result predicted by Equation (31) may occur in the absence of NH_3 if a nucleophilic group participates in catalysis, giving an enzyme-bound XMP derivative that subsequently is attacked by NH_3 to give GMP. The mechanism shown in Equation (31) could also be followed by observing the ^{18}O–^{16}O isotopic shift on ^{13}C NMR if ^{13}C-enriched [^{18}O]XMP were used. The ^{18}O replacement by ^{16}O in XMP is predicted from Equation (31).

9.2. Pyruvate-Phosphate Dikinase

An intriguing enzymatic reaction that involves two different phosphorylated enzyme intermediates is catalyzed by pyruvate-phosphate dikinase. The net reaction involves synthesis of phosphoenolpyruvate from pyruvate, ATP, and P_i as shown in Equation (32).

$$\text{Pyruvate} + \text{ATP} + P_i \rightleftharpoons \text{P-enolpyruvate} + \text{AMP} + PP_i \tag{32}$$

The partial reactions that have been suggested are the result of ^{32}P-labeling studies [Equations (33)–(35); Milner *et al.*, 1978; Milner and Wood, 1972,1976; Reeves *et al.*, 1968].

$$E + \text{ATP} \rightleftharpoons E\text{—}P_\beta P_\gamma + \text{AMP} \tag{33}$$

$$E\text{—}P_\beta P_\gamma + P_i \rightleftharpoons E\text{—}P_\beta + P_\gamma P_i \tag{34}$$

$$E\text{—}P_\beta + \text{pyruvate} \rightleftharpoons E + P_\beta\text{-enolpyruvate} \tag{35}$$

Based on these partial reactions, several PIX experiments suggest themselves as well as several rapid-quench kinetic experiments to establish whether these phosphorylated intermediates are formed rapidly enough to be along the reaction path. ATP labeled in the α,β bridge would be useful to detect PIX in the reaction in Equation (33) as redrawn in Equation (36). If

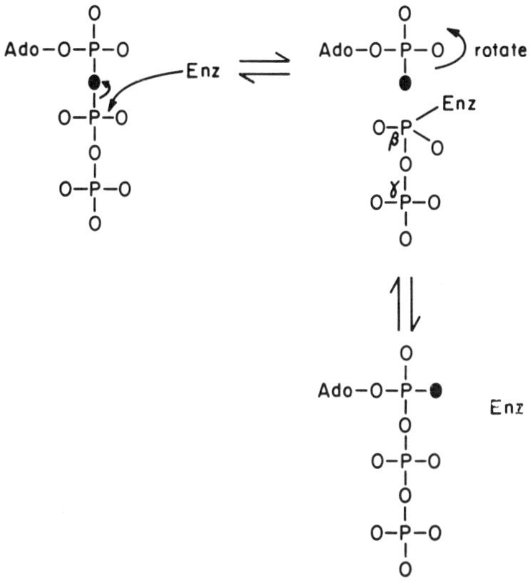

O
|
Ado—O—P—O
|
●
|
O—P—O   —Enz ⇌
|
O
|
O—P—O
|
O

O
|
Ado—O—P—O   rotate
|
●

Enz
O—P$_\beta$—O
|
O
|
O—P$_\gamma$—O
|
O

⇅   (36)

O
|
Ado—O—P—●   Enz
|
O
|
O—P—O
|
O
|
O—P—O
|
O

ATP were labeled in the β,γ bridge position, it could be used to detect PIX in the reaction in Equation (34) shown in more detail in Equation (37).

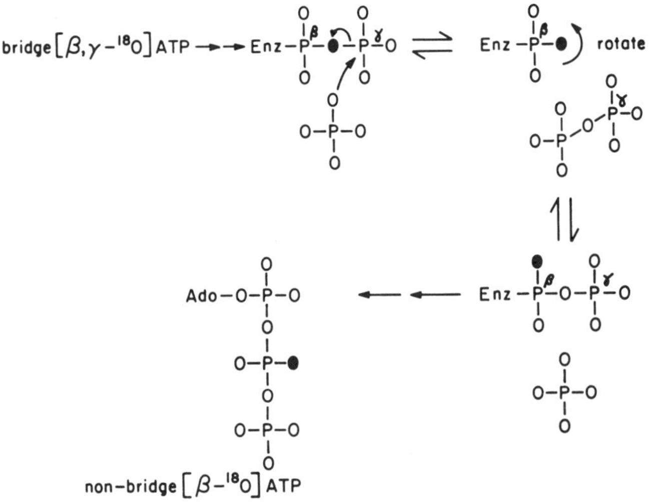

bridge[β,γ-^{18}O]ATP ⟶⟶ Enz—P$_\beta$—●—P$_\gamma$—O ⇌ Enz—P$_\beta$—●   rotate

O
|
O—P—O
|
O

O—P$_\gamma$—O
O—P—O

⇅   (37)

O
|
Ado—O—P—O
|
O
|
O—P—●
|
O
|
O—P—O
|
O
non-bridge [β-^{18}O] ATP

⟵ ⟵ Enz—P$_\beta$—O—P$_\gamma$—O
O
|
O—P—O
|
O

The rate constants for formation of Enz—$P_\beta P_\gamma$ and Enz—P_β could be determined by rapid-quench methods using appropriately labeled [^{32}P]ATP, and these rate constants could be combined with the mathematical expressions for PIX for the reactions in Equations (36) and (37) to establish if these intermediates could be part of the overall mechanism that operates when pyruvate is also present.

9.3. Phosphoenolpyruvate Carboxylase

This enzyme has been isolated from several plant and bacterial sources (Utter and Kolenbrander, 1972) and catalyzes the following reaction:

$$\text{Phosphoenolpyruvate} + HCO_3^- \rightarrow \text{Oxaloacetate} + P_i \qquad (38)$$

The enzyme requires a divalent cation for activity and has a pH optimum of about 8–9. The ^{18}O incorporation from bicarbonate into products in the ratio of 2:1 for oxaloacetate: P_i is the major reason that HCO_3^-, not CO_2, is considered the substrate for the reaction (Maruyama *et al.*, 1966).

Since this reaction is considered irreversible, a concerted mechanism has been proposed [Equation (39); Maruyama *et al.*, 1966].

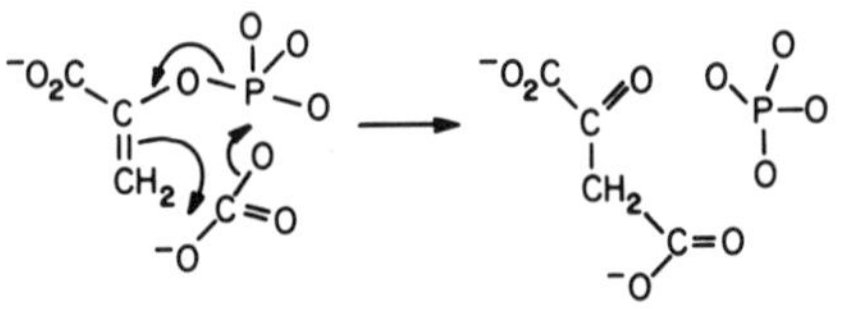

(39)

The reaction as drawn would require unassisted displacement from the sp^2 carbon of HCO_3^-, a process that is not observed. Further, an unprecedented, synchronous, coplanar frontside displacement at phosphorus would be required. Examination of the transition-state structure (or short-lived intermediate) in this process [Equation (40)] reveals that the ‘‘enolate’’ oxygen of

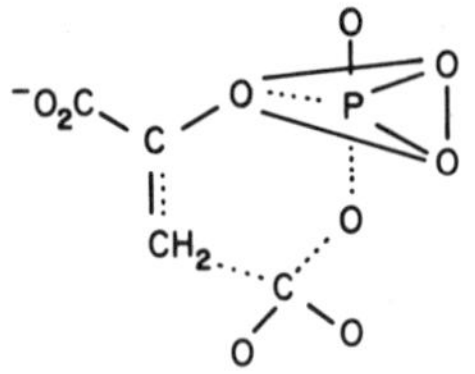

(40)

PEP would have to leave from an equatorial position on pentacoordinate phosphorus, which violates the concepts of reactions of phosphorus compounds developed by Westheimer (1968).

A more likely mechanism is one in which a pentacoordinated phosphorus intermediate is formed that undergoes pseudorotation at phosphorus to expel the enolate oxygen, thus making the reaction a stepwise process.

Another reasonable stepwise mechanism is presented in Equation (41), in

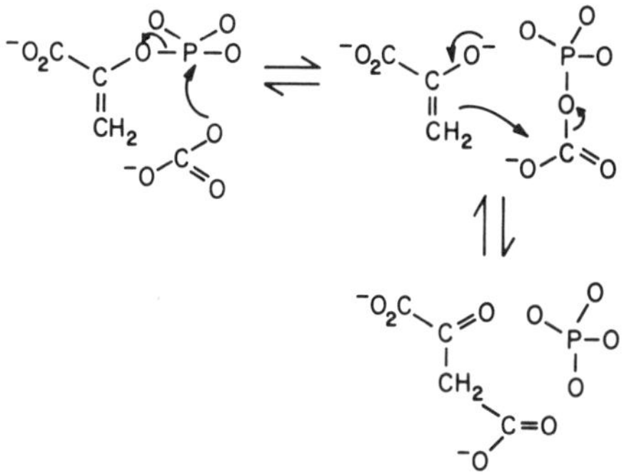

(41)

which HCO_3^- attacks phosphorus, giving carboxyphosphate and enolpyruvate. In a second step, attack of the double bond of enolpyruvate on carboxyphosphate is driven by ketonization to produce the C—C bond and expel P_i. Incorporation of all three oxygens of HCO_3^- would still result from the mechanism in Equation (41) with a 2:1 ratio of oxaloacetate to P_i. Evidence for the stepwise reaction could be obtained using the chiral phosphate strategy.

Using the chiral phosphate approach, a distinction can be made between (1) frontside attack and (2) backside attack on phosphorus [Equation 42)].

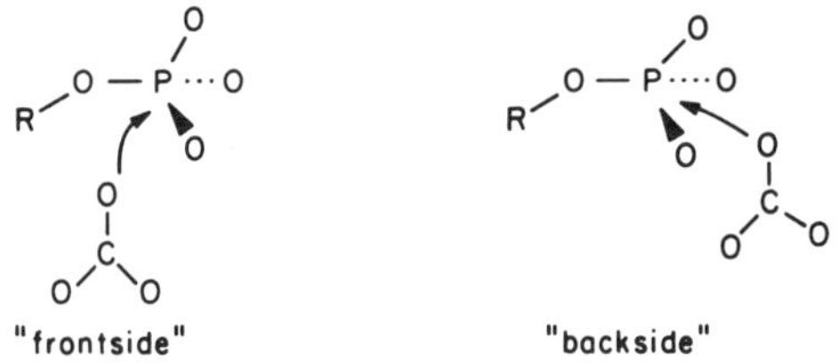

"frontside" "backside" (42)

For the frontside attack by HCO_3^-, two mechanisms will be considered, that is, concerted, and stepwise with pseudorotation. For backside attack two mechanisms, stepwise, and stepwise with pseudorotation, will be depicted. The stereochemical consequences of these mechanisms are given in Equations (43)–(46).

A) Concerted, one step mechanism

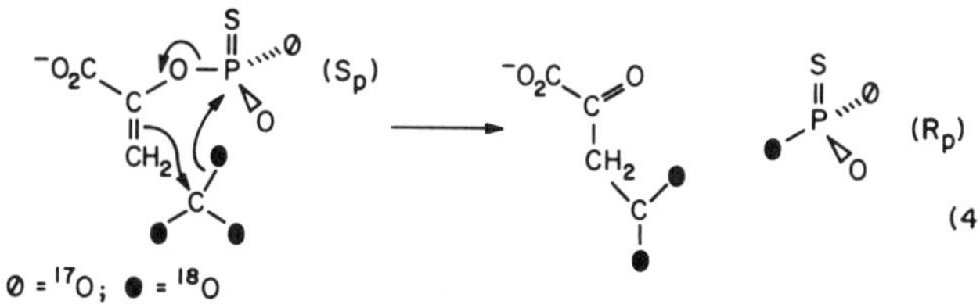

(S_p) → (R_p) (43)

⊘ = ^{17}O; ● = ^{18}O

B) Stepwise mechanism, via pseudorotation

(from S_p) (44) (R_p)

C) Stepwise mechanism, via carboxyphosphate

(S_p) (45) (S_p)

D) Stepwise mechanism, via pseudorotation

(from S_p) (46) (S_p)

The chiral phosphate resulting from frontside attack [Equations (43) and (44)] would have a net retention of configuration (priority rules reverse the absolute configuration), whereas the chiral phosphate from backside attack would have an inversion of configuration at phosphorus relative to starting PEP. Thus the critical position of attack by HCO_3^- on PEP can be resolved by the chiral phosphate approach. If inversion of configuration is found, then the concerted mechanism [Equations (39) and (43)] is eliminated. While this article was in press, Hansen and Knowles (1982) showed that this reaction goes by a stepwise pathway with backside attack.

9.4. Phosphoenolpyruvate Carboxykinase

The enzyme has been isolated from several sources (Foster *et al.*, 1967; Colombo *et al.*, 1978; Brinkworth *et al.*, 1981) and catalyzes the reaction shown in Equation (47).

$$\text{PEP} + CO_2 + \text{GDP} \rightleftharpoons \text{oxaloacetate} + \text{GTP} \qquad (47)$$

The substrate in this reaction seems (based on evidence from earlier work) to be CO_2 and not HCO_3^-, which leads to the mechanism proposed in Equation (48). This mechanism predicts that PIX should be observed in the reverse

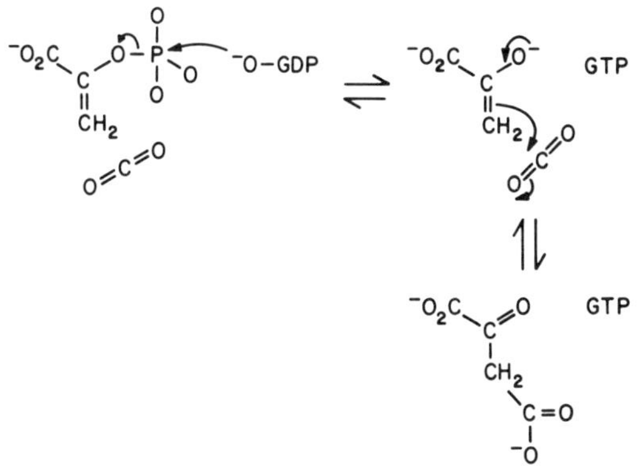

(48)

direction (oxaloacetate plus GTP) if the reaction is reversible at the stage where PEP, CO_2, and GDP are bound. Also, if the reaction proceeds as written, an inversion of configuration at phosphorus would be predicted upon attack of PEP (chiral) by GDP. If instead the first step involves attack by an enzymatic group on PEP, producing a phosphorylated intermediate, retention of configuration would be observed at the P_γ of GTP from this two-step process, after chemical derivation.

9.5. Adenylosuccinate Synthetase

This enzyme catalyzes the formation of adenylosuccinate, a precursor in the formation of AMP. The reaction is given in Equation (49).

$$\text{IMP} + \text{Aspartate} + \text{GTP} \rightleftharpoons \text{Adenylosuccinate} + \text{GDP} + P_i \qquad (49)$$

A mechanism for this reaction has been proposed (Markham and Reed, 1978) that involves formation of two enzyme-bound intermediates [Equation 50)].

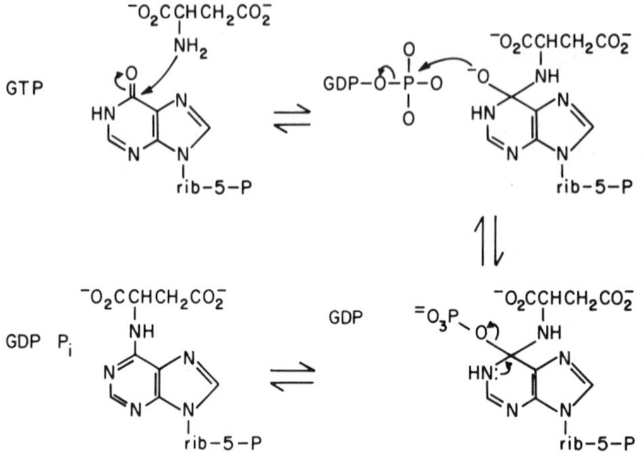

GTP $^{-}O_2CCHCH_2CO_2^{-}$ NH_2 HN N N N rib-5-P ⇌ GDP-O-P-O $^{-}O_2CCHCH_2CO_2^{-}$ NH HN N N N rib-5-P

⇅ (50)

GDP P_i $^{-}O_2CCHCH_2CO_2^{-}$ NH N N N N rib-5-P ⇌ GDP $^{=}O_3P$ O $^{-}O_2CCHCH_2CO_2^{-}$ NH HN N N N rib-5-P

As drawn, this mechanism predicts PIX in GTP only when aspartate is present, since the initial step of the reaction involves formation of aspartyl-IMP. An alternative mechanism for the initial step is written in Equation (51). This scheme has 6-phospho-IMP formed prior to attack by aspartate and

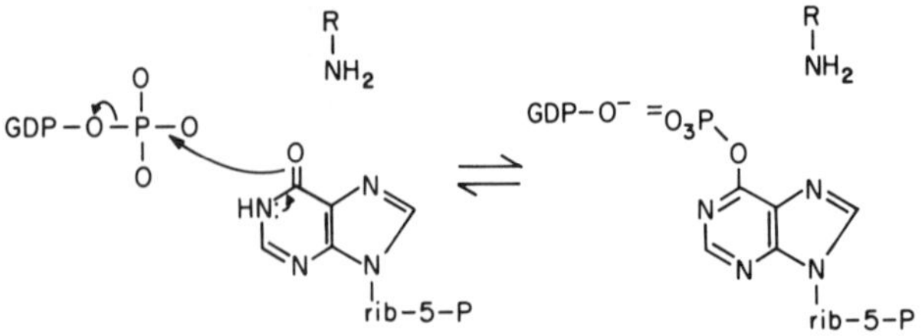

R NH_2 GDP-O-P-O O HN N N N rib-5-P ⇌ GDP-O^- $^{=}O_3P$ O R NH_2 N N N N rib-5-P (51)

predicts PIX in GTP in the absence of aspartate. However, aspartate might be necessary for any reactions to occur. This possibility could be tested by a PIX experiment in the presence of an aspartate analog hadacidin (N-formyl-N-hydroxyglycine) if occupancy of the aspartate site is required for a reaction such as that in Equation (51). In sum, PIX experiments may distinguish between the mechanisms in Equations (50) and (51).

9.6. S-Adenosylmethionine Synthetase

This enzyme catalyzes an unusual reaction of ATP in which the entire tripolyphosphate chain is cleaved [Equation (52)].

$$\text{Ado-P}_\alpha\text{P}_\beta\text{P}_\gamma + \text{L-Methionine} \rightarrow \text{AdoMet} + \text{P}_\alpha\text{P}_\beta + \text{P}_\gamma \qquad (52)$$

The reaction has been studied by Markham *et al.* (1980), and the first step in the reaction involves formation of AdoMet and tripolyphosphate. Subsequently the tripolyphosphate is cleaved to PP_i and P_i. If adenyl-5′-yl imidophosphate (AMP-PNP) is used as a substrate, imidotriphosphate (PPNP) is formed, but it is not cleaved.

At least two experiments could be conducted to establish rates of formation of enzyme-bound intermediates. A PIX experiment with ^{18}O in the 5′ bridge position could be used to determine ratios of rate constants for formation of AdoMet and PPP_i [Equation (53)]. These could be combined

$$\mathrm{R{-}CH_2{-}\bullet{-}P(O)_2{-}O{-}P(O)_2{-}O{-}P(O)_2{-}O}\ (\ddot{S}) \rightleftharpoons \mathrm{R{-}CH_2{-}S^{+}}\ \ \mathrm{\bullet{-}P(O)_2{-}O{-}P(O)_2{-}O{-}P(O)_2{-}O} \rightleftharpoons \mathrm{R{-}CH_2{-}O{-}P(\bullet)(O){-}O{-}P(O)_2{-}O{-}P(O)_2{-}O}\ (\ddot{S}) \qquad (53)$$

with burst kinetics to establish all the rate constants for the initial phase of the reaction. For the second part of the reaction, that is, hydrolysis of PPP_i, incorporation of ^{18}O from water into P_i could easily be detected by ^{31}P NMR. This method could also be used to detect if reversal of this reaction takes place, since reformation of PPP_i may reveal an enrichment of ^{18}O by multiple reversals on the enzyme [Equation (54)]. If reversal of the entire reaction

$$\mathrm{O{-}P_{\alpha}(O)_2{-}O{-}P_{\beta}(O)_2{-}O{-}P_{\gamma}(O)_2{-}O}\ \ \bullet H_2 \rightleftharpoons \mathrm{O{-}P_{\alpha}(O)_2{-}O{-}P_{\beta}(O)_2{-}O}\ \ \mathrm{O{-}P_{\gamma}(\bullet)(O){-}O} \rightleftharpoons \mathrm{O{-}P_{\alpha}(O)_2{-}O{-}P_{\beta}(O)_2{-}O{-}P_{\gamma}(\bullet)(O){-}O}\ \ OH_2 \qquad (54)$$

takes place, enrichment of ^{18}O into the P_{γ} of ATP could also be detected. This enzyme is ripe for study by the methods outlined in this chapter.

ACKNOWLEDGMENTS

It is a pleasure to acknowledge several persons who contributed to some of the work described in this article, in particular, past colleagues F. M. Raushel, T. D. Meek, and M. S. Balakrishnan. Special thanks are also extended to several authors who provided manuscripts prior to publication: G. Lowe, J. A. Gerlt, M. D. Tsai, F. Eckstein, and D. R. Trentham. Partial financial support is acknowledged from the NIH and NSF, and the American

Heart Association is acknowledged for the award of an Established Investigatorship to J. J. Villafranca.

REFERENCES

Abbott, S. J., S. R. Jones, S. A. Weinman, and J. R. Knowles (1978), *J. Am. Chem. Soc.* **100,** 2558.

Abrams, R., and M. Bentley (1959), *Arch. Biochem. Biophys.* **79,** 91.

Anderson, P. M., and A. Meister (1965), *Biochemistry* **4,** 2803.

Anderson, P. M., and A. Meister (1966), *Biochemistry* **5,** 3157.

Balakrishnan, M. S., T. R. Sharp, and J. J. Villafranca (1978), *Biochem. Biophys. Res. Commun.* **85,** 991.

Baraniak, J., K. Lesiak, M. Sockaki, and W. J. Stec (1980), *J. Am. Chem. Soc.* **102,** 4533.

Bernheim, R., and H. Batiz-Hernandez (1967), *Prog. NMR Spectrosc.* **36,** 63.

Bicknell, R., P. M. Cullis, R. L. Jarvest, and G. Lowe (1982), *J. Biol. Chem.* **257,** 8922.

Brinkworth, R. I., R. W. Hanson, F. A. Fullin, and V. L. Schramm (1981), *J. Biol. Chem.* **256,** 10795.

Bryant, F. R., and S. J. Benkovic (1979), *Biochemistry* **18,** 2825.

Burgers, P. M. J., F. Eckstein, and D. H. Hunneman (1979), *J. Biol. Chem.* **254,** 7476.

Cleland, W. W. (1975), *Biochemistry* **14,** 3220.

Coderre, J. A., and J. A. Gerlt (1980), *J. Am. Chem. Soc.* **102,** 6594.

Cohn, M., and A. Hu (1978), *Proc. Natl. Acad. Sci. USA* **75,** 200.

Cohn, M., and A. Hu (1980), *J. Am. Chem. Soc.* **102,** 913.

Colombo, G., G. M. Carlson, and M. A. Lardy (1978), *Biochemistry* **17,** 5321.

Corbridge, D. E. C. (1980), in *Studies in Inorganic Chemistry*, **Vol. 2,** Elsevier, Amsterdam, Chapter 1.

Cornelius, R. D., P. A. Hart, and W. W. Cleland (1977), *Inorg. Chem.* **16,** 2799.

Cullis, P. M., and G. Lowe (1978), *J.C.S. Chem. Commun.*, 512.

Cullis, P. M., and G. Lowe (1981), *J.C.S. Perkin I,* 2317.

Cullis, P. M., R. L. Jarvest, G. Lowe, and B. V. L. Potter (1981), *J.C.S. Chem. Commun.*, 245.

Dunaway-Mariano, D., and W. W. Cleland (1980), *Biochemistry* **19,** 1496.

Eargle, D. H., Jr., V. Licko, and G. L. Kenyon (1977), *Anal. Biochem.* **81,** 196.

Eckstein, F., and R. S. Goody (1976), *Biochemistry* **15,** 1685.

Eckstein, F., W. Bruns, and A. Parmeggiani (1975), *Biochemistry* **14,** 5225.

Foster, D. O., H. A. Lardy, P. D. Ray, and J. Johnson (1967), *Biochemistry* **6,** 2120.

Fukuyama, T. T. (1966), *J. Biol. Chem.* **241,** 4745.

Gass, J. D., and A. Meister (1970), *Biochemistry* **9,** 842.

Gerlt, J. A., and J. A. Coderre (1980), *J. Am. Chem. Soc.* **102,** 4531.

Gerlt, J. A., and W. H. Y. Wan (1979), *Biochemistry* **18,** 4630.

Gray, G. A., and T. A. Albright (1977), *J. Am. Chem. Soc.* **99,** 3243.

Gutowsky, H. S. (1959), *J. Chem. Phys.* **31,** 1683.

Hackney, D., J. Sleep, G. Rosen, R. Hutton, and P. D. Boyer (1978), *NMR Biochem. Symp.* **79,** 285.

Hansen, D. E., and J. R. Knowles (1981), *J. Biol. Chem.* **256,** 5967.

Hansen, D. E., and J. R. Knowles (1982), *J. Biol. Chem.*, in press.

Hartman, S. C., and S. Prusiner (1973), in *The Enzymes of Glutamine Metabolism*, Academic Press, New York, p. 409.

Jaffe, E. K., and M. Cohn (1978), *J. Biol. Chem.* **253,** 4823.

Jarvest, R. L., and G. Lowe (1979), *J.C.S. Chem. Commun.*, 364.

Jarvest, R. L., and G. Lowe (1980), *J.C.S. Chem. Commun.*, 1145.

Jarvest, R. L., and G. Lowe (1981), *Biochem. J.*, **199,** 447.

Jarvest, R. L., G. Lowe, and B. V. L. Potter (1980), *J.C.S. Chem. Commun.*, 1142.

Jarvest, R. L., G. Lowe, and B. V. L. Potter (1981), *Biochem J.* **199,** 427.

Jordan, F., J. A. Patrick, and S. Salamone (1979), *J. Biol. Chem.* **254,** 2384.

Krishnaswamy, P. R., V. Pamiljans, and A. Meister (1962), *J. Biol. Chem.* **237,** 2932.

Lowe, G., and B. V. L. Potter (1981a), *Biochem. J.* **199,** 227.

Lowe, G., and B. V. L. Potter (1981b), *Biochem. J.* **199,** 693.

Lowe, G., and B. S. Sproat (1978a), *J.C.S. Chem. Commun.*, 783.

Lowe, G., and B. S. Sproat (1978b), *J.C.S. Perkin I,* 1622.

Lowe, G., and B. S. Sproat (1980), *J. Biol. Chem.* **255,** 3944.

Lowe, G., B. V. L. Potter, and B. S. Sproat (1979), *J.C.S. Chem. Commun.*, 733.

Lutz, O., A. Nolles, and D. Staschewski (1978), *Z. Naturforsch.* **33A,** 380.

Markham, G. D., and G. M. Reed (1978), *J. Biol. Chem.* **253,** 6184.

Markham, G. D., E. W. Hafner, C. W. Tabor, and H. Tabor (1980), *J. Biol. Chem.* **255,** 9082.

Maruyama, H., R. L. Easterday, H.-C. Chang, and M. D. Lane (1966), *J. Biol. Chem.* **241,** 2405.

Meek, T. D., and J. J. Villafranca (1980) *Biochemistry* **19,** 5513.

Meek, T. D., K. A. Johnson, and J. J. Villafranca (1982), *Biochemistry* **21,** 2158.

Midelfort, C. F., and I. A. Rose (1976), *J. Biol. Chem.* **251,** 5881.

Milner, Y., and H. G. Wood (1972), *Proc. Natl. Acad. Sci. USA* **69,** 2463.

Milner, Y., and H. G. Wood (1976), *J. Biol. Chem.* **251,** 7920.

Milner, Y., G. Michaels, and H. G. Wood (1978), *J. Biol. Chem.* **253,** 878.

Milolajczyk, M., M. Witezak, M. Wieczorek, N. G. Bokij, and Y. T. Struchkov (1976), *J.C.S. Perkin I,* 371.

Pollard-Knight, D., B. V. L. Potter, P. M. Cullis, G. Lowe, and A. Cornish-Bowden (1982), *Biochem. J.* **201,** 421.

Ramsey, N. F. (1952), *Phys. Rev.* **87,** 1075.

Raushel, F. M., and J. J. Villafranca (1979), *Biochemistry* **18,** 3424.

Raushel, F. M., and J. J. Villafranca (1980), *Biochemistry* **19,** 3170.

Raushel, F. M., P. M. Anderson, and J. J. Villafranca (1978), *Biochemistry* **17,** 5587.

Reeves, R. E., R. A. Menzies, and D. S. Hsu (1968), *J. Biol. Chem.* **243,** 5486.

Richard, J. P., H. T. Ho, and P. A. Frey (1978), *J. Am. Chem. Soc.* **100,** 7756.

Rose, I. A. (1978), *Fed. Proc.* **37,** 2775.

Rose, I. A. (1979), *Adv. Enzymol. Relat. Areas Mol. Biol.* **50,** 361.

Sheu, K. R., and P. A. Frey (1977), *J. Biol. Chem.* **252,** 445.

Todhunter, J. A., and D. L. Purich (1975), *J. Biol. Chem.* **250,** 3505.

Tsai, M. D. (1979), *Biochemistry* **18,** 1468.

Tsai, M. D. (1980), *Biochemistry* **19,** 5310.

Tsai, M. D., and T. T. Chang (1980), *J. Am. Chem. Soc.* **102,** 5416.

Tsai, M. D., and K. Bruzik (1982), *J. Am. Chem. Soc.* **104,** 863.

Tsai, M. D., S. L. Huang, J. F. Kozlowski, and C. C. Chang (1980), *Biochemistry* **19,** 3531.

Tsuda, Y., R. A. Stephani, and A. Meister (1971), *Biochemistry* **10,** 3186.

Usher, D., D. A. Richardson, F. Eckstein (1970), *Nature* **228,** 663.

Utter, M. F., and H. M. Kolenbrander (1972), *Enzymes, 3rd Ed.* **6,** 117.

Van Etten, R. L., and J. M. Risley (1978), *Proc. Natl. Acad. Sci. USA* **75,** 4784.

Webb, M. R., and J. F. Eccleston (1981), *J. Biol. Chem.* **256,** 7734.

Webb, M. R., and D. R. Trentham (1980), *J. Biol. Chem.* **255,** 1775.

Webb, M. R., and D. R. Trentham (1981), *J. Biol. Chem.* **256,** 4884.

Webb, M. R., G. C. McDonald, and D. R. Trentham (1978), *J. Biol. Chem.* **253,** 2908.

Webb, M. R., C. Grubmeyer, H. S. Penefsky, and D. R. Trentham (1980), *J. Biol. Chem.* **255,** 11637.

Wedler, F. C., and P. D. Boyer (1972), *J. Biol. Chem.* **247,** 984.

Westheimer, F. H. (1968), *Acc. Chem. Res.* **1,** 70.

Two

Spin-Label Probes of Enzyme Action[1]

Marvin W. Makinen and Lawrence C. Kuo

Department of Biophysics and Theoretical Biology
Cummings Life Science Center
University of Chicago
Chicago, IL 60637

1. **Introduction** **54**
2. **The Molecular Structure of Nitroxide Spin-Labels** **55**
3. **Studies of Spin-Labeled Proteins and Enzymes in Crystals** **61**
   3.1. Hemoglobin, 61
   3.2. α-Chymotrypsin, 68
   3.3. Lysozyme, 73
4. **Application of Cryoenzymology for Structural Characterization of a Spin-Labeled Enzyme Reaction Intermediate** **80**
   4.1. The Cryoenzymologic Approach to the Study of Enzyme Mechanisms, 80
   4.2. The Structure of a Low Temperature Stabilized Spin-labeled Reaction Intermediate of Carboxypeptidase A, 81

Acknowledgments **91**

References **91**

---

[1]In this review, we have avoided the use of abbreviations to designate chemical compounds whenever possible, and only standard IUB-approved abbreviations have been employed to designate amino acids. The abbreviations employed in the text are ClCPL, O-(*trans-p*-chlorocinnamoyl)-L-β-phenyllactate; EPR, electron paramagnetic resonance; Glc, glucose; GlcNAc, 2-acetamido-2-deoxy-D-glucopyranoside, or N-acetylglucosamine; HbO_2, oxygen-liganded hemoglobin; HbCO, carbon monoxide–liganded hemoglobin; metHb, (ferri) or high-spin methemoglobin; TEPOPL, O-3-(2,2,5,5-tetramethylpyrrolinyl-1-oxyl)-propene-2-oyl-L-β-phenyllactate.

1. INTRODUCTION

Enzyme catalyzed reactions occur via the sequential formation of a series of reaction intermediates. A general scheme to illustrate the reaction pathway is given by Equation (1):

$$E + S \underset{k_{-1}}{\overset{k_1}{\rightleftharpoons}} ES \underset{k_{-2}}{\overset{k_2}{\rightleftharpoons}} I_2 \underset{k_{-3}}{\overset{k_3}{\rightleftharpoons}} I_3 \underset{k_{-4}}{\overset{k_4}{\rightleftharpoons}} EP \underset{k_{-5}}{\overset{k_5}{\rightleftharpoons}} E + P \qquad (1)$$

where ES represents the initial collision (Michaelis) complex of the enzyme and substrate; EP, the enzyme–product complex in equilibrium with the free product in solution; and I_2 and I_3 are intermediates of the reaction. The various k's represent the rate constants that govern the corresponding sequential series of reactions in the forward or backward directions. For this general scheme, Mildvan and Cohn (1970) pointed out that the essential requirements to understand the molecular basis of enzyme function are: (1) the rate constants and activation parameters for each of the steps of the reaction; (2) a description of the structures of the reaction intermediates and the electronic and atomic rearrangements involving the enzyme and substrate; and (3) a rationale for the magnitude of the rate constants in terms of structure. Thus, the study of enzyme function at a detailed molecular level requires an integrated, multidisciplinary approach employing chemical, kinetic, and structural methods.

The application of electron paramagnetic resonance methods to characterize the structural properties of enzymic reaction intermediates requires a paramagnetic probe that is not chemically modified by the reaction. No enzyme or protein has a naturally occurring paramagnetic center that satisfies this requirement. The use of intrinsic paramagnetic sites, such as the Fe^{3+} cation in heme (Palmer, 1978) and iron-sulfur (Orme-Johnson and Sands, 1973) proteins, or free-radical centers generated in enzymic reactions (Schepler *et al.*, 1975; Hoffman *et al.*, 1979; de Beer *et al.*, 1979), is often restricted to only one form of the enzyme. Therefore, structural characterization of several or all of the intermediates of an enzyme reaction requires the introduction of paramagnetic probes. This objective can be accomplished by (1) use of synthetic substrates with paramagnetic centers that are not altered chemically during the reaction, (2) substitution of paramagnetic cations for catalytically required diamagnetic metal ions in metalloenzymes or in metal-activated enzymes, or (3) introduction of a bound paramagnetic species that does not participate in catalytic events but is sufficiently close to the active site for monitoring structural relationships of the substrate and active-site residues. All these methods have been employed in magnetic resonance studies. An implicit requirement is that introduction of the spectroscopic probe does not alter the reaction mechanism or distort the structure of the enzyme–substrate complex into a catalytically inactive form. Despite the voluminous literature of magnetic resonance studies of enzymes, this requirement has been rigorously satisfied in few instances.

The necessity to employ paramagnetic probes for structural characterization has led to the application of the free-radical, nitroxide-spin-label method, first introduced successfully into biophysical studies by McConnell and co-workers (Hamilton and McConnell, 1968; McConnell and McFarland, 1970; McConnell, 1971). A variety of magnetic resonance studies of biological macromolecules have been carried out with nitroxide spin-labels, and the studies have been extensively reviewed (Rozantsev, 1970; Dwek, 1973; Berliner, 1976; Likhtenstein, 1976). Therefore, this review is limited to a selective survey of the literature to discuss how spin-label probes have been employed to characterize enzyme complexes and reaction intermediates in catalytically active states and to analyze aspects of spin-label–protein interactions that have limited their application in structural studies. Since the spin-label method leads to the application of a variety of magnetic resonance techniques, we have selected illustrative examples of the three primary approaches to characterize conformational states of proteins and enzymes: (1) assignment of spin-label orientation in protein crystals as a monitor of protein conformational states or spin-label configuration, (2) measurement of the enhancement of nuclear spin-lattice relaxation rates of enzyme residues in the presence of bound spin-labels to determine their radial separation, and (3) measurement of the quenching of the signal amplitude of a spin label in the presence of a paramagnetic ion to estimate the radial separation between the two paramagnetic sites.

Only those studies are discussed in which the application of other structural methods, in particular X-ray diffraction, corroborate the results and interpretations of the spectroscopic studies. We discuss neither the detailed theoretical foundations of the methods nor the mathematical relationships by which quantitative structural description of the paramagnetic species is made, for these aspects have been presented in sufficient detail in numerous reviews (see Berliner, 1976; Burton *et al.*, 1979). Furthermore, specific technical details for the handling of biological samples in spin-label studies and the construction of goniometers for single-crystal work have been extensively described (Jost and Griffith, 1978; Berliner, 1978; Fee, 1978; Chien and Dickinson, 1981). Since the emphasis of the review is on structural characterization of catalytically active states of enzymes, we discuss certain aspects of the combined application of cryoenzymologic methods (Douzou, 1977; Makinen and Fink, 1977) with magnetic resonance techniques (see Section 4) because of the general importance of the results and the potential for applying this combined approach to a variety of enzyme systems.

2. THE MOLECULAR STRUCTURE OF NITROXIDE SPIN-LABELS

Of the nitroxide spin-labels employed in biophysical studies, there are structurally two basic types. These are illustrated in Figure 1 and are correspondingly derivatives of 2,2,6,6-tetramethylpiperidine-1-oxyl and 2,2,5,5-tetramethylpyrrolidine-1-oxyl. A large number of derivatives of these spin-

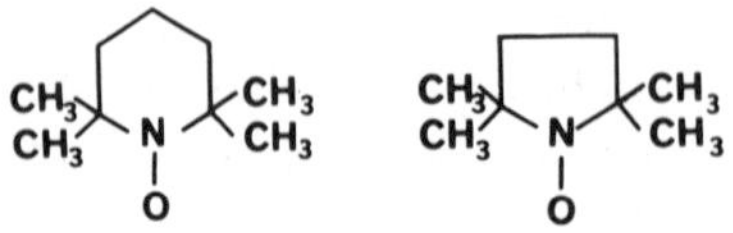

Figure 1. Structural formulas of 2,2,6,6-tetramethylpiperidine-1-oxyl (left) and 2,2,5,5-tetramethylpyrrolidine-1-oxyl (right), showing their respective six- and five-membered ring structures.

labels have been synthesized with strategically placed substituents to gain specificity in binding to macromolecules.[1] The general chemistry of spin-labels in enzyme studies has been discussed by Morrisett (1976). In this section, we point out details of their stereochemical properties that influence their interactions with macromolecules. Hitherto, little attention has been directed toward these aspects of spin-label structure.

Derivatives of 2,2,6,6-tetramethylpiperidine-1-oxyl with a six-membered ring structure exhibit a variety of different molecular configurations. The cyclic compound 2,2,6,6-tetramethyl-4-piperidinol-1-oxyl, for which the X-ray determined bond lengths and angles are illustrated in Figures 2 and 3, is found in a chair-like configuration, as generally associated with cyclohexane-like molecules. However, in crystals of 2,2,6,6-tetramethylpiperidine-1-oxyl (Capiomont *et al.*, 1972) and di-(2,2,6,6-tetramethylpiperidine-1-oxyl)suberate (Capiomont, 1972), the molecules are found in chair–chair inversion movement. Since inversion of the configuration of the six-membered ring is found for nitroxide derivatives in crystals, a

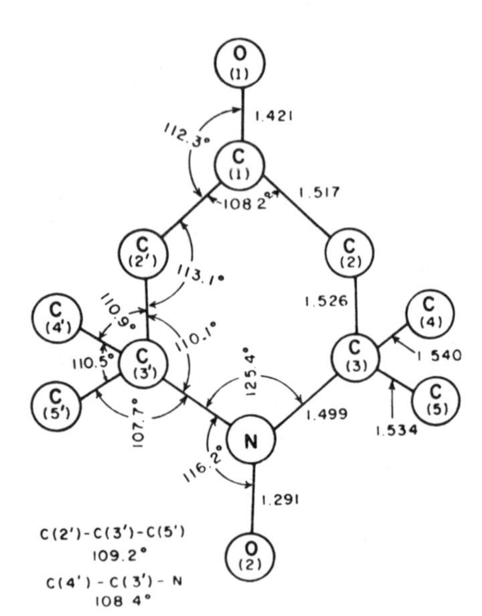

Figure 2. The bond angles (in degrees) and bond lengths (in Ångströms) of 2,2,6,6-tetramethyl-4-piperidinol-1-oxyl determined through X-ray diffraction studies. From Berliner (1970) with permission.

[1]For instance, there are currently 64 different spin-label derivatives available from Molecular Probes (Plano, Texas), a company that specializes in the synthesis of fluorescent and spin-label compounds for biochemical studies.

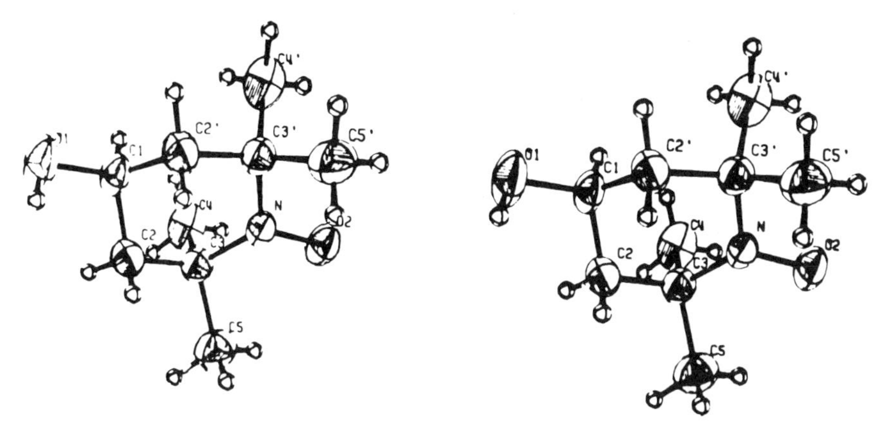

Figure 3. Stereoview of the configuration of the nitroxide molecule in Figure 2. Thermal motions are represented by 50% probability ellipsoids. A "flattened" cyclohexane-like configuration of the ring is evident. From Berliner (1970) with permission.

similar structural change may also occur for spin labels bound to macromolecules in solution. On the other hand, the ketone 2,2,6,6-tetramethyl-4-piperidinone-1-oxyl is found in a twisted configuration in which the C(4)—O(8) and N(1)—O(7) bonds form a twofold axis of rotation (Bordeaux and Lajzerowicz, 1974). Changes in the configuration of the ring from a chair-like to a twisted form alter the geometrical relationships of the principal components of the g-matrix to the C—N—C plane. Such changes may lead to complications in the interpretation of experimental results. Furthermore, temperature changes may alter the relative populations of twisted and chair-like forms.

Differences in the planarity of the C(NO)C group in five- and six-membered nitroxide derivatives govern the nonbonded van der Waals interactions of spin-labels, since the overall molecular dimensions are dependent upon these stereochemical differences (Lajzerowicz-Bonneteau, 1976). In the six-membered piperidinyl spin-labels, the C(NO)C group is not planar, and the N(1)—O(7) bond makes an angle of ~16° with the C—N—C plane. On the other hand, planarity of the C(NO)C group is invariably observed in the five-membered pyrrolinyl and pyrrolidinyl nitroxide spin-labels. The differences in overall van der Waals dimensions, dependent upon C(NO)C planarity, are illustrated in Figure 4. The planarity of the five-membered ring system constrains the C(NO)C group to have a twofold axis of symmetry including the positions of the adjacent methyl groups bonded to the C(2) and C(5) positions. This configuration remains constant with time. In contrast, the "bent" ring structure of piperidinyl-1-oxyl derivatives results in a pyramid-shaped configuration of the four methyl groups and the C(NO)C group. Since the ring structure of piperidinyl derivatives may interconvert among the several observed conformers, a change of configuration results in a change in overall van der Waals dimensions. Such changes may effect

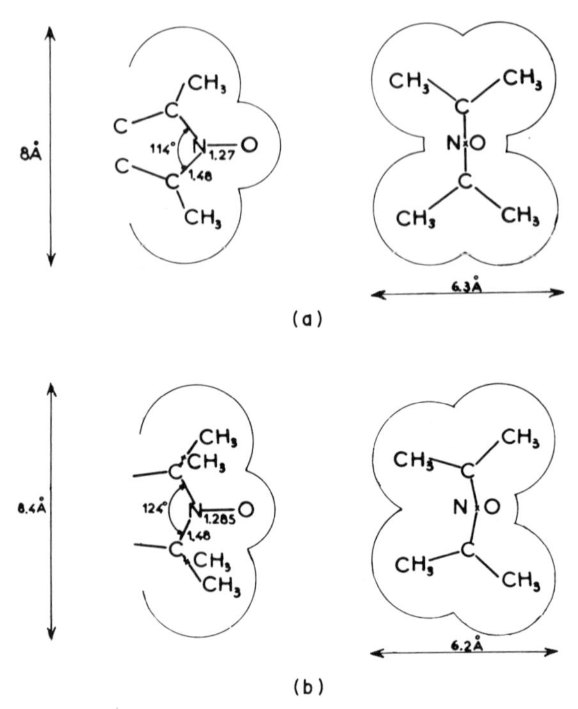

Figure 4. Diagrams of the nitroxide group and its overall van der Waals thickness for (*a*) a five-membered ring with a planar C(NO)C configuration, and (*b*) a six-membered ring with a nonplanar C(NO)C configuration. From Lajzerowicz-Bonneteau (1976) with permission. Copyright: Academic Press, Inc.

steric interactions of the spin-label with macromolecules that are dependent upon their overall dimensions.

A drawing of the molecular structure of the five-membered cyclic compound (+)-3-carboxy-2,2,5,5-tetramethylpyrrolidine-1-oxyl is shown in Figure 5, and a diagram of the bond lengths and angles is provided in Figure 6 (Ament *et al.*, 1973). The atoms C(2), C(5), N(1), and O(1) are nearly coplanar, with the N(1) atom displaced only 0.03 Å from the plane defined by the other three atoms. For the pyrrolidinyl compounds, the ring structure constrained by the planarity of the C(NO)C group gives rise to optically active enantiomers with a chiral center at C(3). The stereochemical relationships are illustrated in Figure 7. In the chemical synthesis of analogs of 2,2,5,5-tetramethylpyrrolidine-1-oxyl, the enantiomers are usually not resolved, and their use in macromolecular studies has been largely made on the basis of racemic mixtures of the $R(+)$ and $S(-)$ isomers. Therefore, spectroscopic data must be considered to be complicated by spin-label–macromolecule interactions dependent upon enantiomeric structural differences when unresolved racemic mixtures are employed. As we discuss below (see Section 3.2), enantiomeric specificity is observed in the rate-

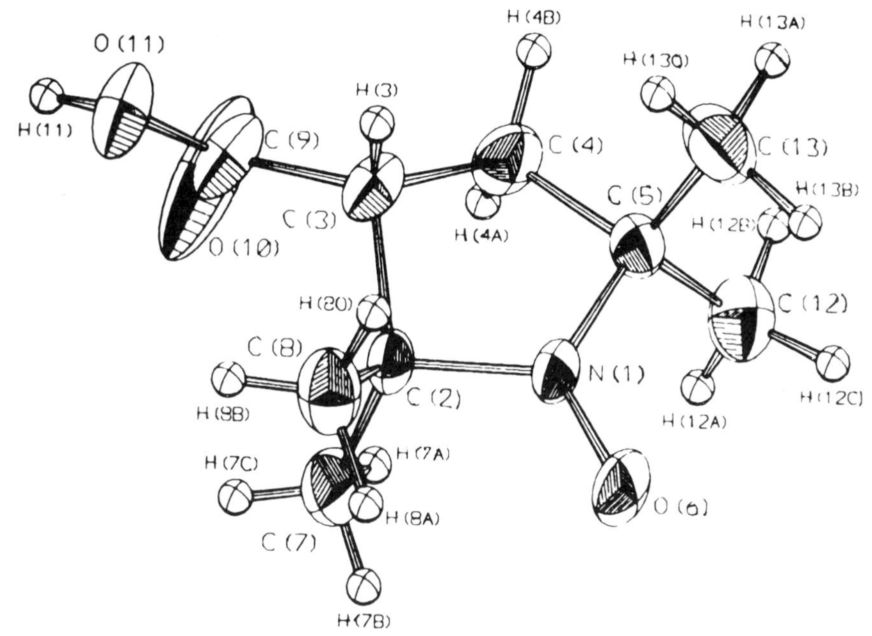

Figure 5. Diagram with thermal ellipsoids of the *R* configuration of (+)-3-carboxy-2,2,5,5-tetramethylpyrrolidine-1-oxyl. Note the position of the 3-carboxy substituent conferring stereoisomerism on this spin label. Reprinted with permission from Ament *et al.* (1973). Copyright 1972 American Chemical Society.

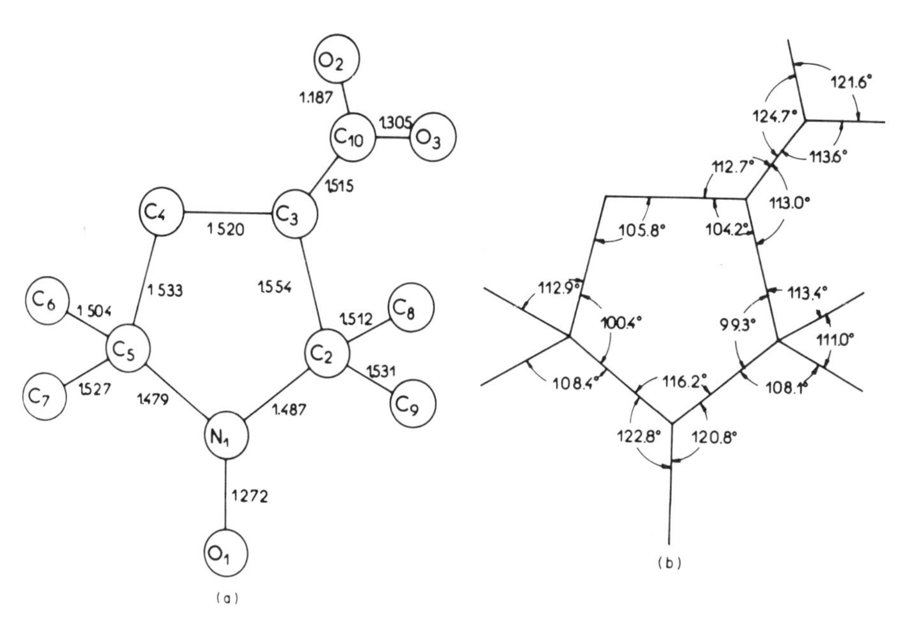

Figure 6. Drawing of the (*a*) bond distances and (*b*) bond angles of (+)-3-carboxy-2,2,5,5-tetramethylpyrrolidine-1-oxyl.

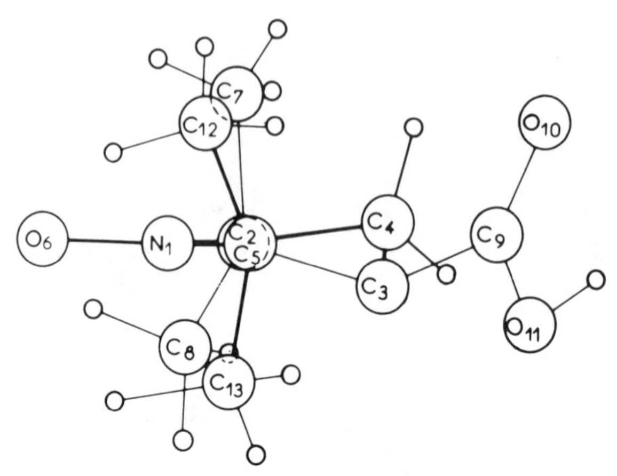

Figure 7. Projection of the molecule in Figures 5 and 6 parallel to its mean plane to illustrate the chiral center at C(3). (The hydrogen atom bonded to C(3) is mistakenly not shown in this diagram.) From Lajzerowicz-Bonneteau (1976) with permission. Copyright: Academic Press, Inc.

limiting step of the hydrolysis of asymmetrical 3-carboxy-2,2,5,5-tetramethylpyrrolidine-1-oxyl-*p*-nitrophenyl ester catalyzed by α-chymotrypsin (Flohr *et al.*, 1975; Bauer and Berliner, 1979). Comparable interactions, dependent upon chiral differences, are likely to occur with other macromolecules.

The presence of enantiomers can be avoided by use of derivatives of the five-membered ring analog 2,2,5,5-tetramethylpyrroline-1-oxyl in which the C(3)—C(4) bond is unsaturated. The structure of this spin-label with a 3-carbamido substituent is illustrated in Figure 8. In this spin-label the entire ring system is planar, and the olefinic bond ensures the absence of chirality at the C(3) position. Figure 9 illustrates the chemical-bonding structure of 3-(2,2,5,5-tetramethylpyrrolinyl-1-oxyl)propen-2-oic acid, which has been

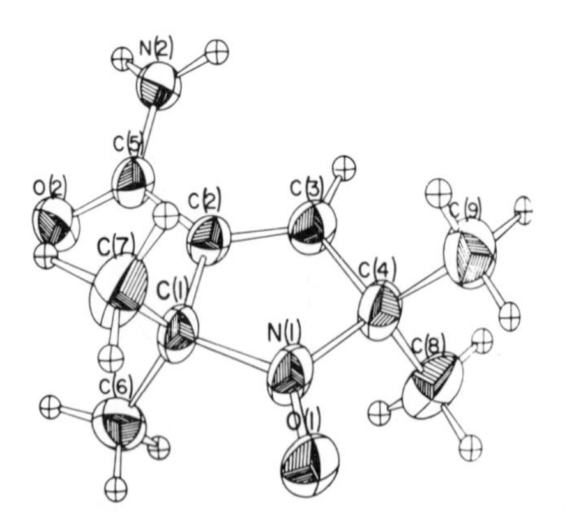

Figure 8. Molecular structure of 3-carbamido-2,2,5,5-tetramethylpyrroline-1-oxyl with thermal ellipsoids drawn at 50% probability. From Turley and Boer (1972) with permission.

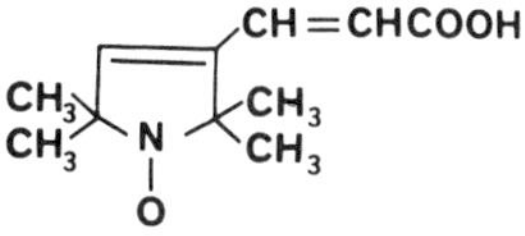

Figure 9. Chemical-bonding structure of 3-(2,2,5,5-tetramethylpyrrolinyl-1-oxyl)-propene-2-oic acid. The unsaturated C=C bond in the ring results in a coplanar configuration of the C(3)-substituent with no chiral center.

described recently as a potentially useful intermediate in the synthesis of spin-label substrates (Koch *et al.*, 1979). This spin-label lacks a chiral center and is structurally similar to arylacryloyl derivatives employed as synthetically derived substrates for a variety of enzymes (Bernhard and Lau, 1971). Moreover, in contrast to the very weak optical absorption properties of aliphatic nitroxides (Janzen, 1971), the conjugated system of double bonds gives rise to absorption changes of moderate intensity near 275 nm that are convenient for spectrophotometric collection of kinetic data (see Figure 27). Recent studies (Kuo *et al.*, 1983) with use of the L-β-phenyllactic acid ester of this spin-label to determine the structure of a reaction intermediate of carboxypeptidase A are discussed below.

3. STUDIES OF SPIN-LABELED PROTEINS AND ENZYMES IN CRYSTALS

3.1. Hemoglobin

The cooperative binding of molecular oxygen by hemoglobin is attributed to differential affinity of the ligand to two forms of the protein, differing in quaternary structure and designated R (relaxed) for the oxy or liganded form and T (tensed) for the deoxygenated form (Monod *et al.*, 1965). The liganded derivatives of hemoglobin have the same R quaternary structure, whereas only deoxyhemoglobin is found in the quaternary T structure. A stereochemical mechanism of cooperative ligand binding has been suggested on the basis of the X-ray-defined tertiary and quaternary structural changes that are associated with ligand binding (Perutz, 1970). More recent studies at higher resolution show that the reactivity of the hemes for ligand binding is regulated by differences in the nonbonded protein-heme and protein–ligand interactions in the R and T states (Baldwin and Chothia, 1979; Moffat *et al.*, 1979). Insight into the nature of the structural changes responsible for cooperative ligand binding was provided by early studies of spin-labeled hemoglobin. At the time of these studies, sufficient resolution had not been achieved in X-ray crystallographic studies of hemoglobin to determine whether the relative configurations of protein residues, particularly with respect to the heme group in each subunit, were altered in oxygen binding and release.

The two spin-labels illustrated in Figure 10 have been those most widely applied to study conformational changes associated with oxygen binding to hemoglobin. Both spin-labels are covalently attached to the sulfhydryl group

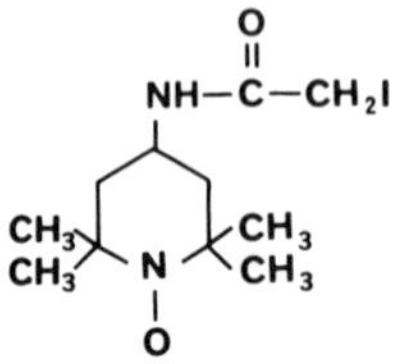

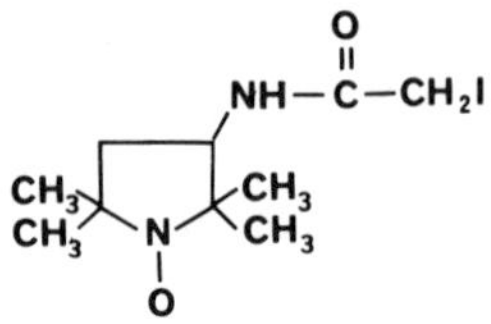

Figure 10. Chemical-bonding structures of N-(2,2,6,6-tetramethyl-4-piperidinyl-1-oxyl)iodoacetamide (upper) and N-(2,2,5,5-tetramethyl-3-pyrrolidinyl-1-oxyl) iodoacetamide (lower).

of Cys-β93, a position that is close to the proximal His-β94 residue coordinated to the heme iron and to residues involved in transmitting cooperative structural changes across the $\alpha_1\beta_2$ interface of the hemoglobin molecule. McConnell and Hamilton (1968) observed that the spectrum of the spin-label N-(2,2,6,6-tetramethyl-4-piperidinyl-1-oxyl)iodoacetamide attached to the Cys-β93 sulfhydryl group indicates two isomeric configurations corresponding to a strongly immobilized and a weakly immobilized probe. This spectrum is illustrated in Figure 11. In single crystals the spectrum of the spin-label indicates similar but not identical configurations for both metHb and HbCO, whereas the spectra of spin-labeled metHb and HbCO in solution are distinctly different. On the other hand, only one configuration of the spin-label is observed for the deoxygenated protein (Ogawa *et al.*, 1968). These studies demonstrated that there must be differences in the detailed tertiary structure of liganded derivatives of the protein in the R quaternary state. Subsequent, more-detailed paramagnetic resonance studies of spin-labeled hemoglobin have shown that the single crystal spectra of low- and high-spin derivatives of liganded Fe^{3+} hemoglobin complexes are not identical, and that the orientations of the principal nitrogen hyperfine axes with respect to the heme group differ perceptibly by up to 10–15° (Chien, 1979).

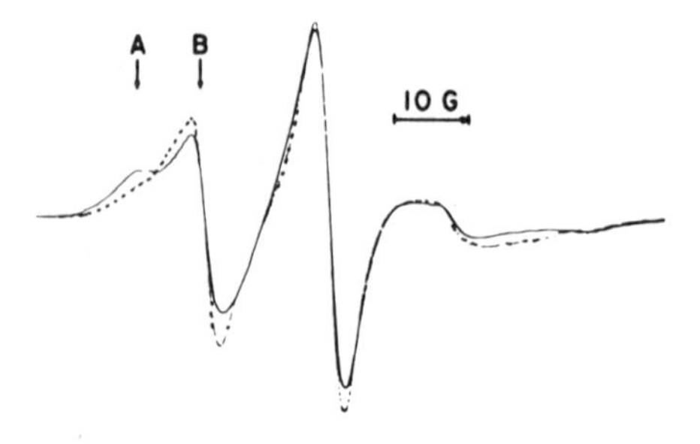

Figure 11. The (first derivative) electron paramagnetic resonance spectrum of spin—labeled horse hemoglobin in the carbonmonoxy (unbroken line) and met (broken line) forms. The spectral features designated *A* and *B* refer to distinct signals from isomeric conformational states of the spin-labeled protein. From McConnell and Hamilton (1968) with permission.

These studies have confirmed that differences in protein tertiary structure obtain for all liganded complexes of hemoglobin, as had been earlier implied through polarized optical studies of hemoglobin single crystals (Makinen and Eaton, 1974).

Detection of the two states of the spin-label designated A and B in Figure 11 indicates an equilibrium between two conformations with a half-life for interconversion $\geqslant 10^{-7}$ sec. These two conformational states have been characterized by X-ray diffraction studies. A difference Fourier synthesis of horse HbCO with the piperidinyl derivative covalently attached to Cys-β93 has been calculated at 3.5-Å resolution (Moffat, 1971). Two conformations of the spin-label identified in the difference electron density map were shown to be consistent with the principal hyperfine axis orientations described in the paramagnetic resonance studies. In addition, numerous distortions of the protein induced by the presence of the spin-label were described, with the largest number localized to both heme environments in the α and β subunits, the (proximal) F helix of the β chain containing the Cys-β93 residue, and the FG region near the $\alpha_1\beta_2$ interface.

Figure 12 illustrates stereoviews of the two configurations of the spin-label attached to the Cys-β93 group in crystals of horse HbCO identified through difference Fourier studies. The structural features belonging to the two conformations A and B described in the paramagnetic resonance spectra were identified by (1) difference electron density features, (2) correspondence with orientations of the nitrogen hyperfine axes of the C(NO)C group determined in spectroscopic studies, and (3) stereochemical constraints of the van der Waals contacts with surrounding amino acid residues. On this basis, the strongly immobilized conformation A was assigned to an orientation of the spin-label that displaces the Tyr-β145 residue from the HG region of the molecule, and the weakly immobilized conformation B was assigned to a configuration of the spin-label on the surface of the hemoglobin molecule. The assignment of conformations A and B was confirmed in these studies, since a molecular model with orientation B could not be constructed into the difference electron density features associated with displacement of the tyrosine residue so that it remained in agreement with the EPR-determined direction cosines of the hyperfine axes. For the strongly immobolized conformation A, a variety of structural perturbations are induced by interactions of the piperidinyl ring with amino acid side chains. The four methyl groups are in van der Waals contact with His-β97, Val-β98, and Asp-β99, whereas the rest of the piperidinyl ring is in contact with Leu-β141 and Ala-β142. The numerous contacts with the protein and spatial displacement of these residues in addition to those of Phe-β103 and Ala-β142 account for the strength of the immobilization. For the conformation B, a range of structures could be accommodated by the difference electron density features, as is consistent with its weakly immobilized nature. Furthermore, the piperidinyl ring induces no significant structural alterations of the protein in this orientation, and the configuration around the C_α atom of the Cys-β93 residue is not altered from that of the native protein.

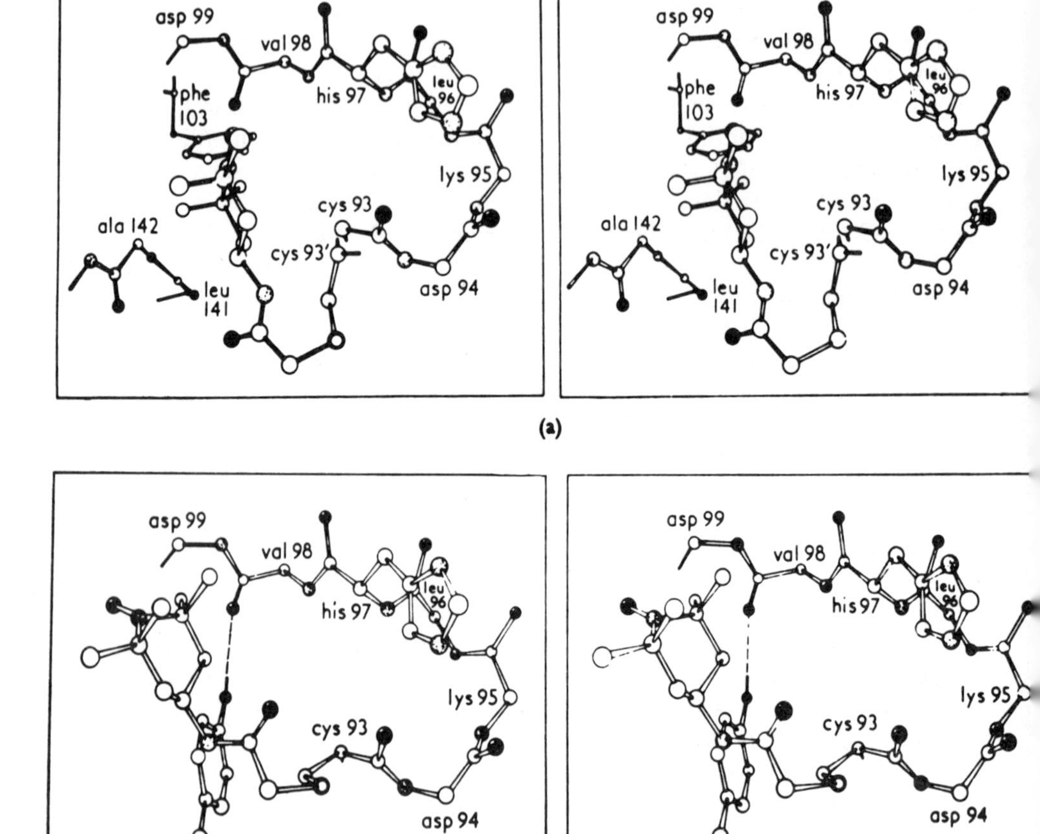

Figure 12. Stereo drawing of parts of the atomic model of spin-labeled horse HbCO viewed in a direction perpendicular to the crystallographic *y* and *z* axes. Carbon atoms, open circles; nitrogen, dotted circles; oxygen, closed circles; and sulfur, double circles. (*a*) Isomeric state A of Figure 11 with a strongly immobilized spin-label group. (*b*) Isomeric state B with a weakly immobilized spin label. From Moffat (1971) with permission. Copyright: Academic Press, Inc., (London) Ltd.

Since the spin-label in conformation A was shown to displace the Tyr-β145 residue, the results implied that the C-terminus of the β chain in the native molecule exists in two conformational states in rapid equilibrium. A decrease in the population of the strongly immobilized conformation in metHb compared with that in HbCO implied that the equilibrium is altered in metHb relative to that in HbCO. This change in the position of the con-

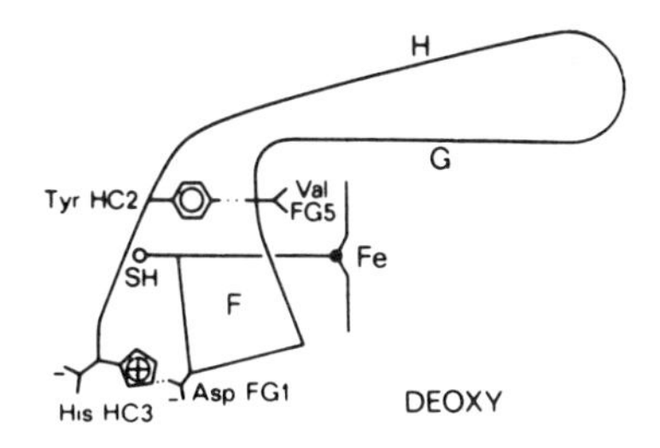

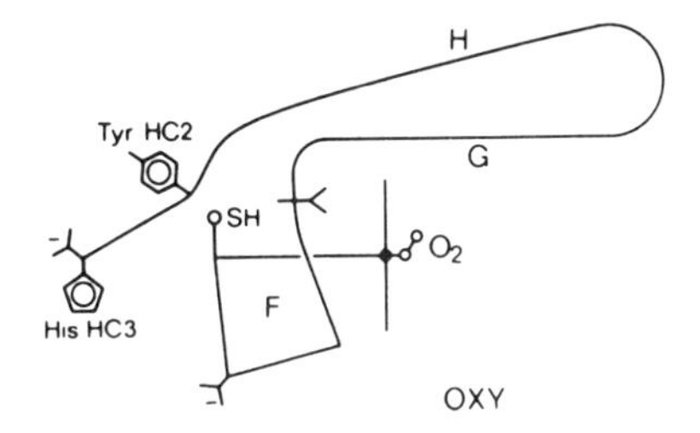

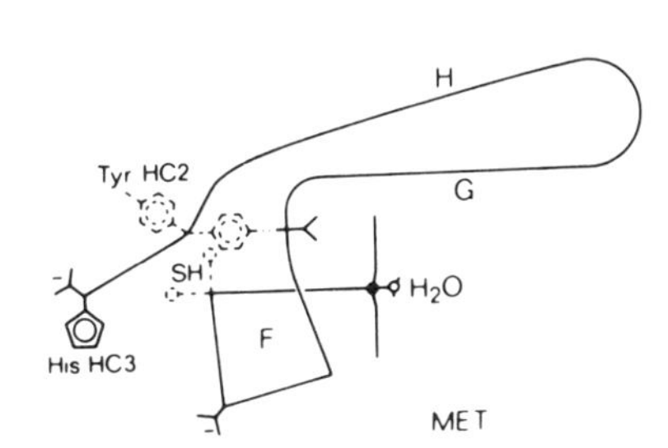

Figure 13. The positions of the side chains of Tyr-β145 and Cys-β93 in deoxy, liganded (corresponding to HbCO or HbO_2), and metHb. In the R quaternary conformation, the equilibrium SH_{out}, $Tyr_{in} \leftrightarrows SH_{in}$, Tyr_{out} occurs and is shifted to the right by low-spin derivatives and to the left by high-spin derivatives. These correspond to states A and B of spin-labeled hemoglobin, respectively. The conformational state in deoxyHb corresponds to the weakly immobilized state. From Heidner *et al.* (1976) with permission. Copyright: Academic Press, Inc. (London), Ltd.

formational equilibrium involving the Tyr-β145 residue was subsequently confirmed in a difference Fourier analysis of native horse HbCO (Heidner *et al.*, 1976). The conformational equilibrium is illustrated in Figure 13. For HbCO the electron density maps show that the β93 sulfhydryl group is in the tyrosine pocket, with the Tyr-β145 residue displaced; for metHb the difference density features indicate occupancy of the pocket by the tyrosine residue with a corresponding altered position of the β93 sulfhydryl group. Thus, in metHb the conformational equilibrium

$$(\text{SH-}\beta 93 \text{ out, Tyr in}) \rightleftharpoons (\text{SH-}\beta 93 \text{ in, Tyr out})$$

has the left-hand side predominating; in HbCO the right-hand side predominates. The occupancy of the pocket by the Tyr-β145 residue prevents the insertion of the spin-label residue into this site, accounting for the decreased population of the strongly immobilized conformation in metHb. In HbCO

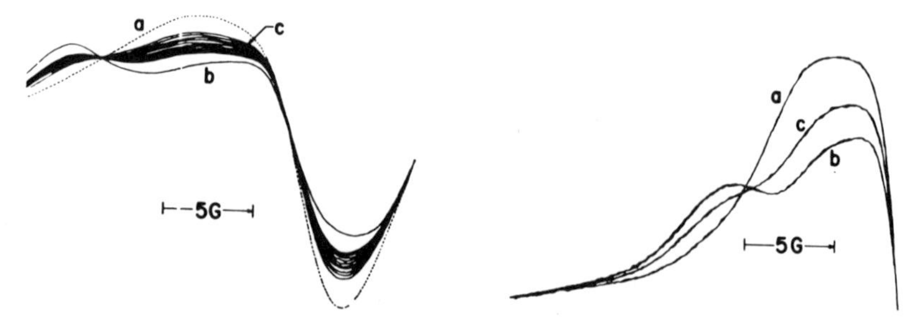

Figure 14. The (first derivative) electron paramagnetic resonance spectrum of human hemoglobin labeled with N-(2,2,5,5-tetramethyl-3-pyrrolidinyl-1-oxyl)iodoacetamide (left) and N-(2,2,6,6-tetramethyl-4-piperidinyl-1-oxyl)iodoacetamide (right). The spectra *a, b,* and *c* refer to deoxygenated, fully oxygenated, and partially oxygenated solutions of the spin-labeled hemoglobins, respectively. From Ogawa *et al.* (1968) with permission.

the displaced tyrosine residue allows a proportionately higher occupancy of the highly immobilized conformation by the piperidinyl ring.

The paramagnetic resonance spectra of spin-labeled HbO_2 in solution are dependent on the percentage saturation of oxygen binding and the chemical structure of the spin-label (Ogawa *et al.*, 1968). The left side of Figure 14 illustrates the changes observed with use of the five-membered iodoacetamide pyrrolidinyl moiety. There is retention of isoclinic (Marriott and Griffith, 1974) points in the high- and low-field regions throughout the course of deoxygenation. On the other hand, with the piperidinyl analog, deviations from isoclinicity are observed upon complete deoxygenation of the solution, as shown in the right side of Figure 14. These results indicate that the two spin-labels exhibit differential sensitivity to protein conformational changes that occur in the $\alpha_1\beta_2$ interface upon deoxygenation. In addition, the loss of isoclinic points in the spectra of the piperidinyl derivative reflects structural changes in the heme environment and in the $\alpha_1\beta_2$ interface that are directly associated with cooperative ligand binding. Use of this spin-label with chemically modified or mutant hemoglobins that lack the property of cooperative ligand binding because of amino acid substitutions in the $\alpha_1\beta_2$ interface region does not result in a loss of isoclinic points upon deoxygenation (Ho *et al.*, 1970; Baldassare *et al.*, 1970).

Through X-ray crystallographic studies, it was first recognized that the side chain residues of amino acids in the heme crevice may sterically hinder the entrance of ligands in binding to the heme (Perutz, 1970). This observation implied differential reactivity of the heme groups in the α and β subunits of hemoglobin. Numerous spectroscopic approaches have been applied to investigate the differences in protein–heme interactions in the α and β subunits that are associated with differences in heme reactivity. One approach has been through the application of the spin-label method. With use of appropriate chemical methods, the propionic acid substituents of the heme group can be covalently attached to spin-label residues (see Figure 15) with insertion of the spin-labeled heme group into apohemoglobin (Asakura

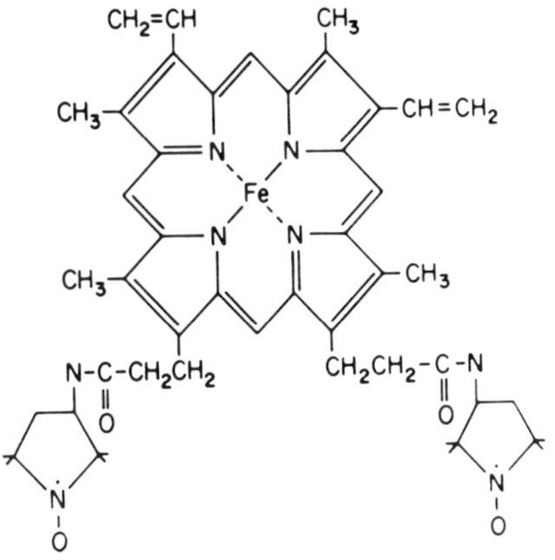

Figure 15. The structure of the di-spin-labeled protoheme synthesized by Asakura *et al.* (1971).

et al., 1969). The paramagnetic resonance spectra of reconstituted mono-spin-labeled heme in hemoglobin are shown in Figure 16; the monosubstituted heme has been selectively incorporated either into the α subunit or into the β subunit (Lau and Asakura, 1979). The spectra give evidence for two components, corresponding to strongly immobilized and weakly immobilized conformations of the spin-label. The relative populations of the

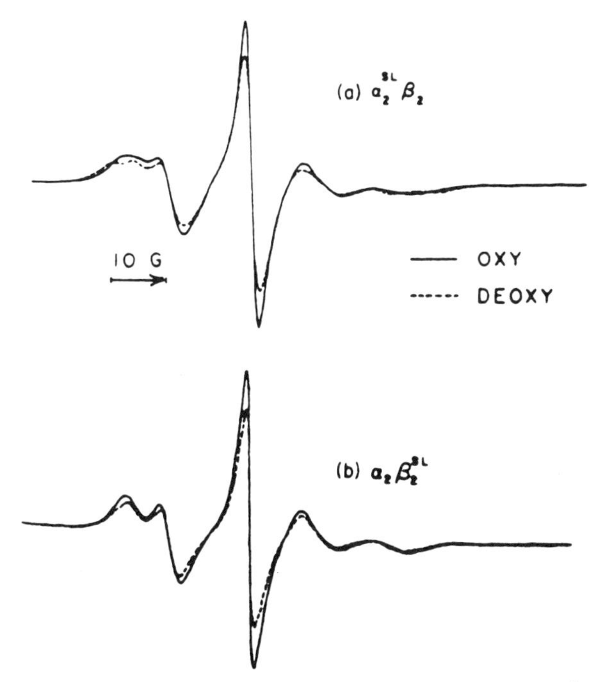

Figure 16. Changes in the paramagnetic resonance spectra upon deoxygenation of mono-spin-labeled heme inserted into apohemoglobin. (*a*) The spin-labeled heme is in the α subunits. (*b*) The spin-labeled heme is in the β subunits. From Lau and Asakura (1979) with permission.

two conformations can be manipulated by changes in the extent of binding of molecular oxygen and by changes in the oxidation and spin state of the heme iron. With respect to changes in ligand and oxidation state, the spectra of the α- and β-hemes are not equivalent. Thus, the spin-labeled heme derivatives provide evidence for nonequivalent structural interactions between the heme and its protein environment.

It is not possible, however, to assign specific configurations of the spin-labels for each of the immobilized states of the spin-labeled heme derivatives, since the paramagnetic resonance spectra must arise not only from different conformational states of the spin-label with respect to the protein but also from unresolved, chirally active enantiomers. Moreover, recent nuclear magnetic resonance studies have shown that reconstitution of myoglobin from apomyoglobin and heme may lead to "disordering" of the complex so that two types of protein–heme complexes are formed, differing by a 180° rotation around an axis through the α and γ methene carbon atoms of the heme (LaMar *et al.*, 1978). The similar heme environments of hemoglobin and myoglobin with nearly identical protein–heme contacts (Perutz, 1965) suggest that comparable disordering is to be expected for spin-labeled heme inserted into apohemoglobin.

3.2. α-Chymotrypsin

Chymotrypsin is a member of the serine protease family that catalyzes the hydrolysis of peptide bonds on the carboxyl side of the aromatic amino acids tryptophan, tyrosine, and phenylalanine. In the hydrolytic reaction catalyzed by this enzyme, the active-site residue Ser-195 is acylated by the substrate to form a covalent acyl-enzyme intermediate, which was initially detected by use of chromophoric substrates (Hartley and Kilbey, 1954; Kèzdy and Bender, 1962). The reaction schematically outlined in Figure 17 can be represented by Equation (2):

$$\mathrm{E} + \mathrm{S} \underset{K_S}{\rightleftharpoons} \mathrm{ES} \xrightarrow{k_2} \underset{+\mathrm{P}_1}{\mathrm{ES}'} \xrightarrow{k_3} \underset{+\mathrm{P}_1}{\mathrm{EP}_2} \rightleftharpoons \mathrm{E} + \mathrm{P}_2 + \mathrm{P}_1 \qquad (2)$$

where ES′ represents the acyl-enzyme intermediate, P_1 and P_2 are the first and second products of hydrolysis, and K_S is the substrate dissociation constant.

The structural relationships of the acyl group to active site residues in the reaction intermediate are illustrated in Figure 18, based on the difference Fourier synthesis of indoleacryloyl-α-chymotrypsin (Henderson, 1970), a derivative that is sufficiently stable at pH 4 for crystallographic investigations. The indoleacryloyl group has acylated the $—CH_2OH$ side chain of Ser-195 in a planar *trans-S* configuration. A hydrogen-bonded water molecule is observed between the N_ε atom of His-57 and the carbonyl oxygen of the acyl moiety. Model building studies suggest that this water molecule

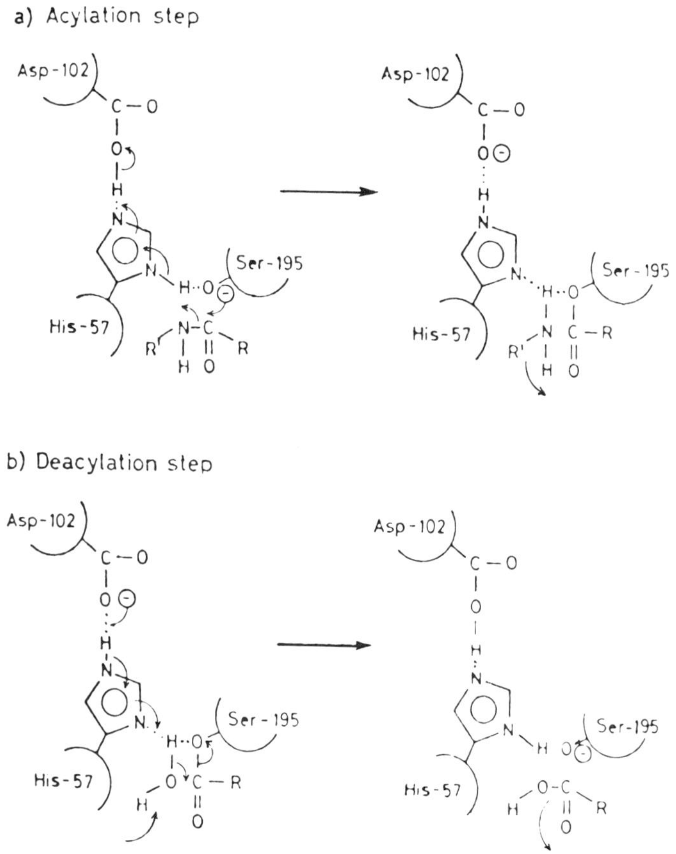

Figure 17. The mechanism of substrate hydrolysis catalyzed by α-chymotrypsin. The catalytically active residues are indicated.

would be in a favorable orientation for nucleophilic attack upon the transformation of the carbonyl carbon atom into a tetrahedral configuration. Studies with the chirally resolved *p*-nitrophenyl esters of 3-carboxy-2,2,5,5-tetramethylpyrrolidine-1-oxyl have helped to confirm these mechanistic details and to show how the relative orientation of the carbonyl group to His-57 determines the hydrolytic reactivity of the acyl-enzyme intermediate (Bauer and Berliner, 1979).

Kinetic data for hydrolysis of the $R(+)$ and $S(-)$ enantiomers of the ester substrate (Flohr *et al.*, 1975) are compared in Table I with those of specific ester substrates. For both spin-label substrates, breakdown of the acyl-enzyme intermediate [ES′ of Equation (2)] is the rate-limiting step, and the values of the rate constants k_2 and k_3 for the $R(+)$ form are 9- and 21-fold greater than those for the $S(-)$ enantiomer. No stereospecific effects are reflected in the value of K_S, which governs the equilibrium of the binding

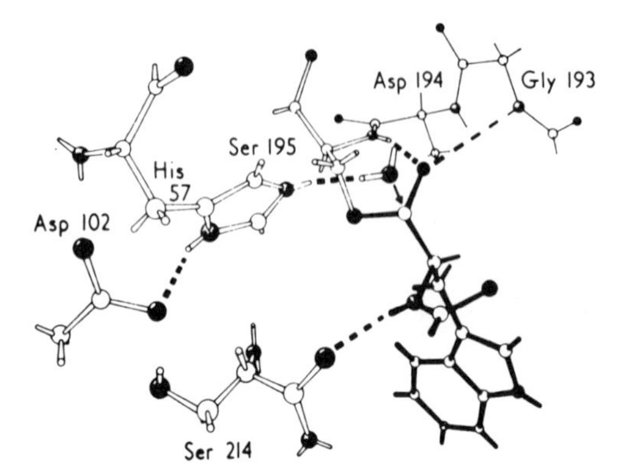

Figure 18. The postulated model of the reactive acyl-enzyme reaction intermediate of α-chymotrypsin, based on the difference Fourier synthesis of 3-(indole)acryloyl-Ser-195-α-chymotrypsin. The arrow shows the possible direction of attack on the carbonyl carbon of the substrate by the water molecule hydrogen bonded to His-57. From Henderson (1970) with permission. Copyright: Academic Press, Inc. (London) Ltd.

step. Under appropriate conditions (pH ~4.2), crystals of α-chymotrypsin can be reacted with either enantiomer of the spin-label ester substrate to form a stable crystalline intermediate in which Ser-195 is acylated by the substrate (Bauer and Berliner, 1979). Each of the spin-labels exhibits a unique orientation in the active site, and the structural relationships of both the spin-labels to active-site residues have been determined on the basis of the orientation of the spin-label, as defined by the hyperfine axes in the crystal through paramagnetic resonance experiments and model building studies.

On the basis of the direction cosines of the nitrogen hyperfine axes to the crystal axes, molecular models of each of the acyl moieties of spin-label substrates were constructed in the active site of the enzyme (Bauer and Berliner, 1979). A comparison of the orientation of the $R(+)$ and $S(-)$ nitroxide acyl groups in the active site of α-chymotrypsin is made in Figures 19 and 20. For both acyl-enzymes the pyrrolidinyl group can be positioned so as to extend outside the hydrophobic pocket. This position suggests that for the acyl-enzyme in solution, the nitroxide ring may have some motional freedom, consistent with the moderate immobilization reflected in spectra of the spin-labeled acyl-enzyme in solution. The most significant difference in the two acyl-enzymes determined by these model building studies is the position of the carbonyl oxygen of the acyl moiety. In the case of the $S(-)$ acyl-enzyme, the carbonyl oxygen is directed toward the N_ε atom of His-57, whereas for the $R(+)$ acyl-enzyme, it is directed away from the N_ε atom (see Figures 19 and 20). The different orientation determined for the $R(+)$ acyl-enzyme suggests that the carbonyl carbon is more favorably oriented for attack by the bridging hydrogen bonded water molecule (see Figure 18). This stereochemical configuration is consistent with the 21-fold greater rate of

Table I. Rate and equilibrium constants for the α-chymotrypsin catalyzed hydrolysis of specific ester substrates

Substrate	K_S (M)	k_2 (s^{-1})	k_3 (s^{-1})	Reference
$R(+)$ spin label[a]	4.1×10^{-4}	37×10^{-2}	52×10^{-4}	Flohr *et al.* (1975)
$S(-)$ spin label[a]	5.1×10^{-4}	4.1×10^{-2}	2.5×10^{-4}	Flohr *et al.* (1975)
N-acetyl-L-Trp ethyl ester[b]	0.21×10^{-4}	35	0.84	Brandt *et al.* (1967)
N-Furoylacryloyl-L-Tyr ethyl ester[b]	0.07×10^{-4}	53	1.5	Himoe *et al.* (1969)
N-β-(3-indole)acryloyl-imidazole	—	—	6×10^{-5}	Rossi and Bernhard (1970)

[a]3-Carboxy-2,2,5,5-tetramethylpyrrolidine-1-oxyl-*p*-nitrophenyl ester. Results are given for kinetic studies at 25°C and pH 7.
[b]Results are reported for kinetic studies at 25°C and pH 5.

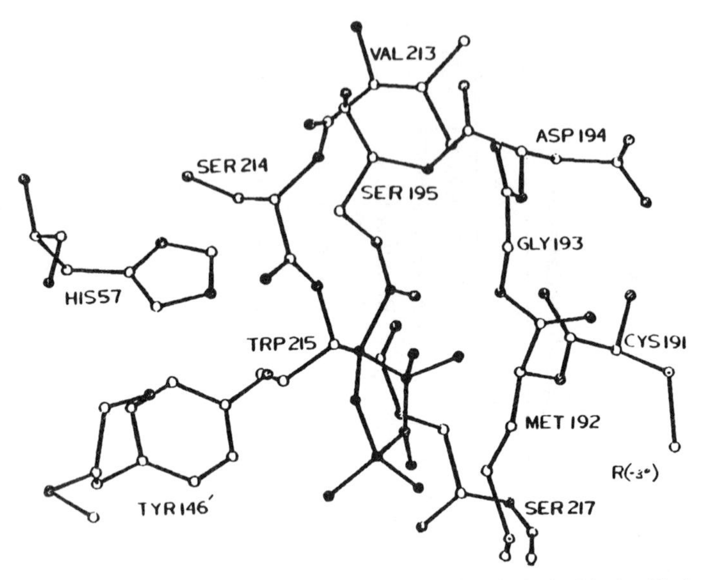

Figure 19. Diagram of the active site of α-chymotrypsin labeled with the $R(+)$ nitroxide acyl group of 2,2,5,5-tetramethylpyrrolidinyl-1-oxyl. The diagram is based on Figure 10 of Bauer and Berliner (1979). (○) C; (◐) N; (●) O; (⊙) S.

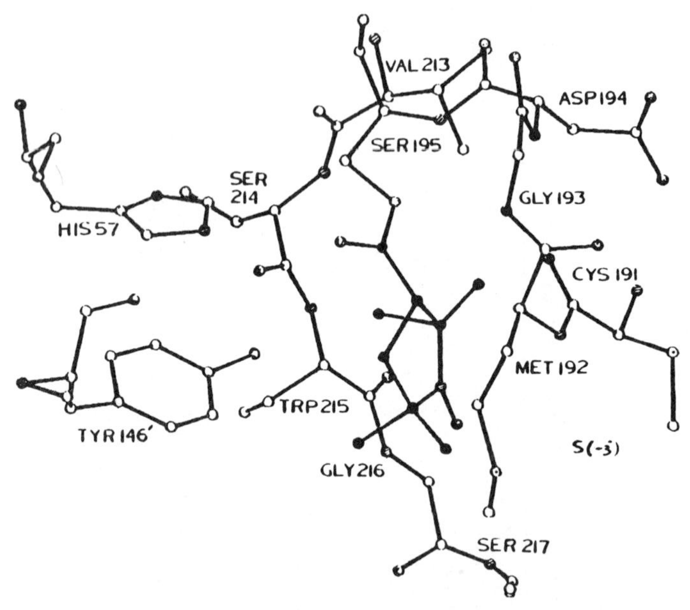

Figure 20. Diagram of the active site of α-chymotrypsin labeled with the $S(-)$ acyl nitroxide of 2,2,5,5-tetramethylpyrrolidinyl-1-oxyl. The diagram is based on Figure 11 of Bauer and Berliner (1979). Atoms are labeled as in Figure 19.

deacylation determined in kinetic studies (Flohr *et al.*, 1975). On the other hand, the orientation of the less reactive $S(-)$ enantiomer is more similar to that found for the acyl residue in the kinetically sluggish derivative of indoleacryloyl-α-chymotrypsin at pH 4.

Although a variety of other spin-label derivatives have been employed for acylating the active site of serine proteases (see Morrisett, 1976), detailed stereochemical information has not been derived through ancillary model building or X-ray structural studies. The results of Bauer and Berliner (1979) illustrate particularly well how a detailed spectroscopic analysis of the structural relationships of spin-label–enzyme interactions can lead to valuable mechanistic conclusions about enzyme catalyzed reactions.

3.3. Lysozyme

The active site cleft of hen egg white lysozyme has six subsites, conventionally labeled A–F, each accommodating a saccharride unit of an oligomer of N-acetylglucosamine (GlcNAc) for substrate hydrolysis. The structural relationships of substrate binding to active site residues are illustrated in Figure 21. The results of chemical, crystallographic, and nuclear magnetic resonance studies indicate that specific hydrogen bonding relationships of the acetamido side chain of the saccharide unit in subsite C are required to orient the substrate into a catalytically productive configuration in subsites D and E (Imoto *et al.*, 1972). Bond cleavage between the residues in subsites D and E is catalyzed by the side chains of Glu-35 and Asp-52. The specific binding relationships in subsite C that are thought to account for the energetic requirements of geometric distortion of the substrate in subsite D obtain not only in hydrolysis of naturally occurring substrates derived from chitin or N-acetylmuramic acid but also with synthetic, mixed-disaccharide inhibitors such as β-(D-glucopyranosyl)-D-(2-acetamido-2-deoxyglucopyranoside) and mixed-disaccharide substrates such as the aromatic linked glycosides of β-(p-nitrophenyl-β-D-glucopyranosyl)-D-(2-acetamido-2-deoxyglucopyranoside). These binding requirements induced by interactions of the N-acetyl group with residues in subsite C and the relative positioning of substrates, inhibitors, and mixed-disaccharide substrates are schematically illustrated in Figure 22.

The spin-label inhibitor analogs of lysozyme illustrated in Figure 23 have been synthesized with strategically positioned N-acetyl groups to mimic the binding relationships of GlcNAc saccharide units in the active site. Berliner (1971) synthesized the spin-label inhibitor 4-N-acetamido-2,2,6,6-tetramethylpiperidine-1-oxyl and characterized its binding relationships to lysozyme by difference Fourier methods. From three-dimensional intensity data collected to only 6 Å resolution, three binding sites of the inhibitor were located, two in the active site cleft in subsites A and C, and one in a hydrophobic pocket near Trp-123. The difference Fourier synthesis calculated on the basis of low resolution data did not allow an assignment of

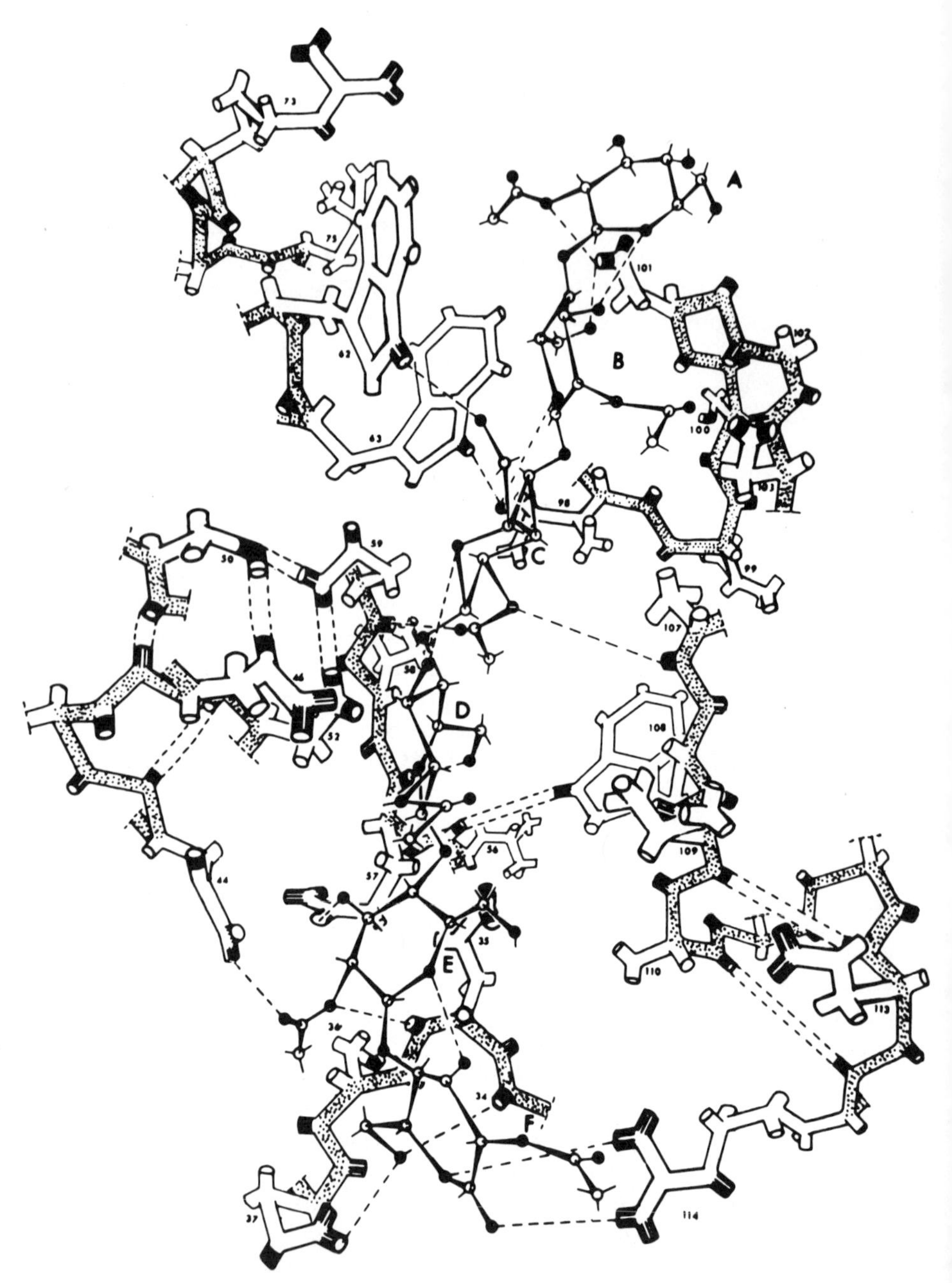

Figure 21. Illustration of the arrangement of an N-acetylchitohexaose $(GlcNAc)_6$ molecule bound in the active site of lysozyme. The carbohydrate residues in subsites *A*, *B*, and *C* are observed in the binding of N-acetylchitotriose. Residues in subsites *D*, *E*, and *F* occupy positions inferred from model building. From Johnson *et al.* (1968) with permission.

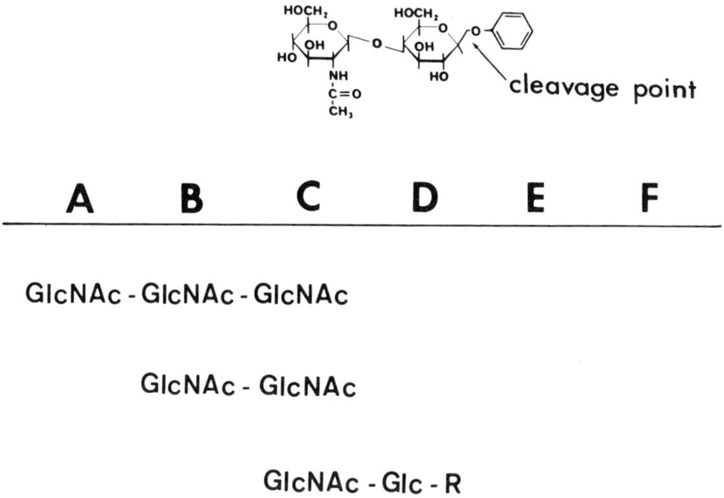

cleavage point

A B C D E F

GlcNAc - GlcNAc - GlcNAc

GlcNAc - GlcNAc

GlcNAc - Glc - R

Figure 22. Comparison of binding modes of oligosaccharides and mixed glycosides to lysozyme. The enzyme is regarded as having an active site with six subsites, each accommodating a GlcNAc ring as shown in Figure 21. Subsites *A*, *B*, and *C* form tight complexes with GlcNAc, whereas others do not. The relative positions of GlcNAc units in the subsites *A–F* are illustrated in the lower part of the diagram. *GlcNAc-GlcNAc-GlcNAc* represents the oligosaccharide chitotriose $(GlcNAc)_3$ and *GlcNAc-Glc-R* represents the aryl-linked mixed-disaccharide substrate β-(phenyl)-4-O-(2-acetamido-2-deoxy-β-D-glucopyranosyl)-D-glucopyranoside. The structure of this mixed-disaccharide substrate and position of bond cleavage is illustrated in the upper part of the diagram.

the binding configurations of the spin-label but showed that inhibitor molecules are to be found in each site. Hydrogen bonding interactions of the N-acetyl group with the carbonyl oxygen of Ala-107 and the peptide NH of Asn-59 in subsite C were assumed to be the stabilizing factors for orienting the inhibitor in a configuration similar to that expected for a GlcNAc residue. However, a different orientation stabilized by hydrophobic interactions of the four methyl groups of the spin-label with active site residues leaving the N-acetyl group to protrude into the solvent channels of the crystal could not be excluded in interpreting the difference Fourier map.

The highest occupancy of the three binding sites was for the inhibitor in the hydrophobic pocket near Trp-123. This is not in the active site cleft and bears no direct relevance to the binding of natural substrates. An approximate fit of a molecular model to the difference electron density features for the site near Trp-123 suggested that the piperidinyl ring is parallel to the indole ring. The binding of 4-N-acetyl-2,2,6,6-tetramethylpiperidine-1-oxyl to the Trp-123 residue of lysozyme in solution has been confirmed by nuclear magnetic resonance studies in which the bound nitroxide served as a paramagnetic probe to enhance the nuclear spin-lattice relaxation of the C-2 proton of His-15 (Wien *et al.*, 1972). Binding of the spin-label in subsites C and A, although of lower occupancy in the crystal than that near Trp-123,

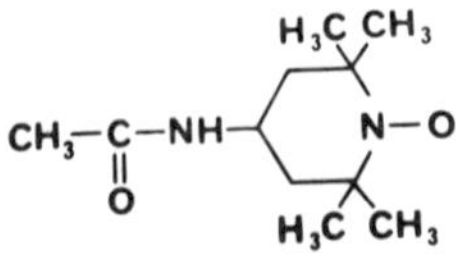

Figure 23. Structural formulas of nitroxide spin-label compounds specific for binding to lysozyme. (*I*) Structure of N-(2,2,6,6-tetramethyl-4-piperidinyl-1-oxyl)acetamide; (*II*) structure of the inhibitor β-(2,2,5,5-tetramethyl-3-carbinol-pyrrolidinyl-1-oxyl)-4-O-(2-acetamido-2-deoxy-D-glucopyranosyl)-(2-acetamido-2-deoxy-β-D-glucopyrano-side) employed in the studies of Wien *et al.* (1972); and (*III*) the structure of β-(2,2,6,6-tetramethyl-4-piperidinyl-1-oxyl)-O-(2-acetamido-2-deoxy-D-glucopyranoside) employed in the studies of Makinen (1976). The monosaccharide tetramethylpyrrolidinyl-nitroxide inhibitor employed by Wien *et al.* (1972) is not pictured.

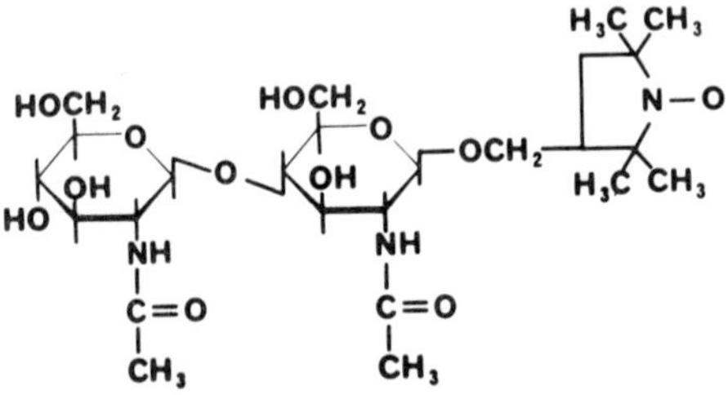

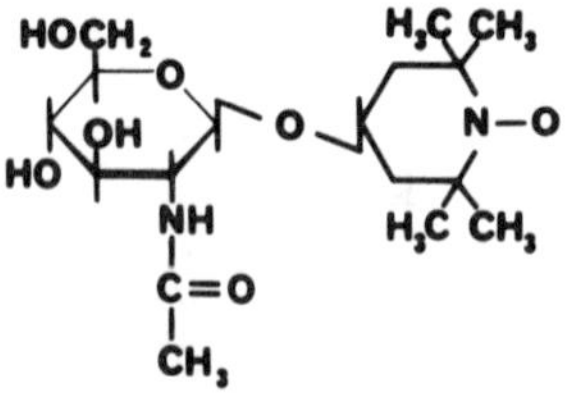

must also occur for the molecule in solution since the spin-label is an inhibitor of the enzyme.

Two glycoside inhibitors with the five-membered spin-label group as the aglycone residue were also synthesized to probe active site binding relationships (Wien *et al.*, 1972). The disaccharide spin-label inhibitor is illustrated as Structure II in Figure 23. On the basis of the binding configurations of GlcNAc and $(GlcNAc)_2$ defined by X-ray studies (Imoto *et al.*, 1972), the monosaccharide analog should occupy subsites C and D, whereas the disaccharide derivative should occupy subsites B, C, and D. In both cases, it is expected that the aglycone spin-label moiety occupies subsite D with the N-oxyl group near the catalytic side chains of Asp-52 and Glu-35. The broadening (~1 Hz) of the C-2 proton of His-15 and the estimated binding constants of 185 and 150 M^{-1} for the monosaccharide and disaccharide inhibitors, respectively, led to calculated distances of 20 and 19 Å between the nitrogen atom of the nitroxide and the C-2 proton of His-15 (Wien *et al.*, 1972). The measured distance on the molecular model of the enzyme is approximately in the 15–21 Å range, in good agreement with the *assumed* binding configuration of both spin-label inhibitors.

However, the results of a combined crystallographic and spectroscopic investigation of the binding of the spin-label β-(2,2,6,6-tetramethyl-4-piperidinyl-1-oxyl)-O-(2-acetamido-2-deoxy-D-glucopyranoside) to lysozyme have demonstrated that the inhibitor binds in an anomalous manner because of the hydrophobic interactions of the methyl groups (Makinen, 1976). This inhibitor is illustrated in Figure 23 as Structure III. A stereoview of the binding configuration of the inhibitor in the active-site cleft of hen egg white lysozyme determined by difference Fourier methods is illustrated in Figure 24. The acetamido group of the carbohydrate unit forms hydrogen bonds with Ala-107 and Asn-59 in subsite C, as do oligomers of GlcNAc in catalytically active configurations. However, the carbohydrate residue is rotated so that the reducing end with the spin-label aglycone residue is directed toward subsite B. The methyl groups make hydrophobic contacts with Trp-62 and Trp-63 in the active-site cleft. These amino acid residues normally make hydrogen bonded contacts with the carbohydrate OH groups through their pyrrolic N-H units. In the normal binding configuration of GlcNAc in subsite C, the methyl group of the acetamido side chain is directed toward the indole ring of Trp-108 to form a hydrophobic contact (Imoto *et al.*, 1972). In the rotated configuration of the GlcNAc residue illustrated in Figure 24, this interaction is no longer possible. It is probable that energetically the simple additive effect of two methyl–indole hydrophobic interactions formed in subsite B with the tetramethylpiperidinyl ring is greater than that of the methyl group of the acetamido group in subsite C with Trp-108 alone. These hydrophobic interactions may account for stabilization of the unexpected binding configuration.

The binding configuration of this spin-label in the active site cleft of lysozyme is unusual and highly anomalous (Makinen, 1976). In this study, the spin-label inhibitor was synthesized via the lysozyme catalyzed transglycosylation reaction, comparable to the coupling of cyclohexanol to oligomers of GlcNAc (Rupley *et al.*, 1968). The observation that the enzyme can be employed to synthesize the inhibitor via the transglycosylation reaction indicates that the 2,2,6,6-tetramethyl-4-piperidinol ring is sterically compatible with subsite D. Because of the 16° angle that the N—O bond makes with the C—N—C plane (Berliner, 1970), the aglycone residue derived from the 2,2,6,6-tetramethyl-4-piperidinol-1-oxyl moiety exhibits a "flattened" chair-like configuration (Figure 3) not unlike that expected for a distorted carbohydrate moiety of the substrate in subsite D during active

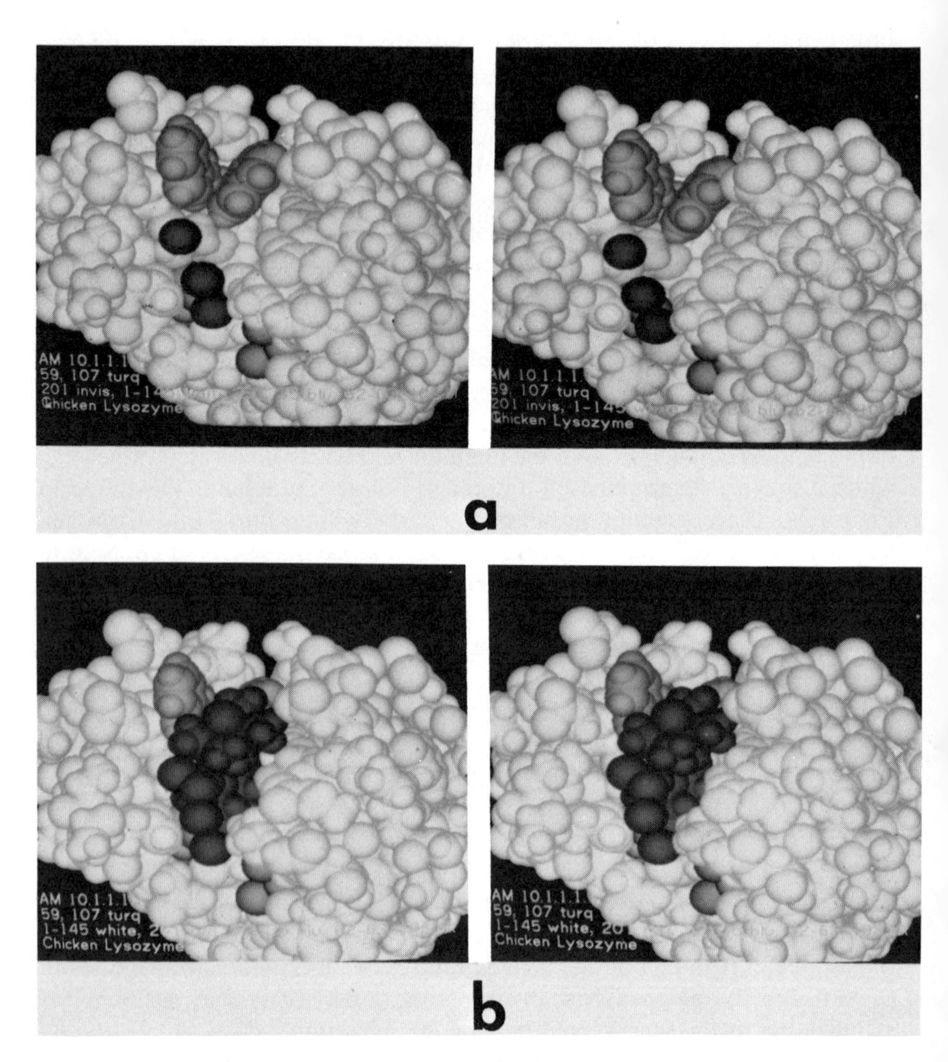

Figure 24. Stereoview of the inhibitor *III* of Figure 23 bound in the active site of lysozyme according to the X-ray defined coordinates of the inhibitor molecule (Makinen, 1976, and unpublished results). (*a*) The active-site cleft is shown for the enzyme as illustrated through the molecular-graphics surface-display method of Feldmann *et al.* (1978) using the 2.0-Å refined coordinates of hen egg white lysozyme (Diamond, 1974). The protein residues appear in white and the active-site residues important in binding to substrates and to the nitroxide spin label *III* of Figure 23 are shaded. The residues Trp-62 and Trp-63 in subsite B are seen as the light gray features in the upper part of the figure, and the catalytically active residues Glu-35 and Asp-52 in subsite D appear in the lower part of the figure. The three dark, spherical features in the front part of the active site are hydrogen-bonded water molecules that are displaced upon substrate binding. (*b*) The binding configuration of β-(2,2,6,6-tetramethyl-4-piperidinyl-1-oxyl)-O-(2-acetamido-2-deoxy-D-glucopyranoside) is illustrated as defined through difference Fourier studies. For interpretation of the difference Fourier map, an idealized structure was constructed from

catalysis (Ford *et al.*, 1974). In addition, a space-filling CPK model of the inhibitor is structurally compatible with the active site of the enzyme with the acetamido group in subsite C serving to orient the aglycone residue into subsite D, as would be anticipated for oligomers and mixed glycosidic derivatives of GlcNAc. This model building study thus suggested that binding of the spin-label glycoside should be compatible with the steric constraints of the active site. In contrast, CPK models of the pyrrolidinyl analog used by Wien *et al.* (1972) cannot be sterically accommodated by the active site with the nitroxide moiety in subsite D (Makinen, 1976).

Together with the anomalous binding of 4-acetamido-2,2,6,6-tetramethyl-piperidine-1-oxyl to Trp-123 as shown by X-ray studies, the difference Fourier studies of the complex of the spin-label glycoside with lysozyme illustrated in Figure 24 show that spin-labels have marked hydrophobic interactions with protein residues. These interactions may lead to severe structural perturbations and prevent spin-label substrate analogs from binding in configurations analogous to those of catalytically active substrates. It is likely that the pyrrolidine analogs employed by Wien *et al.* (1972) also bind in a configuration analogous to that shown in Figure 24. The association equilibrium binding constants of the two tetramethylpyrrolidinyl-glycosides of GlcNAc and $(GlcNAc)_2$ with lysozyme are 185 and 150 M^{-1}, respectively (Wien *et al.*, 1972), whereas those for naturally occurring monosaccharide and disaccharide inhibitors alone are 30 and 1000 M^{-1} (Imoto *et al.*, 1972). Binding of two GlcNAc units into the active-site cleft in a catalytically active configuration should be reflected by a significant increase in the binding constant over that for a monosaccharide derivative. This is not the case with the two pyrrolidinyl glycosides employed by Wien *et al.* (1972). Furthermore, the distance from His-15 to the pyrrolidinyl ring when placed into subsite B to correspond to the anomalous binding configuration of the spin label in Figure 24 estimated from the molecular model of lysozyme is compatible (Makinen, 1976) with the distance of ~20 Å measured in nuclear magnetic resonance studies by Wien *et al.* (1972). Since such anomalous binding configurations have no counterpart in the interaction of naturally occurring substrates and inhibitors with enzymes, the results emphasize the caution that must be placed in the application of spin-label derivatives to protein structural studies.

---

molecular fragments of 2,2,6,6-tetramethyl-4-piperidinol-1-oxyl (Berliner, 1970) and GlcNAc (Mo and Jensen, 1975) joined according to the stereochemistry of the glycosidic linkage in β-aryl-linked glucopyranosides (Brehm and Moult, 1975; Makinen and Isaacs, 1978). The atomic coordinates were defined by a least-squares fit of this idealized structure to the difference electron-density map. The inhibitor is seen as the dark gray shaded residue in the central part of the figure. The nitroxide group is bound in subsite B, and the tight interactions of the methyl groups with Trp-62 and Trp-63 are evident. The nitroxide nitrogen atom was determined to be 3.7 Å distant from the indole NH group by electron nuclear double resonance methods (Makinen, 1976).

4. APPLICATION OF CRYOENZYMOLOGY FOR STRUCTURAL CHARACTERIZATION OF A SPIN-LABELED ENZYME REACTION INTERMEDIATE

4.1. The Cryoenzymologic Approach to the Study of Enzyme Mechanisms

Enzyme catalyzed reactions, as illustrated by Equation (1), consist of a series of sequential steps. Each step is characterized by a rate constant k_i and enthalpy of activation. The relationship between these two parameters, as derived by transition-state theory, is shown in Equation (3):

$$k_i = \frac{kT}{h} e^{\Delta S\ddagger/R} \; e^{-\Delta H\ddagger/RT} \tag{3}$$

The step with the smallest rate constant is the rate-determining step. For most enzymic reactions, the rate-determining step is associated with an enthalpy of activation in the range of 12–20 kcal/mole. On this basis, lowering of the temperature from +25 to −50 or −100°C decreases the velocity of the rate-limiting step by a factor of 2×10^3 to 3×10^{10}, and the reaction can be brought consequently to a negligible rate of turnover in fluid media. Under these conditions, the intermediates I_i preceding the rate-determining step can be accumulated, and those that are formed under the condition $k_i \gg (k_{-i} + k_{i+1})$ are the most readily stabilized.

The major value of cryoenzymology is the potential to accumulate reaction intermediates for structural characterization, often in stoichiometric concentrations. This approach can be employed to prepare reaction intermediates with half-lives sufficiently long for investigation by ordinary spectroscopic and X-ray diffraction methods. This method, thus, stands in contrast to the rapid-freeze-quenching experiment (Bray, 1961) employed to prepare temperature-labile derivatives for EPR studies. The rapid-freeze-quenching method yields only low concentrations of reaction intermediates since the reaction is initiated under near steady-state conditions at normal temperatures. The successful introduction of cryoenzymology in biochemical studies is due primarily to the studies of Douzou and co-workers (Douzou *et al.*, 1970; Douzou, 1974, 1977). A useful compendium of the physical properties of cryosolvents suitable for proteins has been provided by Douzou *et al.* (1976), and the kinetic relationships that define the accumulation of intermediates have been discussed by Fink (1976). Recent reviews summarize results of cryoenzymologic studies (Fink, 1977; Douzou, 1979; Fink and Cartwright, 1981), and a detailed discussion of the physical chemistry of cryosolvent systems applied to enzymes in crystals and in solution is given by Makinen and Fink (1977).

Since structural characterization is an important objective of cryoenzymologic studies, it is essential to ensure that the catalytic reaction pathway in cryosolvent mixtures remains unchanged from that in pure aqueous medium and that the enzyme–substrate species accumulated under low-temperature conditions are identical to the catalytically active intermediates of the enzyme reaction at normal temperatures. Cryosolvents frequently exhibit an inhibitory influence on enzyme reactions and may alter the chemical pathway or the conversion of substrates to products. According to whether the cryosolvent binds to the free enzyme or to a reaction intermediate, different inhibition patterns may be observed, depending upon the relationship of the formation of the cryosolvent-bound species to the rate-limiting step. The influence of cryosolvent inhibitor binding may lead then to the detection of "new" spectral species that are mistaken as catalytic reaction intermediates. The assignment of "new" spectral species accumulated at subzero temperatures as the true catalytic intermediates of the reaction rests upon rigorous application of kinetic criteria. Such criteria probably have been most extensively applied to the esterolytic reaction catalyzed by carboxypeptidase A (Makinen *et al.*, 1979).

4.2. The Structure of a Low Temperature Stabilized Spin-Labeled Reaction Intermediate of Carboxypeptidase A

The mechanism of action of carboxypeptidase A has been conjectural despite a wide variety of kinetic, structural, and chemical studies. Neither had the roles of catalytic residues of the enzyme been defined unambiguously for both ester and peptide hydrolysis, nor had intermediates formed subsequent to substrate binding been detected by application of classical rapid kinetic techniques. With use of cryoenzymologic methods, temporal resolution of the reaction between carboxypeptidase A and a specific ester substrate was first achieved by Makinen and co-workers (1976, 1979). These studies demonstrated that hydrolysis of the specific ester substrate O-(*trans-p*-chlorocinnamoyl)-L-β-phenyllactate (ClCPL) proceeds via formation of the covalent mixed-anhydride (acyl-enzyme) intermediate schematically illustrated in Figure 25. By lowering the temperature to $-60°C$, the intermediate is accumulated with a half-life of $\sim$120 min, sufficient for spectroscopic characterization. Figure 26 illustrates the difference (ultraviolet) absorption spectrum of the low temperature stabilized reaction intermediate formed with the specific ester substrate ClCPL and with both the native enzyme and the paramagnetic, Co^{2+}-reconstituted form of carboxypeptidase A. Of considerable importance in these studies has been the demonstration that Co^{2+}-carboxypeptidase A catalyzes the esterolytic reaction by the same mechanism as the native enzyme (Makinen *et al.*, 1979).

To structurally characterize the low temperature stabilized reaction intermediate of carboxypeptidase A, the spin-label ester O-3-(2,2,5,5-tetramethylpyrroline-1-oxyl)-propene-2-oyl-L-β-phenyllactate (TEPOPL) was synthesized (see Figure 9). This spin-label ester, structurally analogous to the substrate ClCPL, has been shown by kinetic studies to be a specific

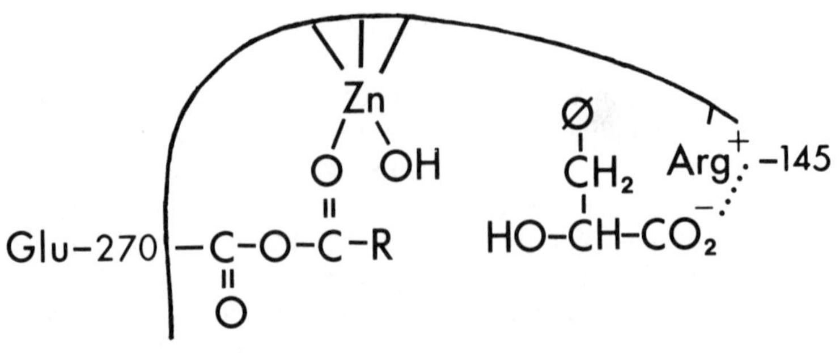

Figure 25. Structure of the mixed-anhydride (acyl-enzyme) reaction intermediate of carboxypeptidase A formed with specific ester substrates. The active site metal ion is coordinated by Glu-72, His-69, and His-196. The aromatic side chain and free terminal COOH-group of the substrate are bound in the active site by the hydrophobic pocket of Arg-145, respectively, and are presumably not released until breakdown of the acyl-enzyme intermediate. The chemical bonding structure is based on the results of chemical and kinetic studies of the enzyme reaction under cryoenzymologic conditions (Makinen *et al.*, 1979; Kuo and Makinen, 1982; Makinen *et al.*, 1983).

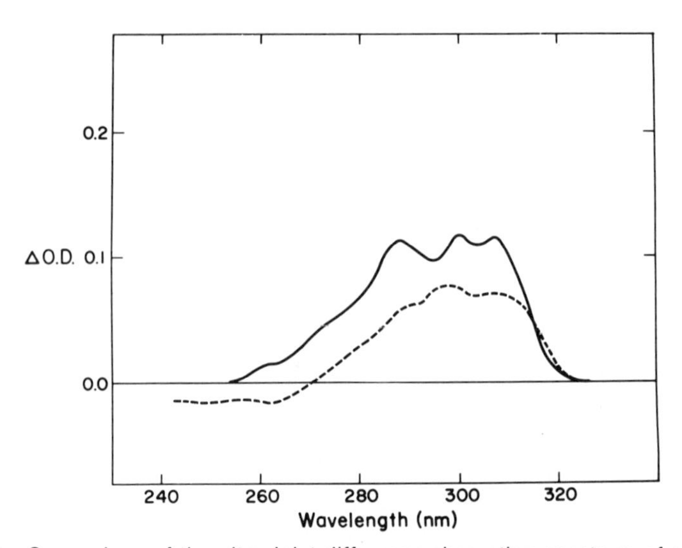

Figure 26. Comparison of the ultraviolet difference absorption spectrum of equimolar mixtures of carboxypeptidase A and the specific ester substrate ClCPL at −60°C versus free enzyme and substrate at ambient temperature. The solid line corresponds to the spectrum of the intermediate formed with the native enzyme; the broken line corresponds to the spectrum of the intermediate of the Co^{2+}-reconstituted enzyme. The fluid cryosolvent mixture consists of a buffered solution of ethylene glycol, methanol, and water in 40:20:40 (v/v) ratio. From Makinen *et al.* (1979) with permission.

substrate of carboxypeptidase A. Moreover, the mechanism of hydrolysis of the spin-label is identical to that of ClCPL (Kuo *et al.*, 1983). Kinetic parameters governing the hydrolysis of TEPOPL are compared to those of other specific substrates in Table II. Because of the prominent ultraviolet absorption intensity of the spin label, a difference spectrum of the corresponding low temperature stabilized reaction intermediate near −60°C is obseved as with ClCPL. The difference spectrum illustrated in Figure 27 shows a shift in maximum intensity to shorter wavelength and a decrease in absorption intensity in the long wavelength region. These changes are indicative of a structural perturbation and shortening of the conjugated system of double bonds in the spin-label substrate.

In Figure 28 are illustrated the (first derivative) EPR spectra of the spin-labeled reaction intermediate stabilized near −60°C formed with both the native enzyme and with Co^{2+}-carboxypeptidase A. The decrease in the peak-to-peak amplitude in the spectrum of the spin-label in the active site of the Co^{2+}-enzyme is due to the dipolar interaction of the Co^{2+} ion and the nitroxide spin-label. The apparent quenching of the signal intensity provides a means to estimate the radial separation of the paramagnetic metal ion and the nitroxide group. The pertinent relationships derived by Leigh (1970) define the linewidth δH of the nitroxide spin-label in the presence of the paramagnetic metal ion

$$\delta H = C(1 - 3\cos^2\theta_r)^2 + \delta H_0 \quad (4)$$

where δH_0 is the linewidth in the absence of the paramagnetic ion, θ_r is the azimuthal angle of the position vector between the two sites and the applied magnetic field, and C is a constant defined by the relationship

$$C = \frac{2\pi g \beta \mu^2}{hr^6} \tau_1 \quad (5)$$

In Equation (5), τ_1 is the electronic spin-lattice relaxation time of the metal ion, r is the radial separation between the two sites, and μ is the (effective) magnetic moment of the metal ion complex. The other quantities have their usual definitions. Since the values of C are calculated according to the percentage decrease in the signal amplitude of the nitroxide free radical (Leigh, 1970; Dwek, 1973), the distance between the two paramagnetic sites can be estimated if the values of μ and τ_1 are known. Proton relaxation enhancement studies (Kuo *et al.*, 1983) show that for Co^{2+}-substituted carboxypeptidase A, $\tau_1 \sim 8.2 \times 10^{-12}$ sec at −60°C, whereas magnetic susceptibility studies (Rosenberg *et al.*, 1975) have shown that the value of μ is 4.77 B.M. When these quantities are substituted into Equation (5), it can be calculated directly from the quenching of the spin-label signal amplitude in Figure 28 that the Co^{2+} ion is located 7.7 ± 0.2 Å from the nitroxide group. This distance is considerably shorter than that which would be anticipated for the spin-label substrate in a fully extended *trans* configuration of the olefinic bond of the propene-2-oyl side chain, and

Table II. Kinetic parameters governing the carboxypeptidase A catalyzed hydrolysis of specific ester and peptide substrates

Substrate	k_{cat} (s^{-1})	$K_{m(app)}$ (M)	Reference
TEPOPL	2.3	0.96×10^{-4}	Kuo *et al.* (1983)
ClCPL	73	0.98×10^{-4}	Makinen *et al.* (1979)
O-Hippuryl-L-β-phenyllactate	527	0.73×10^{-4}	Bunting *et al.* (1974)
Hippuryl-L-Phe	52	3.1×10^{-4}	Kuo *et al.* (1983)
Cbz-Gly-Gly-L-Phe[a]	133	0.25×10^{-4}	Auld and Vallee (1970)

[a]Cbz = carbobenzoxy-.

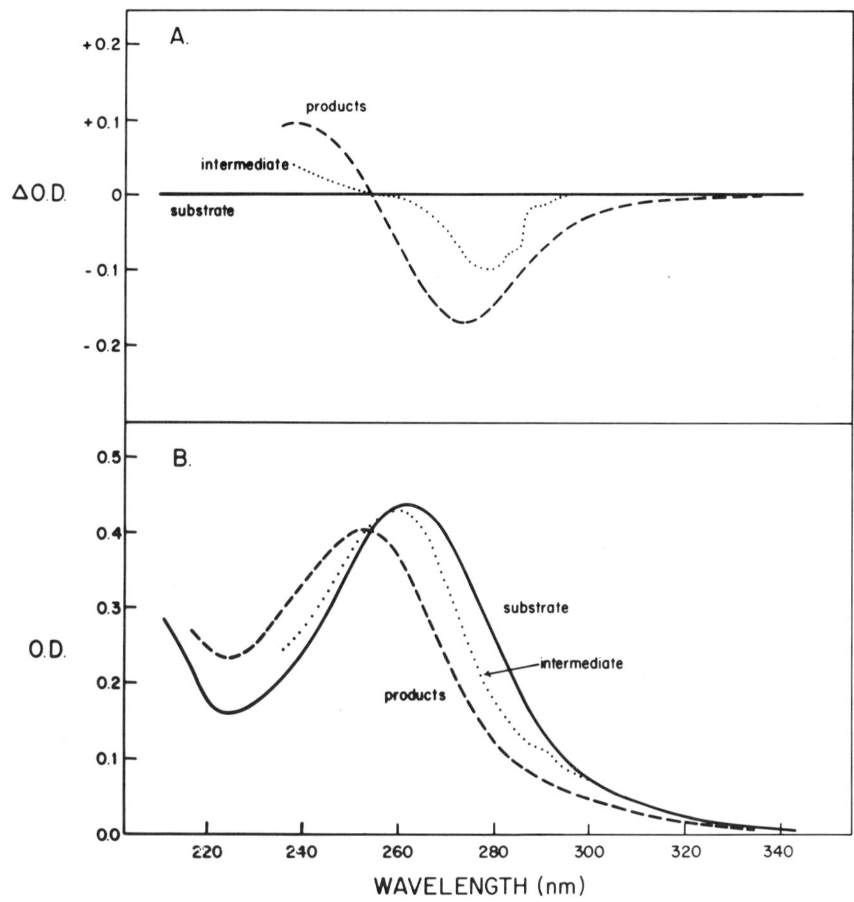

Figure 27. (*A*) Comparison of the ultraviolet difference absorption spectrum of an equimolar mixture of carboxypeptidase A and the spin-label ester substrate TEPOPL at −60°C versus free enzyme and substrate at ambient temperature. (*B*) The absolute absorption spectrum of the spin-label moiety in the low-temperature-stabilized reaction intermediate is constructed on the basis of the difference spectrum and the absorption spectrum of the free substrate. The shift in maximum absorption intensity to lower wavelength and loss of intensity in the long wavelength region signify a distortion of the cross-conjugated system of double bonds in the spin-label substrate upon formation of the reaction intermediate. From Kuo *et al.* (1983) with permission. Copyright: Academic Press, Ltd. (London).

model-building studies indicate that the intact spin-label substrate cannot be sterically accommodated by the active-site cleft without considerable rotation around the C=C double bond. To account for the distance of 7.7 Å estimated on the basis of magnetic resonance studies, the configuration of the substrate that can be accommodated by the active site and that is compatible with the estimate of 7.7 Å was determined with the aid of molecular-graphics methods (Kuo *et al.*, 1983).

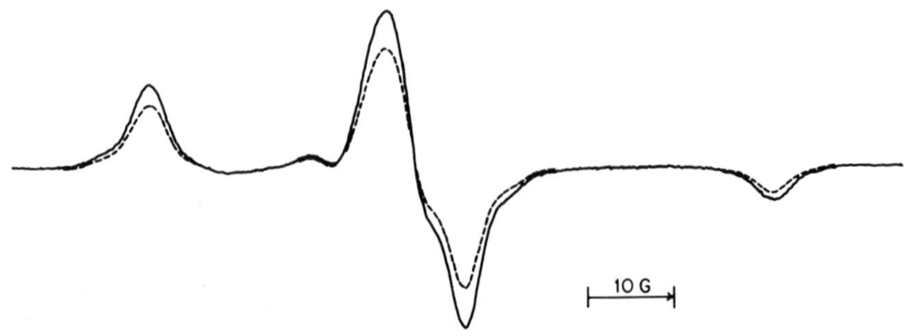

Figure 28. First derivative EPR absorption spectra of the acylenzyme intermediate formed with carboxypeptidase A and TEPOPL. The reaction intermediate was prepared directly in the EPR quartz sample tube maintained at −60°C. The solid line is the spectrum with native carboxypeptidase A; broken line is the spectrum with Co^{2+}-substituted carboxypeptidase A. The presence of the active site Co^{2+} ion results in a 25% reduction in the peak-to-peak amplitude of the spin-label spectrum. This corresponds to a distance of 7.7 Å between the Co^{2+} ion and the nitroxide group according to Equation (5), as derived for the dipolar interaction of two spins in a rigid-lattice matrix (Leigh, 1970). From Kuo *et al.* (1983) with permission. Copyright: Academic Press, (London) Ltd.

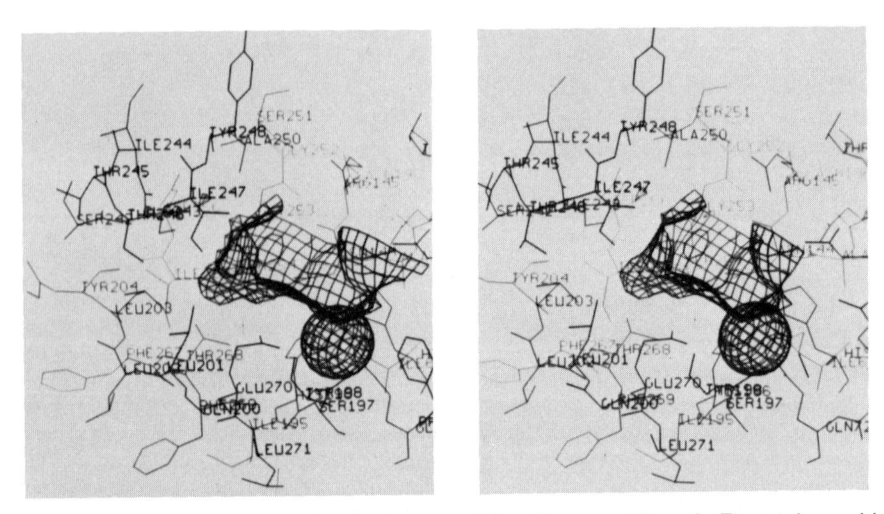

Figure 29. Stereodrawing of the active site of carboxypeptidase A. The amino acid residues in the immediate vicinity of substrate binding are illustrated as skeletal stick figures, whereas the van der Waals surface of the active site pocket into which the COOH-terminal residue of the substrate must fit is shown by the (chicken-wire-like) contours calculated on the basis of the molecular-graphics display system of Marshall *et al.* (1979). To produce this surface contour, a pseudo electron density map is generated for the nonhydrogen atoms of protein residues. The radii of these atoms are deliberately increased by an amount equal to the van der Waals radius of a "typical" nonhydrogen atom of the substrate. This contoured surface then defines the region within which a substrate molecule can bind to the enzyme through nonbonded interactions. The active site Zn^{2+} ion is represented by a sphere of 2.0 Å radius. The abrupt upper border of the surface contours indicates the interface of the active-site cleft with bulk solvent. From Kuo *et al.* (1983) with permission. Copyright: Academic Press, (London) Ltd.

In Figures 29 and 30 are illustrated, respectively, the active-site surface of carboxypeptidase A and two configurations of the spin-label ester in the active site of the enzyme. The substrate is shown as a skeletal stick figure, and the calculated "extended" van der Waals surface of all the active-site residues is represented by the (chicken-wire-like) contours. The requirement for allowed, nonbonded contacts between the substrate and protein residues is that the nonhydrogen atoms of the substrate must lie within the van der Waals surface. For the substrate molecule with a fully extended *trans* configuration of the side-chain olefinic group, there are severe violations of the requirements for nonbonded contacts. This means simply that the substrate cannot bind in the active site in a fully extended configuration. In Figure 30 is illustrated a configuration of the spin-label in the active site achieved by rotation around the C=C bond of the propene-2-oyl substituent and rotation of the bulky groups of the substrate around the C—C bonds adjacent to the olefinic double bond. In this rotated configuration, there are no violations of the requirements for nonbonded contacts, and the fit of the spin-label is sterically compatible with the active site surface. The distance between the nitroxide nitrogen atom and the metal ion for this configuration

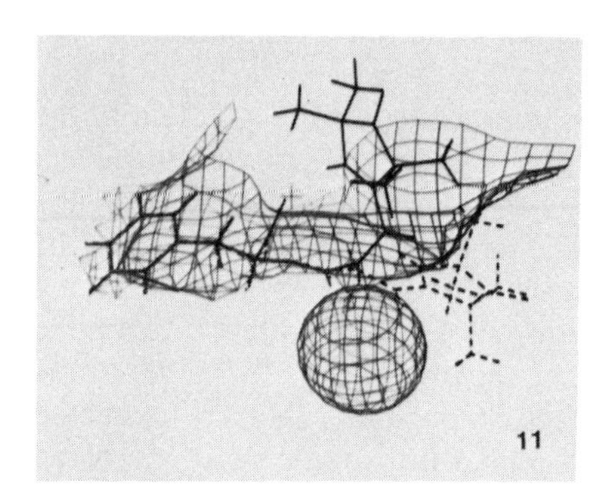

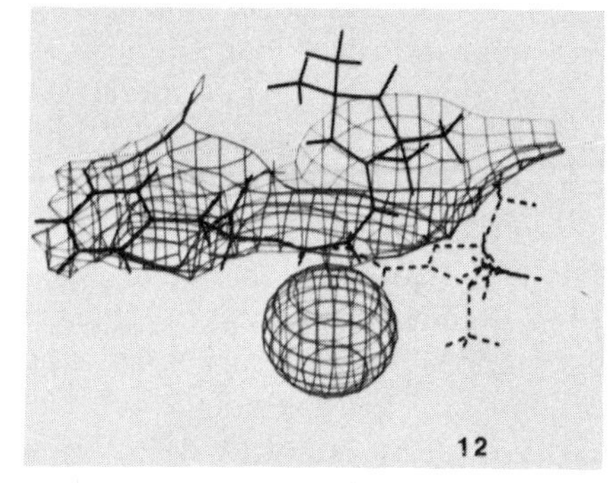

Figure 30. Stereodrawing of the "best fit" configuration (solid skeletal stick figure) of the spin-label ester substrate TEPOPL in the active site of carboxypeptidase A. The configuration of the spin-label is defined so that the carbonyl oxygen is coordinated to the metal ion and the terminal L-β-phenyllactic acid residue is in a salt-linked position with Arg-145, as shown for the inhibitor glycyl-L-tyrosine in difference Fourier studies (Quiocho and Lipscomb, 1971). The skeletal stick figure represented by broken lines corresponds to the fully extended *trans* configuration of the substrate. The solid lined figure corresponds to the torsionally distorted configuration that is sterically compatible with binding in the native site. To obtain the best fit with only allowed nonbonded contacts, the C=C double bond of the propene-2-oyl side chain must be rotated by 89° from the fully extended *trans* configuration of the molecule. The C—C bonds adjacent to the double bond of the propene-2-oyl have also been rotated to position the pyrrolinyl moiety within the extended van der Walls surface contours. The tetramethylpyrrolinyl group in this rotated configuration points toward the opening of the active site cleft into the bulk solvent. The extended *trans* configuration would experience considerable steric crowding with both side-chain residues and polypeptide backbone atoms, indicating that it cannot be accommodated by the active site. Based on Figure 10 of Kuo *et al.* (1983).

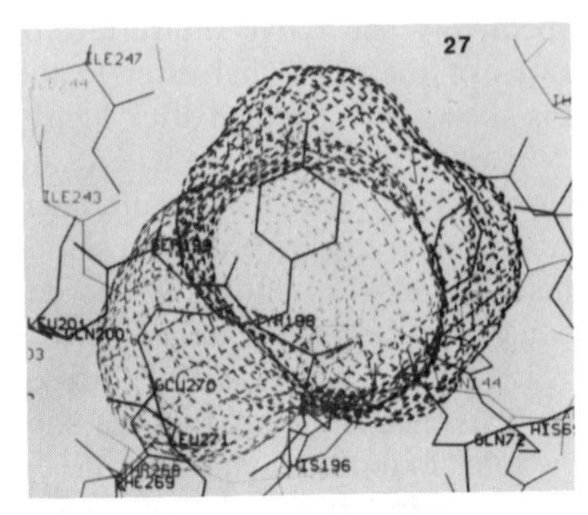

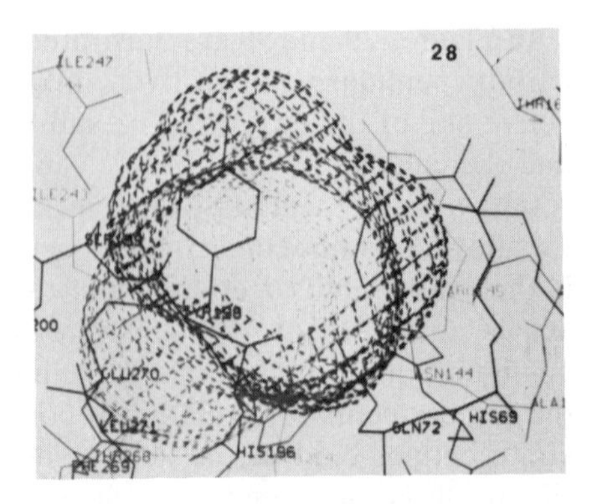

Figure 31. Stereodrawing of the "inside" of the extended van der Waals surface contour of TEPOPL in its fully extended *trans* configuration to illustrate the steric crowding with protein side-chain residues. In this case the enzyme active site residues are represented as skeletal stick figures, and the stereodrawings are recorded with the frontal end of the surface contour zoomed past the image focal plane of the computer-controlled visual display system. By inspection of all regions within the van der Waals surface of the extended *trans* configuration of the substrate, the protein residues found penetrating through the surface contours are Glu-72, Ser-197, Tyr-198, and Phe-279.

is 7.8 Å, in excellent agreement with the radial separation estimated through paramagnetic resonance studies.

The molecular graphics studies have been extended further to identify the protein residues that sterically prevent the spin-label from binding in an extended *trans* configuration (Kuo *et al.*, 1983). Figure 31 illustrates a view inside the calculated van der Waals surface of the spin-label substrate. The side chains of the four amino acid residues Glu-72, Ser-197, Tyr-198, and Phe-279 found within this volume are the origin of the forbidden, nonbonded interactions. Glu-72 serves as a ligand to the metal ion; the other three residues are located on the protein surface at positions that are expected to have steric contacts with the second and third amide residues of an oligopeptide substrate (Quiocho and Lipscomb, 1971). The steric interactions of the spin-label with these residues must, therefore, distort the substrate molecule, leading to rotation of the bulky groups around its C—C bonds to allow binding in the active site.

In Figure 32 are illustrated two reaction schemes for the hydrolysis of TEPOPL catalyzed by carboxypeptidase A. The appropriate rate equations are also presented. In the first scheme only bound substrates having a distorted configuration compatible with the steric requirements of the active site allow *further* reaction steps to proceed, whereas other configurations are nonproductive and eventually dissociate from the enzyme. The effect of nonproductive binding is to lower both k_{cat} and $K_{m(app)}$. According to this model, the ratio $k_{cat} : K_{m(app)}$ will be unaffected. It follows that a smaller $K_{m(app)}$ than that for ClCPL should obtain for TEPOPL, since k_{cat} for the nitroxide is smaller by more than an order of magnitude; this is not observed

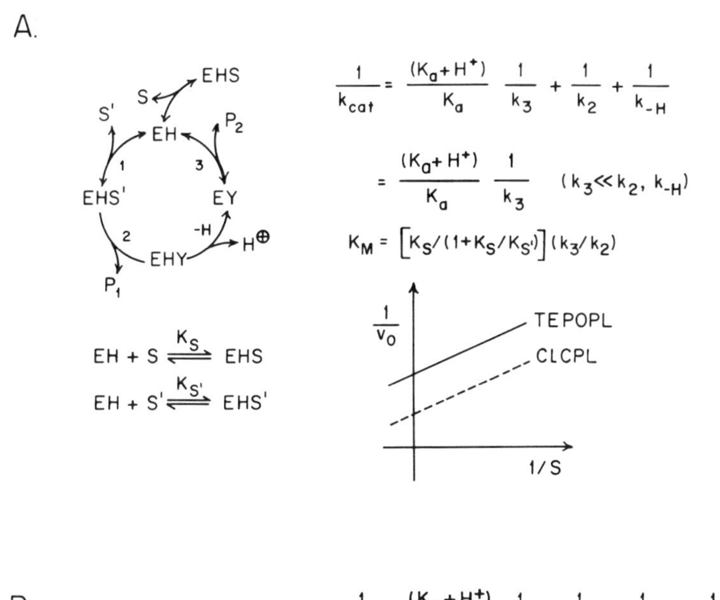

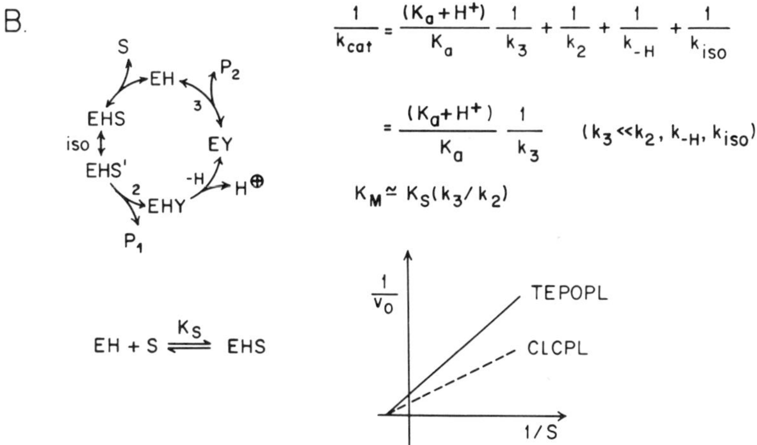

Figure 32. Mechanistic schemes for two possible reaction pathways for the hydrolysis of TEPOPL catalyzed by carboxypeptidase A. (*A*) Only a bound substrate having a distorted configuration compatible with the steric requirements of the active site is productive. Substrates having other conformations bind the enzyme but are nonproductive. Both k_{cat} and $K_{m(app)}$ are expected to be smaller compared with the parameters for hydrolysis of the analogous ester ClCPL. Therefore, on the basis of kinetic data, this mechanism can be ruled out. (*B*) An isomerization step is invoked, and the nitroxide substrate is distorted upon binding the enzyme. The overall rate constant for the reaction includes the rate constant of the isomerization step as an obligatory part of the catalytic process. The value of $K_{m(app)}$ for TEPOPL should be comparable to that for ClCPL. This mechanism is consistent with both kinetic data and structural studies defining the configuration of the spin-label in the active site.

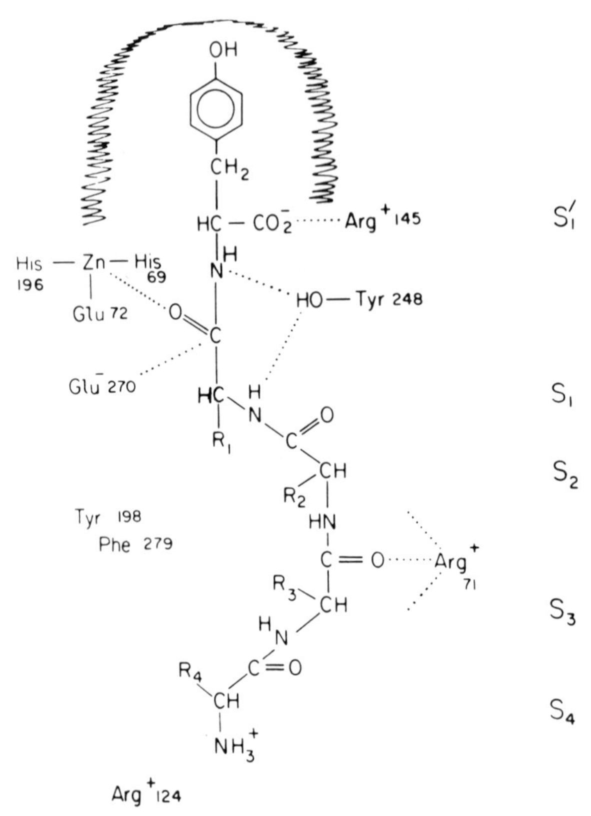

Figure 33. Interactions of an oligopeptide substrate of carboxypeptidase A as derived from experimental results on glygyl-L-tyrosine binding and model building (Quiocho and Lipscomb, 1971). The residues Arg-71, Tyr-198, and Phe-279 are denoted as secondary recognition sites of substrate binding by model building studies. From Quiocho and Lipscomb (1971) with permission. Copyright: Academic Press, Ltd. (London).

in kinetic studies (see Table II). The second mechanism invokes an isomerization step in which the nitroxide ester is distorted following initial binding. In this scheme distortion of the substrate occurs prior to the acylation step and is an obligatory part of the catalytic reaction pathway. The value of $K_{m(\mathrm{app})}$ is not affected, since molecules with different configurations bind to the enzyme productively, but the rate of isomerization, which must be faster than that of rate-limiting deacylation, is included in the overall kinetic rate constant. This scheme accounts for the results of both structural and kinetic studies (Kuo *et al.*, 1983).

The results of this study suggest that geometric distortion of substrates is an obligatory process in the catalytic action of carboxypeptidase A. The active site of carboxypeptidase A can be considered to consist of three distinct regions—the hydrophobic pocket with Arg-145 for binding the C-terminal residue; the bond-cleavage site, consisting of the metal ion and the side chain of Glu-270; and four amino acid residues constituting the sites of secondary substrate recognition. These are schematically illustrated in Figure 33. Although splitting of the scissile bond occurs in the bond-cleavage site by the action of the catalytic metal ion and Glu-270, hydrolysis is dependent upon the interactions of the substrate molecule with the hydrophobic pocket on one hand and the distal protein surface on the other. Substrate binding is mediated largely through hydrophobic interactions of the aromatic side chain of the C-terminal residue and electrostatic interactions of the free carboxylate group with Arg-145. Torsional distortion of a substrate, as shown through the study of Kuo *et al.* (1983), is promoted by the sites of secondary substrate recognition. Interactions of the substrate with residues of the distal surface of the active site exert considerable steric strain on the second and third residues of an oligopeptide substrate upon binding into the active site.

The effect causes rotation around the scissile bond of the substrate with subsequent bond cleavage. Also, the requirement for torsional distortion of the substrate explains the difference in rate-limiting steps for peptide and ester hydrolysis catalyzed by carboxypeptidase A (Kuo *et al.,* 1983). These studies with use of a specific spin-label ester substrate have yielded one of the most detailed descriptions thus far of the molecular basis of enzyme action.

ACKNOWLEDGMENTS

We thank Mr. R. J. Feldmann, Professor C. D. Barry, and Dr. J. McAlister for assistance in the preparation of illustrations in molecular graphics studies. L. C. K. was a predoctoral student supported by a Public Health Service Training Grant of the National Institutes of Health (GM 07183). This work was supported by grants from the National Institutes of Health (GM 21900 and HL 16005).

REFERENCES

Ament, S. S., J. B. Wetherington, J. W. Moncrief, K. Flohr, M. Mochizuki, and E. T. Kaiser (1973), *J. Am. Chem. Soc.* **95,** 7896.

Asakura, T., H. R. Drott, and T. Yonetani (1969), *J. Biol. Chem.* **244,** 6626.

Asakura, T., J. S. Leigh, Jr., H. R. Drott, T. Yonetani, and B. Chance (1971), *Proc. Natl. Acad. Sci. USA* **68,** 861.

Auld, D. S., and B. L. Vallee (1970), *Biochemistry* **9,** 602.

Baldassare, J., S. Charache, R. P. Jones, and C. Ho (1970), *Biochemistry* **9,** 4707.

Baldwin, J., and C. Chothia (1979), *J. Mol. Biol.* **129,** 175.

Bauer, R. S., and L. J. Berliner (1979), *J. Mol. Biol.* **128,** 1.

Berliner, L. J. (1970), *Acta Cryst.* **B26,** 1198.

Berliner, L. J. (1971), *J. Mol. Biol.* **61,** 189.

Berliner, L. J. (1976), *Spin Labelling. Theory and Applications*, Academic Press, New York, p. 592.

Berliner, L. J. (1978), *Meth. Enzymol.* **49,** 418.

Bernhard, S. A., and S. J. Lau (1971), *Cold Spr. Harb. Symp. Quant. Biol.* **36,** 75.

Bordeaux, D., and J. Lajzerowicz (1974), *Acta Cryst.* **B30,** 790.

Brandt, K. G., A. Himoe, and G. P. Hess (1967), *J. Biol. Chem.* **242,** 3973.

Bray, R. C. (1961), *Biochem. J.* **81,** 196.

Brehm, L., and J. Moult (1975), *Proc. Roy. Soc. Lond.* **B188,** 425.

Bunting, J. W., J. Murphy, C. D. Myers, and G. G. Cross (1974), *Canad. J. Chem.* **52,** 2648.

Burton, D. R., S. Forsen, G. Karlstrom, and R. A. Dwek (1979), *Prog. NMR Spectrosc.* **13,** 1.

Capiomont, A. (1972), *Acta Cryst.* **B28,** 2298.

Capiomont, A., D. Bordeaux, and J. Lajzerowicz (1972), *C. R. Acad. Sci. Paris* **275C,** 317.

Chien, J. C. W. (1979), *J. Mol. Biol.* **133,** 385.

Chien, J. C. W., and L. C. Dickinson (1981), in *Biological Magnetic Resonance*, Vol. 3, L. J. Berliner and J. Reuben, Eds., Plenum Press, New York, pp. 155–212.

De Beer, R., D. van Ormondt, M. A. van Ast, R. Banen, J. A. Duine, and J. Frank (1979), *J. Chem. Phys.* **70,** 4491.

Diamond, R. (1974), *J. Mol. Biol.* **82,** 371.

Douzou, P. (1974), in *Methods of Biochemical Analysis*, Vol. 22, D. Glick, Ed., Academic Press, New York, pp. 401–512.

Douzou, P. (1977), *Cryobiochemistry*, Academic Press, New York, p. 286.

Douzou, P. (1979), *Quart. Rev. Biophys.* **12,** 521.

Douzou, P., G. Hui Bon Hoa, P. Maurel, and F. Travers (1976), in *Handbook of Biochemistry and Molecular Biology*, 3rd ed. G. D. Fasman, Ed., Chemical Rubber Publishing, Cleveland, OH, pp. 520–540.

Douzou, P., R. Sireix, and F. Travers (1970), *Proc. Natl. Acad. Sci. USA* **66,** 787.

Dwek, R. A. (1973), *Nuclear Magnetic Resonance in Biochemistry: Application to Enzyme Systems*, Clarendon Press, Oxford, p. 395.

Fee, J. A. (1978), *Meth. Enzymol.* **49,** 512.

Feldmann, R. J., D. H. Bing, B. C. Furie, and B. Furie (1978), *Proc. Natl. Acad. Sci. USA* **75,** 5409.

Fink, A. L. (1976), *J. Theor. Biol.* **61,** 419.

Fink, A. L. (1977), *Acc. Chem. Res.* **10,** 233.

Fink, A. L., and S. J. Cartwright (1981), *CRC Crit. Rev. Biochem.* **11,** 145.

Flohr, K., R. M. Paton, and E. T. Kaiser (1975), *J. Am. Chem. Soc.* **97,** 1209.

Ford, L. O., L. N. Johnson, P. A. Machin, D. C. Phillips, and R. Tjian (1974), *J. Mol. Biol.* **88,** 349.

Hamilton, C. L., and H. M. McConnell (1968), in *Structural Chemistry and Molecular Biology*, A. Rich and N. Davidson, Eds., W. W. Freeman, San Francisco, pp. 115–149.

Hartley, B. S., and B. A. Kilbey (1954), *Biochem. J.* **56,** 288.

Heidner, E. J., R. C. Ladner and M. F. Perutz (1976), *J. Mol. Biol.* **104,** 707.

Henderson, R. (1970) *J. Mol. Biol.* **54,** 341.

Himoe, A., K. G. Brandt, R. J. DeSa, and G. P. Hess (1969), *J. Biol. Chem.* **244,** 3483.

Ho, C., J. J. Baldassare, and S. Charache (1970), *Proc. Natl. Acad. Sci. USA* **66,** 722.

Hoffman, B. M., J. E. Roberts, T. G. Brown, C. H. Kang, and E. Margoliash (1979), *Proc. Natl. Acad. Sci. USA* **76,** 6132.

Imoto, T., L. N. Johnson, A. C. T. North, D. C. Phillips, and J. A. Rupley (1972), in *The Enzymes*, Vol. 7, 3rd ed., P. D. Boyer, Ed., Academic Press, New York, p. 665.

Janzen, E. G. (1971), *Top. Stereochem.* **6,** 177.

Johnson, L. N., D. C. Phillips, and J. A. Rupley (1968), *Brookhaven Symp. Biol.* **21,** 120.

Jost, P. C., and O. H. Griffith (1978), *Meth. Enzymol.* **49,** 369.

Kèzdy, F. J., and M. L. Bender (1962), *Biochemistry* **1,** 1097.

Koch, T. R., L. C. Kuo, E. G. Douglas, S. Jaffer, and M. W. Makinen (1979), *J. Biol. Chem.* **254,** 12310.

Kuo, L. C., and M. W. Makinen (1982), *J. Biol. Chem.* **257,** 24.

Kuo, L. C., J. M. Fukuyama, and M. W. Makinen (1983), *J. Mol. Biol.* **163,** p. 63.

Lajzerowicz-Bonneteau, J. (1976), in *Spin-Labeling*, L. J. Berliner, Ed., Academic Press, New York, pp. 239–249.

LaMar, G. N., D. L. Budd, D. B. Viscio, K. M. Smith, and K. C. Langry (1978), *Proc. Natl. Acad. Sci. USA* **75,** 5755.

Lau, P. W., and T. Asakura (1979), *J. Biol. Chem.* **254,** 2595.

Leigh, J. S. Jr. (1970), *J. Chem. Phys.* **52,** 2608.

Likhtenstein, G. I. (1976), *Spin-Labeling Methods in Molecular Biology*, Wiley, New York, p. 258.

Makinen, M. W. (1976), D. Phil. thesis, University of Oxford.

Makinen, M. W., and W. A. Eaton (1974), *Nature* **247,** 62.

Makinen, M. W., and A. L. Fink (1977), *Ann. Rev. Biophys. Bioeng.* **6,** 301.

Makinen, M. W., J. M. Fukuyama, and L. C. Kuo (1982), *J. Am. Chem. Soc.* **104,** 2667.

Makinen, M. W., and N. W. Isaacs (1978), *Acta Cryst.* **B34,** 1584.

Makinen, M. W., K. Yamamura, and E. T. Kaiser (1976), *Proc. Natl. Acad. Sci. USA* **73,** 3882.

Makinen, M. W., L. C. Kuo, J. J. Dymowski, and S. Jaffer (1979), *J. Biol. Chem.* **254,** 356.

Marriott, T. B., and O. H. Griffith (1974), *J. Magn. Reson.* **13,** 45.

Marshall, G. R., C. D. Barry, H. E. Bosshard, R. A. Dammkoehler, and D. A. Dunn (1979), in *Computer Assisted Drug Design*, E. C. Olson and R. E. Christofferson, Eds., American Chemical Society, Washington, D. C., pp. 205–226.

McConnell, H. M. (1971), *Ann. Rev. Biochem.* **40,** 227.

McConnell, H. M., and C. L. Hamilton (1968), *Proc. Natl. Acad. Sci. USA* **60,** 776.

McConnell, H. M., and B. G. McFarland (1970), *Quart. Rev. Biophys.* **3,** 91.

McConnell, H. M., W. Deal, and R. T. Ogata (1969), *Biochemistry* **8,** 2580.

Mildvan, A. S., and M. Cohn (1970), *Adv. Enzymol.* **33,** 1.

Mo, F., and L. H. Jensen (1975), *Acta Cryst.* **B31,** 2867.

Moffat, K. (1971), *J. Mol. Biol.* **55,** 135.

Moffat, K., J. F. Deatherage, and D. W. Seybert (1979), *Science* **206,** 1035.

Monod, J., J. Wyman, and J. P. Changeux (1965), *J. Mol. Biol.* **12,** 88.

Morrisett, J. D. (1976), in *Spin-Labeling*, L. J. Berliner, Ed., Academic Press, New York, pp. 273–338.

Ogawa, S., H. M. McConnell, and A. Horwitz (1968), *Proc. Natl. Acad. Sci. USA* **61,** 401.

Orme-Johnson, W. H., and R. H. Sands (1973), in *Iron-Sulfur Proteins*, Vol. 2, W. Lovenberg, Ed., Academic Press, New York, pp. 195–238.

Palmer, G. (1978), in *The Porphyrins*, Vol. 4, D. Dolphin, Ed., Academic Press, New York, pp. 313–353.

Perutz, M. F. (1965), *J. Mol. Biol.* **13,** 646.

Perutz, M. F. (1968), *J. Crystal Growth* **2,** 64.

Perutz, M. F. (1970), *Nature* **228,** 726.

Quiocho, F. A., and W. N. Lipscomb (1971), *Adv. Prot. Chem.* **25,** 1.

Rozantsev, E. G. (1970), *Free Nitroxyl Radicals*, Plenum Press, New York.

Rosenberg, R. C., P. K. Bernstein, and H. B. Gray (1975), *J. Am. Chem. Soc.* **97,** 21.

Rossi, G. L., and S. A. Bernhard (1970), *J. Mol. Biol.* **49,** 85.

Rupley, J. A., V. Gates, and R. Bilbrey (1968), *J. Am. Chem. Soc.* **90,** 5633.

Schepler, K. L., W. R. Dunham, R. H. Sands, J. A. Fee, and R. H. Abeles (1975), *Biochem. Biophys. Acta* **397,** 510.

Turley, J. W., and F. P. Boer (1972), *Acta Cryst.* **B28,** 1641.

Wien, R. W., J. D. Morrisett, and H. M. McConnell (1972), *Biochemistry* **11,** 3707.

Three

DNA Backbone Conformation and Dynamics

Heisaburo Shindo

Tokyo College of Pharmacy
Horinouchi, Hachioji
Tokyo 192-03, Japan

1. **Introduction** **95**
2. **NMR Methods** **96**
   2.1. The Chemical-Shift Tensor, 97
   2.2. NMR Relaxation Parameters, 98
3. **Phosphodiester-Backbone Conformation of DNA** **99**
   3.1. DNA in Solution, 99
   3.2. Phosphodiester Orientation in DNA Fibers, 103
      3.2.1. The B Form of DNA, 105
      3.2.2. The A Form of DNA, 108
4. **DNA Dynamics in Solution** **113**
   4.1. ^{31}P NMR Studies, 114
   4.2. Proton NMR Studies, 117
   4.3. ^{13}C NMR Studies, 120
   4.4 Internal Motion and Flexibility of DNA, 125
5. **Concluding Remarks** **126**

References **127**

1. INTRODUCTION

Deoxyribonucleic acid (DNA) is one of the most important of all biological substances because of its genetic function. Essential features of the DNA structure are rather simple but contain a formidable amount of genetic information in a one-dimensional array. The Watson-Crick base pairing and stacking interactions between bases along the chain force the DNA molecule to form a stable double-stranded helical structure, which was once thought to be independent of base composition and sequences.

There are, however, many lines of evidence that indicate that variations in the static and dynamic nature of the secondary structure may play important roles in DNA function and its organization *in vivo* and *in vitro*. Indeed, X-ray diffraction studies and physical properties have suggested that the secondary structure of DNA in solution may be affected by base sequences (see the reviews of Wells *et al.*, 1977, 1980), although it is not known how and to what extent this occurs. If verified by further experiment, these results could be important in understanding how a regulatory protein recognizes specific DNA sequences.

Hydrodynamic properties of high-molecular-weight DNA are adequately described by a wormlike chain model. The stiffness of such a chain is measured by its persistence length. It is apparent that considerable bending and twisting motions are associated with local fluctuation of DNA segments (Barkley and Zimm, 1979; Olson, 1980). Recent NMR results (Tritton and Armitage, 1978) have suggested that a rapid internal motion with a rate on the order of a nanosecond is occurring in DNA. Any bending or twisting motion may cause some strains along the chain; this strain may frequently be localized and occasionally give rise to transient hydrogen-bond breakage and unstacking of the Watson-Crick base pairs. The frequency of such a transient hydrogen-bond breakage may be low (McGhee and Von Hippel, 1977), because such strains may usually be released through local flexibility or conformational changes of the DNA duplex.

In this review, the discussion is divided into two parts: one on static conformations of DNA and the other on the dynamic properties of DNA in solution. In the first category, I discuss how NMR data can be correlated with the secondary structure of DNA, and ^{31}P NMR studies of DNA in solution and the fiber state are emphasized. In the other part of the discussion, it is shown on the basis of NMR data that double-stranded DNA is much more flexible than previously thought. Yet the analysis of NMR data and the models proposed therefrom are rather poor at present. Therefore, it is important to review NMR studies in a consistent manner and also to discuss some limitations of these studies for elucidating the molecular motions in DNA.

2. NMR METHODS

Nuclear magnetic resonance has been widely used for elucidating molecular structure and dynamics in many biological systems. One of the advantages of this method is the ability to independently examine different atoms or atomic groups within a molecule; specific sites in a molecule can be examined with use of probes such as ^{1}H, ^{31}P, ^{13}C, and ^{15}N nuclei, and the results can be integrated to obtain a complete picture of the molecule. The NMR parameters usually determined are chemical shift, linewidth, spin–spin coupling constant, and relaxation parameters (T_1, T_2, and NOE; see Section 2.2). The former three may be correlated with the molecular structure, and the latter characterizes the dynamic nature of molecules.

2.1. The Chemical-Shift Tensor

The chemical shift of a given nucleus arises as a result of the magnetic shielding due to electron clouds surrounding the nucleus. Electron orbitals are not usually spherically symmetrical. Therefore, chemical shift is generally dependent on the anisotropy of the electron distribution, or molecular structure. In fact, it is expressed by a tensor quantity. For example, Figure 1 shows a typical powder pattern of ^{31}P NMR observed for Na-DNA in the solid state. As is seen in Figure 1, the chemical shift of the phosphodiester of DNA spans over about 200 ppm. The principal values of the chemical-shift tensor can be measured to be −82, −22, and 113 ppm for σ_{11}, σ_{22}, and σ_{33}, respectively. The orientation of the principal axis relative to the molecular frame must be experimentally determined (Kohler and Klein, 1976; Herzfeld *et al.*, 1978). For simplicity, we will adopt the axis system based on the symmetry of the phosphodiester (Kohler and Klein, 1976). The principal-axis system used here is as follows: the x axis for σ_{11} is perpendicular to the plane made by the O═P—O⁻ bonds, the y axis for σ_{22} lies on the bisector of the O═P—O⁻ bonds, and the z axis is normal to both the above axes. The orientation of the shift tensor is depicted in Figure 2. An orientation of the principal axis (i.e., a phosphodiester group) relative to the magnetic field must be correlated with a certain chemical shift. In the case of polycrystalline or amorphous materials, there are all possible orientations of

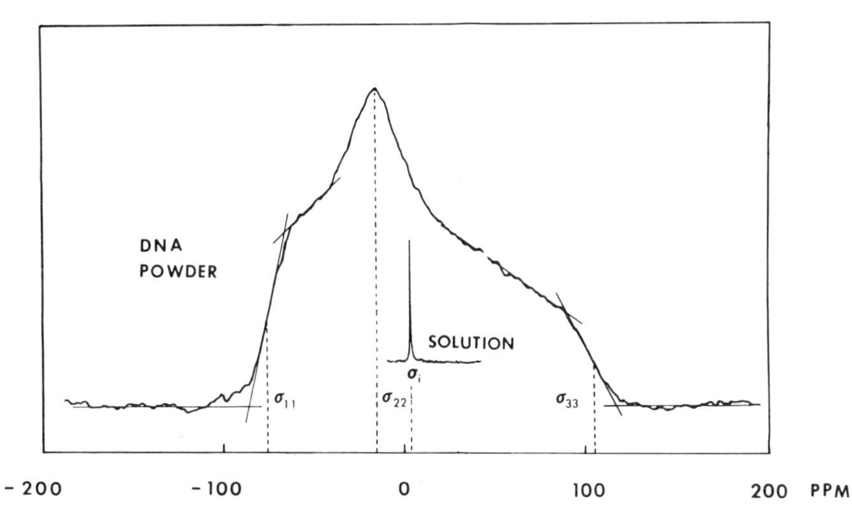

Figure 1. Proton dipolar decoupled ^{31}P NMR powder pattern (24.3 MHz) of DNA. The principal values of the chemical-shift tensor are indicated. A sharp resonance in the inset represents the ^{31}P spectrum of DNA in solution.

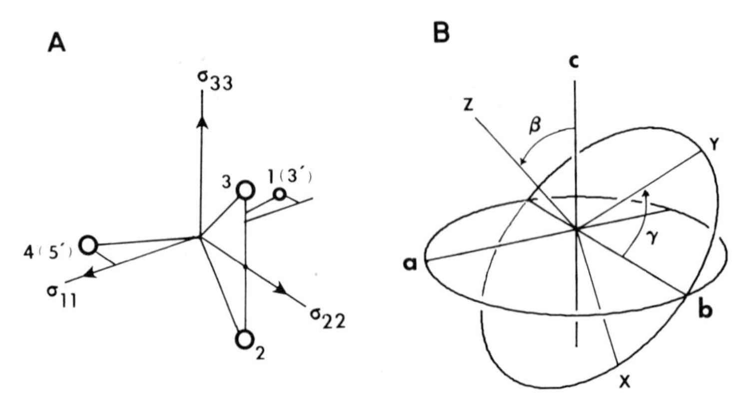

Figure 2. (*A*) Orientation of the chemical-shift tensor relative to the phosphodiester atomic coordinates, and (*B*) representation of the principal axis system (x, y, z) of the shift tensor relative to the helical axis or fiber axis of DNA by Euler angles β and γ.

the group, and hence the spectral pattern seen in Figure 1 results. The chemical-shift shielding is general for all nuclei, although the magnitude and orientation of the chemical shift anisotropy vary depending on the observing nucleus and the chemical-bond characteristics.

By contrast, if the DNA molecule, or more correctly the phosphodiester group, undergoes a rapid tumbling motion in solution, the chemical-shift tensor is averaged out, and consequently a narrow spectral line, which is characterized by the isotropic chemical shift σ_i, is observed (see a sharp line in the center of Figure 1). Of course, information about the spatial orientation of the molecule is precluded in isotropic chemical-shift values, and the variation in the isotropic chemical shift is obviously smaller by an order of 10^2 than the magnitudes of the chemical-shift tensor. However, chemical shifts and spin–spin coupling constants obtained from high-resolution NMR spectra can provide useful information about conformational states of a molecule in solution.

2.2. NMR Relaxation Parameters

When a nuclear spin system is instantaneously perturbed by any means, it approaches an equilibrium state with certain time constants under the given conditions. In a simple case, the process is characterized by two relaxation times T_1 and T_2; the former is defined as the time required for a Boltzmann distribution of the spin system and the latter as the time required for interchange of the spin states. The mechanisms by which the system can relax toward the equilibrium are often very complicated. Any fluctuation in the local magnetic field contributes to the relaxation processes in the spin system. Spin dipole–dipole interaction, spin–quadrupole interaction, and chemical-shielding anisotropy are the typical examples producing such local fluctuating fields. Two relaxation parameters may be related to all these local field strengths and their fluctuation amplitude, whereas the Overhauser effect (nuclear Overhauser enhancement: NOE), which appears as signal enhancement or reduction, is caused by dipole–dipole interaction alone. The observed relaxation parameters including the NOE can be correlated with several motional modes that are occurring in a molecule. The simplest one is a single mode having a single correlation time τ_c, and another is general motions having more than two distinct correlation times. The rationale of multimolecular motions must be treated with caution. Some cases have been discussed extensively and applied to ^{13}C NMR data of peptides and protein systems (see the review of London, 1980). The relaxation mechanism for protonated carbon nuclei is rather simple because dipole–dipole interaction between ^{13}C and ^{1}H is the predominant relaxation mechanism. On the other hand, relaxation mechanisms for ^{31}P and ^{1}H nuclei are more complicated, since mechanisms other than dipole–dipole interaction and interactions between nonbonded atoms become important for these nuclei in macromolecules.

3. PHOSPHODIESTER-BACKBONE CONFORMATION OF DNA

The three-dimensional structure of DNA has been deduced from X-ray diffraction studies on fibers (Langridge *et al.*, 1960a, Marvin *et al.*, 1961; Fuller *et al.*, 1965). Generally, DNA in aqueous solution has been assumed to have the same conformation present in the fibrous state at high relative humidities, namely the B form. But, natural DNA, with varying sequences, may possess conformational variation undetected by the X-ray fiber diffraction technique. The X-ray diffraction method includes model building in such a way that the most probable structure is that which gives the best fit to the diffraction data. Therefore, individual atomic coordinates determined by the diffraction method tend to provide an average conformation of DNA, and this prevents the model from showing sequence dependence of the helical structure of DNA.

In fact, it has been shown that the ^{31}P NMR characteristics of DNA in solution, while corresponding overall to a B form of DNA (Shindo *et al.*, 1979; Simpson and Shindo, 1980), do indicate a nonuniform backbone conformation in natural DNA (Shindo *et al.*, 1980). The nonuniformity of the backbone conformation has several important structural and functional implications.

3.1. DNA in Solution

Gorenstein (1975) extensively surveyed ^{31}P NMR isotropic chemical shifts of phosphodiester groups and found some correlation with the torsion or dihedral angles of the O—P—O bonds. Furthermore, empirical quantum mechanical calculations supported a dependence of ^{31}P chemical shift on the torsion angle of phosphodiesters (Gorenstein and Kar, 1977). Hence it is quite reasonable to assume that a change in the ^{31}P chemical shift of a phosphodiester group reflects a change in its conformation, although the quantitative relationship is not yet known.

Gueron and Shulman (1975) reported the ^{31}P NMR spectrum of $tRNA^{Phe}$ in which several resonance lines are significantly shifted from the main cluster in the spectrum. They interpreted these peaks as arising from phosphodiester groups having abnormal conformations in the molecule. The multiplicity in the ^{31}P NMR spectra for the above-mentioned nucleic acids could be expected because structural variety in these molecules was already known conformation'' for poly(deoxyribonucleotides) having an alternating se- calculate the spectrum of $tRNA^{Phe}$ from the dihedral angles of the phosphodiesters determined by the X-ray crystal structure (Sussman *et al.*, 1978). The agreement was not complete.

The first indication of conformational variation in DNA emerged from the X-ray diffraction studies of the tetranucleotide $(dAdT)_2$ (Viswamitra *et al.*, 1978). From this result, Klug *et al.* (1979) proposed "an alternating B conformation" for poly (deoxyribonucleotides) having an alternating sequence. Subsequently, Patel *et al.* (1979) showed two well-separated resonances in the ^{31}P NMR spectrum of oligo(dGdC) duplexes at an extremely high salt concentration (4 M) at which the Pohl-Jovin transition (Pohl and Jovin, 1972) is known to occur; this observation provided strong evidence of an alternating form of DNA. At almost the same time, Shindo *et al.* (1979) independently reported the observation of two resonances of equal intensity in the ^{31}P NMR spectrum of 145 base pair (bp) poly(dAdT)·poly(dAdT) in mild salt conditions (0.01 and 0.1 M), which they attributed to ApT and TpA sequences. From further studies of Cohen *et al.* (1981), this characteristic doublet in the spectrum was found to be general for alternating polynucleotide duplexes, such as d(IC), d(AT), and d(A^5BrU) but not for d(GC) and d(AU) in mild salt condition (~0.1 M) (Figure 3). Shindo *et al.* (1979) emphasized from this and other evidence that the backbone conformation of natural DNA is significantly nonuniform. The X-ray crystal study by Wang *et al.* (1979) of the hexamer $(dCdG)_3 \cdot (dCdG)_3$ gave rise to the discovery of the left-handed helical structure of the so-called Z form of DNA. These results apparently brought DNA structure–function relationships into a new phase.

Deoxyribonucleic acid fragments having well-defined sequences are available by synthesis or by extraction from natural products. Figure 4 shows the ^{31}P spectrum of the self-complementary oligonucleotide d(GGAATTCC) in the duplex state, in which five resolved peaks are observed between 3.9 and 4.4 ppm upfield from the internal reference of trimethylphosphate. Although it was not possible to assign these resonances to individual internucleotide phosphates, Patel and Canuel (1979) mentioned that the 0.6 ppm variation in the ^{31}P chemical shift of the seven internucleotide phosphodiester bonds

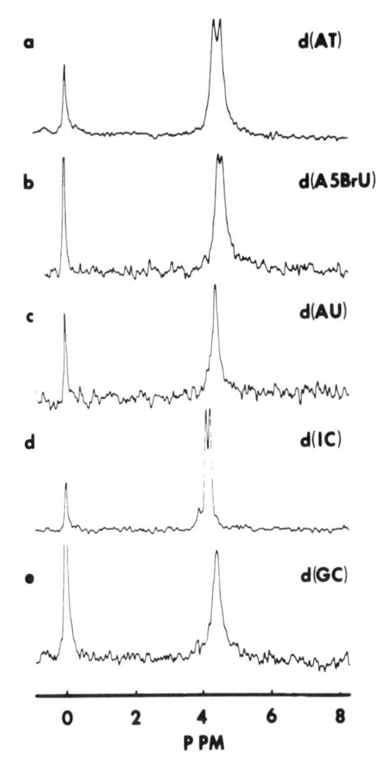

Figure 3. ^{31}P NMR spectra (109.3 MHz) of sonicated polynucleotides with alternating base sequences in solution (0.1 *M* NaCl, 25°C). (Cohen *et al.*, 1981; permission of *Biochemistry.*)

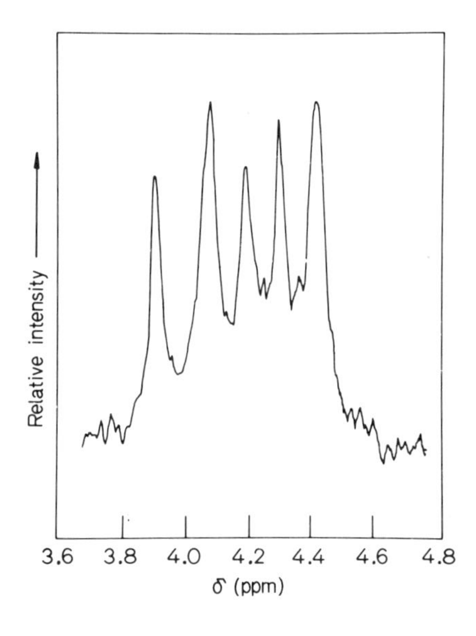

Figure 4. ^{31}P NMR spectrum (145.7 MHz) of deoxyribonucleotide d(GGAATTCC) in the duplex state. (Patel and Canuel, 1979; permission of *Eur. J. Biol.*)

indicates the existence of sequence-dependent variations in the torsional angles about the P—O bonds in the double-helical state. Similarly, a multiplet in the ^{31}P NMR spectrum was recently observed for the dodecamer d(CGCGAATTCGCG) (Patel *et al.*, 1982).

Deoxyribonucleic acid fragments of a well-defined length (145 bp) can be obtained from nucleosome core particles reconstituted from the core histones and synthetic polydeoxynucleotides (Simpson and Kunzler, 1979). All the DNA fragments obtained from this method have a very narrow distribution in their length (145 ± 5 bp). Furthermore, in contrast to both high-molecular-weight and short oligonucleotides, these fragments are short enough to allow resolution in the NMR spectrum but long enough that terminal end effects can safely be ignored.

Figure 5 shows high-resolution ^{31}P NMR spectra (109.3 MHz) of the 145

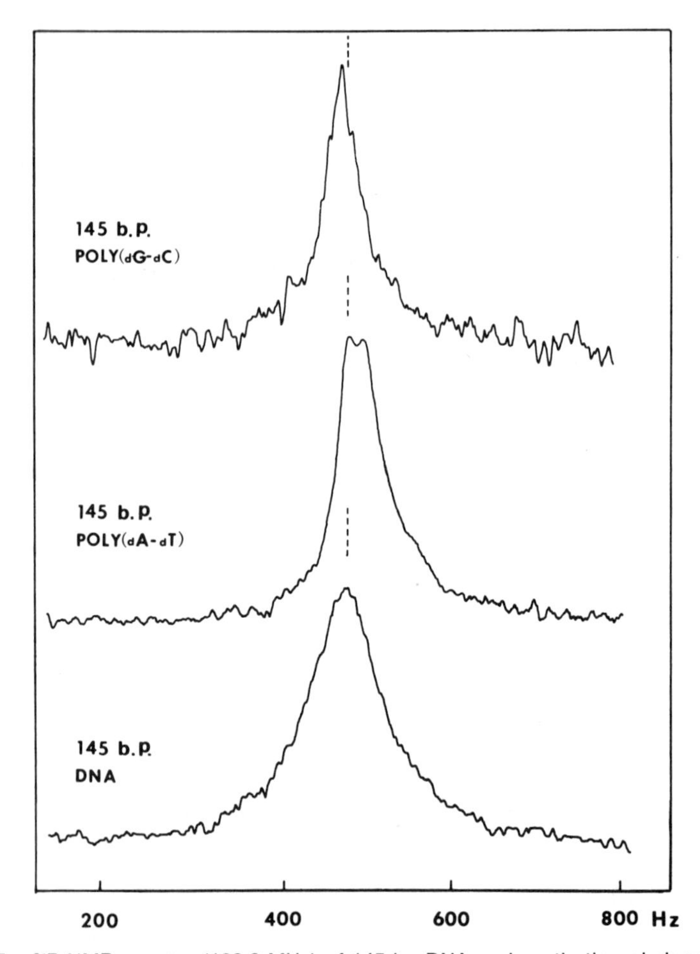

Figure 5. ^{31}P NMR spectra (109.3 MHz) of 145 bp DNA and synthetic polydeoxyribonucleotides isolated from nucleosome core particles.

bp DNAs (10 mM Tris-HCl, 0.5 mM EDTA, pH 7.5) at 20 ± 2° (Simpson and Shindo, 1980). Two important features are seen from the comparison of the spectra: first the linewidth for natural DNA is about twice that for poly(dGdC)·poly(dGdC), and second, poly(dAdT)·poly(dAdT) exhibits a doublet in the spectrum. It is reasonable to assume that dynamic properties, such as correlation times, are very similar for all fragments, since the dimensions of these fragments are almost identical, namely 145 bp long. The linewidth due to anisotropic motion of DNA fragments was previously estimated by the least-squares fits to the relaxation data and was found to be 46 ± 6 Hz at 109.3 MHz. This value is nearly identical to the observed value (43 ± 3 Hz) for poly(dGdC)·poly(dGdC); therefore, no chemical-shift dispersion is measurable within experimental error for this polymer. Thus it can be concluded that all the phosphodiester groups in poly(dGdC)·poly(dGdC) are magnetically equivalent and that its linewidth results from relaxation mechanisms alone (i.e., dipole–dipole interaction and chemical-shift anisotropy). In contrast to the alternating d(GC) polymer, poly(dAdT)·poly(dAdT) exhibits two partially resolved resonances with equal intensity. The linewidth for individual peaks was measured to be 38 and 44 Hz, each of which is nearly identical to the linewidth for the d(GC) polymer.

These results lead us to conclude that extra line broadening observed for natural DNA reflects a variation in the backbone conformations of natural DNA in solution and that such a variation occurs in a sequence-dependent manner.

3.2. Phosphodiester Orientation in DNA Fibers

The Watson-Crick structure of DNA was determined based on the X-ray diffraction images from highly oriented DNA fibers. Assuming that the backbone conformation has a perfect helical symmetry, the ^{31}P NMR spectrum from the fiber oriented parallel to the magnetic field must be a singlet, because in this case all of the phosphodiesters are equivalent, as is seen in the top view of the DNA duplex (Figure 6A). On the other hand, the spectrum from the perpendicularly oriented fiber must be a doublet, corresponding to the side views of phosphodiesters around the helical axis (Figure 6B).

If the DNA structure, and thereby the atomic coordinates of the phosphodiester group, is known, we can calculate theoretical spectra for any orientations of the fiber relative to the magnetic field. An example is shown in Figure 7, where we used atomic coordinates of the DNA structure determined by Langridge *et al.* (1960b) and the principal axis system of the chemical-shift tensor previously discussed (see Section 2.1). As mentioned above, the parallel spectrum is a singlet, the perpendicular spectrum is a doublet, and the spectrum at 45° rotation of the fiber is a triplet (theoretically a quintet, but the center peak is a superposition of two). These calculated chemical shifts may be used for evaluation of the backbone conformation by comparison with actual spectra.

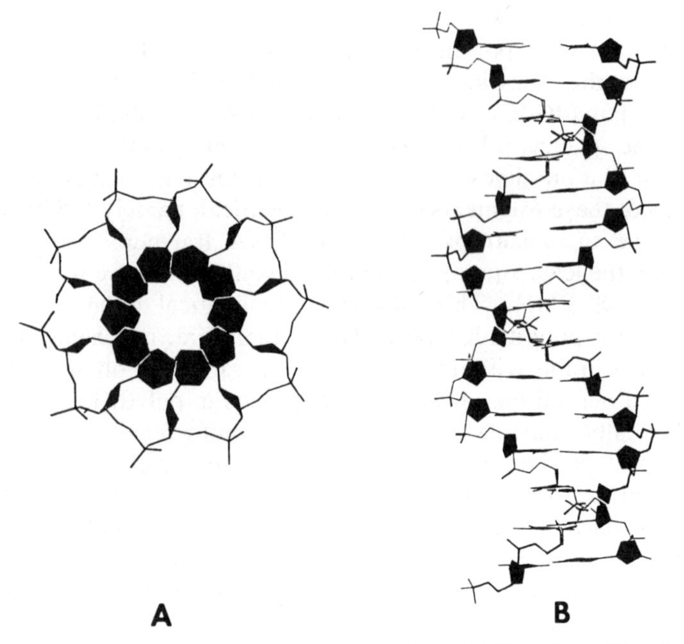

Figure 6. The stick model of the B form of DNA. (*A*) Top view; (*B*) side view.

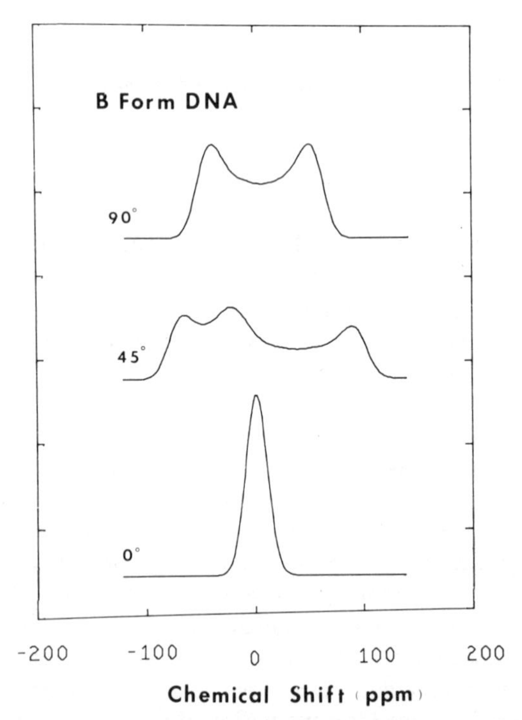

Figure 7. Theoretical spectral patterns of DNA fibers oriented parallel ($\phi = 0°$), perpendicular ($\phi = 90°$), and $\phi = 45°$ relative to the magnetic field.

3.2.1. The B Form of DNA. X-ray diffraction patterns obtained from fibers of salmon sperm DNA and the sodium salt of poly(dAdT)·poly(dAdT) at 98% relative humidity indicate that the molecules are highly oriented along the fiber axis. Helical parameters of the B form of poly(dAdT)·poly(dAdT) were found to be a pitch of 32.3 and a rise per residue of 3.3 Å, corresponding to 9.8 residues per turn. Therefore, it was concluded that preparations of synthetic DNA have the normal B conformation (Shindo and Zimmerman, 1980). Such DNA fibers were inserted into small parallel holes in a Teflon holder (Figure 8), and the relative humidity in the sample was kept at 98%. The orientation of the fibers relative to the magnetic field was measured by a goniometer on the top of the sample tube. Figure 9 compares ^{31}P NMR spectra from the B form of Na-DNA and poly(dAdT)·poly(dAdT) in parallel and perpendicular orientations of the fiber relative to the magnetic field. Both the parallel and the perpendicular spectra for Na-DNA are broad singlets, although a doublet is expected in the perpendicular spectrum based on the spectra calculated assuming the Watson-Crick structure. In fact, a rapid rotational motion with a rate of higher than 3×10^3 sec^{-1} presumably takes place about the helical axis, so that the doublet collapses into a singlet. Two important observations can be made from the spectra for Na-DNA: first, the linewidth of the parallel spectrum is abnormally broad, as much as 41 ppm, and second, the linewidth of the parallel spectrum is broader by 10 ppm than that of the perpendicular spectrum. The chemical-shift value is shifted upfield when the DNA fiber is rotated from the parallel to the perpendicular orientation relative to the magnetic field. It should be noted that the lithium salt of DNA, which does not give the A form, exhibited almost identical NMR spectra to those given by the sodium salt of DNA (Shindo *et al.*, 1980). This result strongly supports the conclusion the spectrum observed for the B form is not due to contamination with the A form.

Remarkably, the B form of poly(dAdT)·poly(dAdT) fiber does indeed exhibit a doublet pattern just as observed in solution. The spectrum for the parallel orientation provides solid evidence for the occurrence of two distinct orientations of the phosphodiester groups in the poly(dAdT) duplex. The perpendicular spectrum is somewhat asymmetrical and is shifted upfield, as was observed for Na-DNA. This spectrum should be a doublet, in principle, even with rapid rotational motion about the helical axis, but the separation may not be sufficiently large.

It appears that poly(dAdT)·poly(dAdT) has two distinct orientations of the phosphodiester group within the molecule. The line broadening of each observed component of the doublet is considered to be induced mainly by imperfect orientation of the DNA molecules along the fiber axis and other minor factors. The apparent linewidth of individual peaks is estimated to be ~18 ppm using the total linewidth of 39 ppm and the peak separation of 21 ppm. Thus the extra line broadening of 23 ppm (= 41 − 18) observed for natural DNA must be assigned to an average chemical-shift distribution, corresponding to a variation in the orientation of the phosphodiester groups in natural DNA.

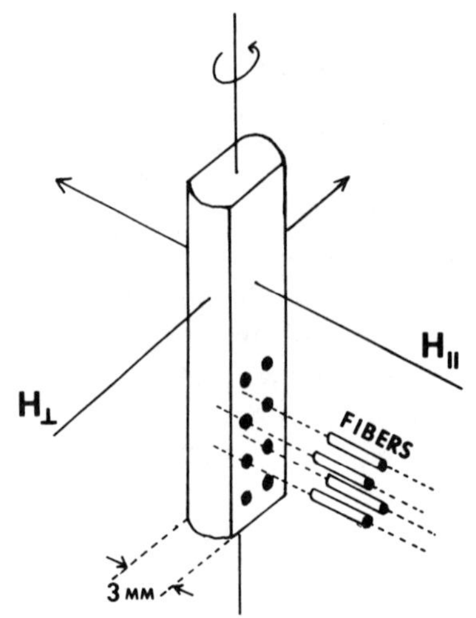

Figure 8. DNA fiber holder that can be rotated in the magnetic field.

This hypothesis is also supported by the reasoning of why the parallel spectrum is broader by 10 ppm than the perpendicular spectrum of natural DNA. For simplicity, consider the largest principal value, $\sigma_{33} = 113$ ppm, of the chemical-shift tensor and the DNA structure for the B conformation determined by Langridge *et al.* (1960b). According to this model, the angle β between the principal axis of σ_{33} and the helical axis is nearly perpendicular,

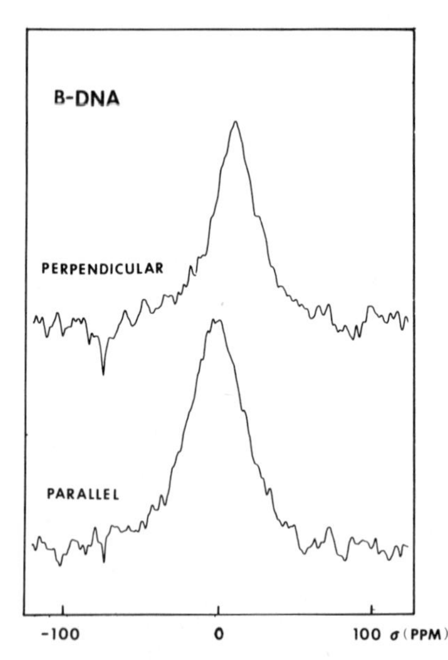

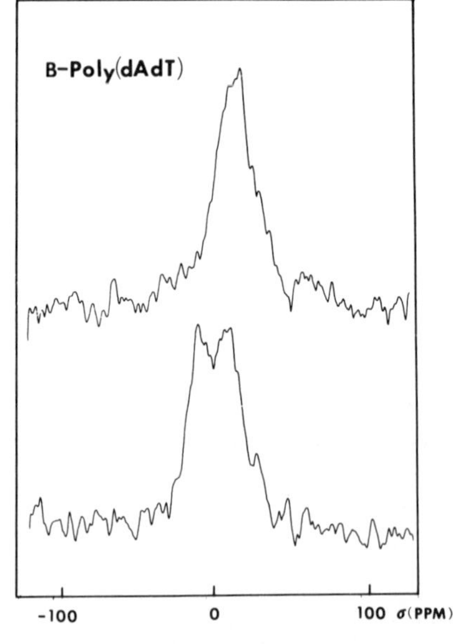

Figure 9. Proton dipolar decoupled ^{31}P NMR spectra (24.3 MHz) observed for the B form of DNA fibers and poly(dAdT)·poly(dAdT) at 98% relative humidity. The fiber orientations relative to the magnetic field are indicated.

Table I. Comparison of the observed and calculated ^{31}P chemical shifts for the B form of DNA and poly(dAdT)·poly(dAdT) fibers

	Chemical Shift (ppm)	
	Parallel	Perpendicular
Observed for Na-DNA	− 6	8
Langridge *et al.* (1960a)	− 5.8	9.0
Arnott and Hukins (1972)	− 41.1	26.5
Observed for poly(dAdT)	− 17	2 to 17 (nonresolved)
	4	
Klug *et al.* (1979)	− 25.6	3.3
	4.4	18.2

that is, $\beta = 119.7°$. If the value of β varies over the range $\pm$ 10°, associated with the variations in orientation of the phosphodiesters, a variation of chemical shifts along the fiber axis is given by $(\sigma_{33} - \sigma_i)[\cos(\beta + 10) - \cos(\beta - 10)]$ (= 30 ppm). On the other hand, the variation in chemical shifts along the axis normal to the fiber axis becomes $(\sigma_{33} - \sigma_i)[\sin(\beta + 10) - \sin(\beta - 10)]$ (= 19 ppm). From these formulations, it is found that line broadening induced by a variation in the phosphodiester orientations is enhanced in the parallel spectrum.

It is now clear for DNA and poly(dAdT)·poly(dAdT) that there is a close correspondence of NMR characteristics between conformations in solution and in the fiber at high relative humidities; the conformation of DNA in solution must be identical to or very similar to the B form of DNAs in the fiber. Furthermore, it can be concluded that irregularity in the backbone conformation generally exists in natural DNA and probably occurs in a sequence-dependent manner.

The orientation of the phosphodiester group with respect to the helical axis cannot be determined for the B form of DNA by ^{31}P NMR, because the perpendicular spectrum from this fiber is averaged out due to a rapid rotational motion of DNA about the helical axis. Thus we can only test DNA structural models proposed by other methods, by comparing the calculated and observed chemical shifts for the parallel or the perpendicular spectrum. For example, the chemical shift values were calculated from the atomic coordinates of the phosphodiester groups; they are compared with the observed values in Table I.

As mentioned previously, the uncertainty of the calculated values may reach ± 14 ppm at most, because of the experimental errors in the chemical-shift anisotropy (± 5 ppm) and also because of the assumption of the orientation of the principal axis system. Thus the following conclusions can be drawn: (1) the NMR data for the B form of DNA can be described well in terms of the phosphodiester coordinates given by Langridge *et al.* (1960b) but not those of Arnott and Hukins (1972); (2) the structure of poly(dAdT)·poly(dAdT) given by Klug *et al.* (1979) is in quite good agreement with the NMR data.

3.2.2. The A Form of DNA. According to the X-ray fiber diffraction, the A form of DNA is usually obtained in fibers containing appropriate salt at lower relative humidities. It is characterized as follows: a pitch of 28 Å, 11 residues per turn, and bases tilted by 16° relative to the helical axis (Fuller *et al.*, 1965). The A form of DNA should give typical fiber patterns in ^{31}P NMR spectra since it is crystalline, and molecular motion is expected to be much more restricted than that of the B form of DNA, as described in the previous section. It is therefore possible in principle to determine the orientation of the phosphodiester group relative to the helical axis within the limit of the spectral resolution. It seems likely, however, that a number of extraneous factors such as varying fractions of amorphous materials and variation in molecular packing within the crystal might be contributing to the observed spectra for the A form of DNA.

An X-ray diffraction image from a fiber of the sodium salt of poly (dAdT)·poly(dAdT) at 79% relative humidity (Shindo and Zimmerman, 1980) was found to be essentially identical to that of the A form of natural DNA as earlier observed by Davies and Baldwin (1963). Figure 10 shows the ^{31}P NMR spectra from the A form of this polymer oriented parallel and perpendicular to the magnetic field. The parallel spectrum seems to display a single line with 39 ppm (or 950 Hz) linewidth at −26 ppm chemical shift from a reference signal of trimethylphosphate. The perpendicular spectrum exhibits a trapezoidal spectral pattern. A deformation of a theoretical doublet (see Figure 7) to the observed trapezoidal shape may be caused by two factors: the presence of rotational motion about the helical axis of the molecule, and imperfect orientation of the molecules along the fiber axis. Small but predictable changes in the perpendicular spectrum were observed when the temperature was lowered. It was shown that the observed trapezoidal pattern is mainly induced by molecular disorientation, whereas molecular motion is greatly diminished for the A form of DNA, compared with that for the B form. The observations of the singlet for the parallel spectrum and the deformed doublet for the perpendicular spectrum of poly(dAdT)·poly(dAdT) lead us to conclude that this polymer has a single uniform backbone conformation in contrast to two distinct conformations in the B form of this polymer.

Nall *et al.* (1981) recently studied the structure of the A form of calf thymus DNA by ^{31}P NMR spectroscopy at high field strength (97.2 MHz) and obtained excellent spectra. Figure 11 shows the orientation dependence of their ^{31}P NMR spectra of highly oriented fibers at 75% relative humidity and 25°C. Simulated spectra are compared; they were calculated by assuming that A-DNA molecules are virtually rigid and that A-DNA has a single conformation. Agreement between the simulated and observed spectra is excellent. Thus, by contrast to the B form, ^{31}P NMR studies of A-DNA and

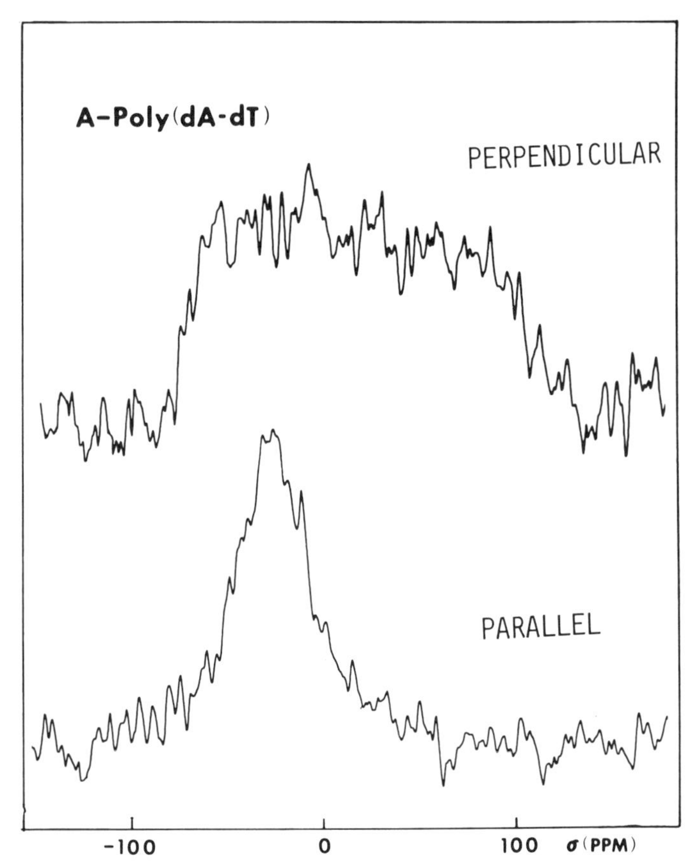

Figure 10. Proton dipolar decoupled ^{31}P NMR spectra (24.3 MHz) observed for the A form of poly(dAdT)·poly(dAdT) fibers at 79% relative humidity.

the A form of synthetic polydeoxynucleotides revealed that A-DNA tends to have a single conformation of the phosphodiester backbone.

In the case in which a molecule is immobile and has a uniform conformation, the orientation of the principal axis system of the chemical-shift tensor can be determined from the NMR data. Since the principal axis system is closely related to the symmetrical axis of the phosphodiester group (see Figure 2*B*), the orientation of the diester group can be determined. Figure 12 shows contour maps of the chemical shifts as a function of the orientation of the chemical-shift tensor relative to the helical axis, namely as functions of Euler angles β and γ. Here $\sigma_{\parallel}$ represents the chemical shift for the fiber oriented parallel to the magnetic field and $\sigma_{\perp}$(min) and $\sigma_{\perp}$(max) correspond to the chemical shifts of the doublet in the perpendicular spectrum for a given set of β and γ. All the contour maps have twofold symmetry with respect to $\beta = 90°$ and $\gamma = 0°$ and $\gamma = 90°$, reflecting the helical and dyad symmetry of the molecule. When the three observed chemical shifts, $\sigma_{\parallel} =$

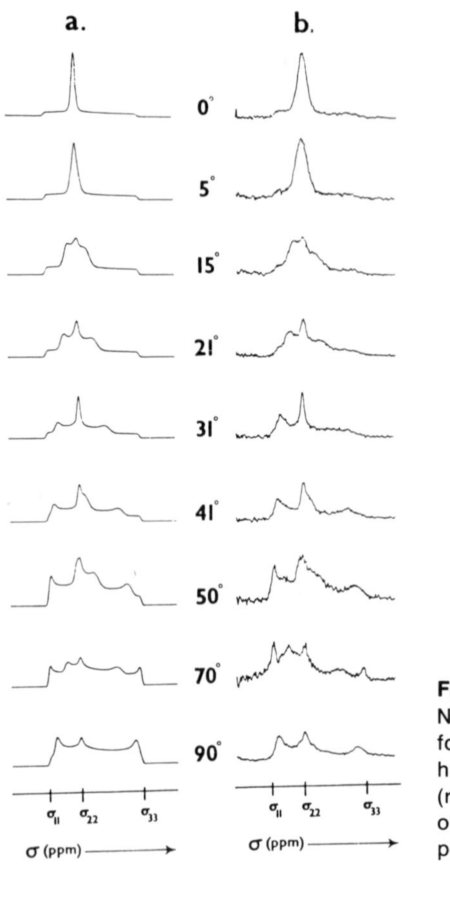

Figure 11. Proton dipolar decoupled ^{31}P NMR spectra (97.2 MHz) observed for the A form of calf thymus DNA at 75% relative humidity. Simulated (left) and observed (right) spectra are compared at various orientations of the fibers. (Nall *et al.*, 1981; permission of *Biochemistry.*)

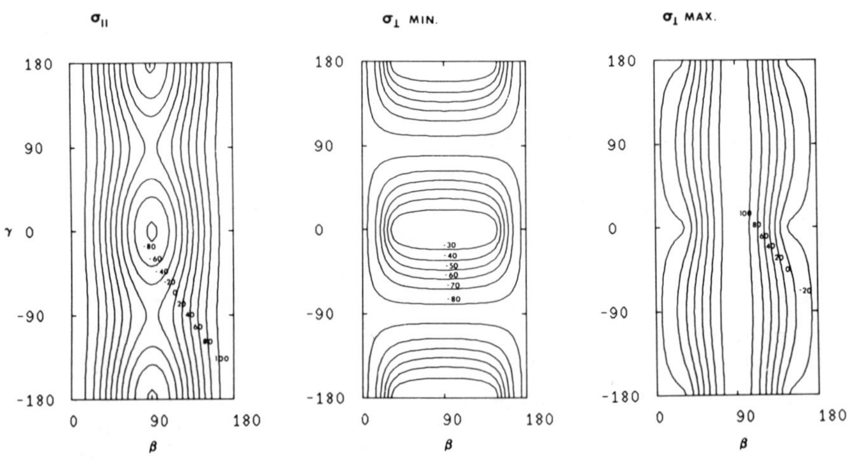

Figure 12. Contour maps of chemical-shift values $\sigma_{\parallel}$, $\sigma_{\perp}$ (max), and $\sigma_{\perp}$ (min) as functions of Euler angles β and γ.

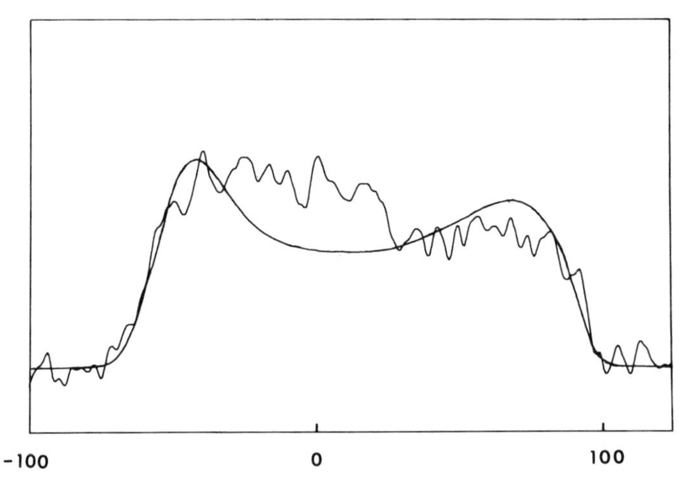

Figure 13. Observed and simulated perpendicular spectra of the A form of poly(dAdT)·poly(dAdT) fibers (79% relative humidity, 2°C).

-26 ± 3, $\sigma_{\perp}(\text{min}) = -59 \pm 5$, and $\sigma_{\perp}(\text{max}) = 97 \pm 7$ ppm for the A form of poly(dAdT)· poly(dAdT) were used, four sets of solutions for β and γ between 0° and 180° were obtained, of which only one set (i.e., $70 \pm 5°$ and $52 \pm 5°$ for β and γ, respectively) is an independent solution. The perpendicular spectrum observed for the A form of this polymer at 2°C is simulated by using the above solution (Figure 13). Here, because of imperfect orientation of molecules along the fiber axis, a standard deviation of DNA molecules from the fiber axis was assumed to be 10° (Shindo and Zimmerman, 1980). Agreement between the calculated and observed spectra is fairly good. Disagreement in the central region of the spectrum may be due mainly to an amorphous part in the fiber and partially to reorientation of DNA molecules about the helical axis.

The orientation of the phosphodiester determined by ^{31}P NMR of the A form of natural DNA (Nall *et al.*, 1981) is compared with those observed for the A-poly(dAdT) duplex in Table II. The orientation parameters β and γ are in good agreement with each other within the limit of experimental errors. The table also contains the parameters obtained by the other methods such as infrared dichroism and X-ray diffraction methods.

Pilet and Brahms (1972) determined the orientation of the phosphodiester group in DNA from infrared dichroism, using the bands at 1240 and 1090 cm^{-1} for the A-DNA. They found that the orientations relative to the helical axis for the O—O vector and for the bisector of the O=P—O^- bonds were 65 and 45°. These angles can easily be converted into our notation (Figure 2) and calculated to be 65 and 51°. Thus results obtained from infrared dichroism are in good agreement with those from ^{31}P NMR studies of poly(dAdT)·poly(dAdT). Of two representative structural models proposed for the A form of DNA, that of Fuller *et al.* (1965) gives the better agreement

Table II. Orientation of the phosphodiester group in the A form of DNA and poly(dAdT)·poly(dAdT) as obtained from ^{31}P NMR and other methods

Material	Method	β^a	γ^a	Reference
Na-Poly(dAdT)·poly(dAdT)	^{31}P NMR	70	52	Shindo *et al.* (1981)
Na-DNA	^{31}P NMR	75	62	Nall *et al.* (1981)
Na-DNA	X-ray	65	60	Fuller *et al.* (1965)
Na-DNA	X-ray	77	79	Arnott and Hukins (1972)
Na-DNA	X-ray	71	70	Arnott *et al.* (1980)
Na-DNA	IR dichroism	65	51	Pilet and Brahms (1972)

[a] Euler angles β and γ represent the angles relative to the helical axis (see Figure 2).

with the NMR data than that of Arnott and Hukins (1972). The later revised model of Arnott and Chandrasekaran (1980) gives better agreement. We assumed the orientation of the chemical-shift tensor to be lying on the symmetrical axis of the phosphodiester group (Kohler and Klein, 1976; see Figure 2). The chemical-shift tensor may deviate from this symmetrical axis, depending on the torsion angle, bond distances, and crystalline field. As far as we are concerned, for the phosphodiesters in DNA, the principal axis system may be invariant relative to the molecular axis system.

It is important to estimate the accuracy of the proposed DNA structure by methods other than X-ray diffraction. In this respect, ^{31}P NMR has become a very useful technique. Of course, other nuclei, ^{13}C, ^{1}H, and ^{15}N, can also be used for investigation of the DNA structure in a similar manner; selective labeling, such as with ^{13}C and ^{2}H, of particular sites in DNA would be especially useful.

4. DNA DYNAMICS IN SOLUTION

Various techniques have been extensively used for investigating chain stiffness or long-range segmental motion of DNA in solution. The hydrodynamic nature and viscoelastic properties of high-molecular-weight DNA are known to be well described by a wormlike chain model in which the stiffness of the DNA chain is measured by its persistence length. To a good approximation, the dynamic properties of a long DNA molecule can be envisioned as a random-flight chain with segmental units whose jumping rate is known to be in the range of microseconds.

Short-range segmental motion, that is, local fluctuation of DNA, is less well characterized, but it is of general interest in connection with dye intercalation, enzyme reactions with DNA, and protein–DNA interactions. When various dyes, including spin-labeled dyes, were used, fluorescence (Wahl *et al.*, 1970) and electron spin resonance measurements (Yamaoka and Noji, 1979) indicated a motion having correlation times of 20 to 50 nsec, nearly independent of the dye. This suggested that significant local motions take place within DNA. Barkley and Zimm (1979) have proposed an elastic model for DNA and have shown that such rapid motions are responsible for local twisting and bending motions at the monomer level in DNA.

Such a local flexibility has gained further support from recent ^{31}P and ^{1}H NMR studies. Internal rotation of the phosphodiester group and twisting motion of the deoxyribose ring having correlation times of a few nanoseconds or less were proposed based on NMR relaxation data (Tritton and Armitage, 1978; Hogan and Jardetzky, 1979). However, it was found that the r.m.s. fluctuation amplitude for such local motions estimated from the theory of Barkley and Zimm (1979) is too small to yield significant effects on NMR parameters. Furthermore, it seems apparent in view of complex relaxation processes in addition to anisotropic motion of DNA that most

NMR analyses are crude approximations. It is necessary to reconsider below further fundamental properties in order to confront the analysis of NMR relaxation data of DNA in solution.

4.1. ^{31}P NMR Studies

Relaxation theory has been successfully applied to account for ^{13}C NMR data of peptides and proteins (see London, 1980), and, as a consequence, useful information about molecular motion has been obtained. The relaxation behavior of protonated ^{13}C atoms is relatively easily analyzed, since the ^{13}C–^{1}H dipole interaction is predominant for the relaxation processes. This situation is not usually met in the case of ^{31}P NMR in macromolecules. Both chemical-shift anisotropy (CSA) and the dipole–dipole (DD) interaction mechanisms must be considered in these cases, and the contribution from CSA is proportional to the square of the magnetic field strength. In this case, total relaxation times are given by the sum of each contribution as follows:

$$\frac{1}{T_1} = \frac{1}{T_1^{DD}} + \frac{1}{T_1^{CSA}} \tag{1}$$

$$\frac{1}{T_2} = \frac{1}{T_2^{DD}} + \frac{1}{T_2^{CSA}} \tag{2}$$

As mentioned in Section 2.2, spin-lattice relaxation time T_1, linewidth $1/\pi T_2$, and the NOE can be correlated with the modes and correlation times of the molecular motions. In order to evaluate the molecular motional modes from the relaxation data, the orientation and magnitude of the ^{31}P—^{1}H dipole vector and the CSA must be known. For example, the dipole vectors may be calculated based on the DNA structure determined by X-ray diffraction studies (e.g., Arnott and Hukins, 1972), and the magnitude of the chemical-shift tensor is evaluated from the relaxation data measured at different frequencies (e.g., Gueron and Shulman, 1975) or it may be obtained from the powder spectrum of solid-state DNA (see Section 2.1).

If DNA is long enough, its overall motion may be treated as that of a wormlike flexible rod; its motion would be described in terms of the isotropic motion of a DNA segment of the persistence length. In this case, problems arise from the fact that a flexible molecule consisting of repeating units exhibits a distribution of correlation times, as was shown by Scheafer (1973). On the other hand, when the DNA size is a persistence length or less, its motion is anisotropic and hence described in terms of the ellipsoid or rodlike model. The latter case is illustrated in Figure 14, together with the parameters for the ^{31}P—^{1}H dipole vectors. (The parameters r_1 and β_1 in the original paper (Shindo, 1980) were incorrect. They should be 3.03 Å and 52.0°, respectively, as in Figure 14.) Here, $D_\perp$ and $D_\parallel$ are the diffusion constants for the rotation of the long axis and the short axis, respectively. The value of

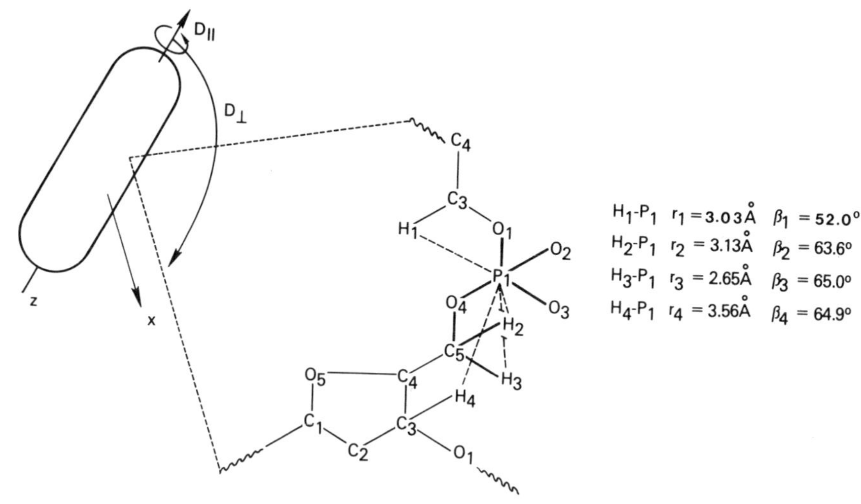

Figure 14. Schematic drawing of anisotropic motion of DNA, and structural parameters of the $^{31}P-^{1}H$ vectors calculated from the atomic coordinates of B-DNA. The angle between the vector and the helical axis is indicated by β_i.

$1/6D_{\perp}$ is equal to the correlation time τ_c for the rotation of the long axis, which becomes longer as the size of DNA increases.

In Figure 15, spin-lattice relaxation time T_1, linewidth, and the NOE at 109.3 MHz are plotted as a function of correlation time τ_c for the above three cases. The solid line represents isotropic motion having a single correlation time (i.e., $D_{\perp} = D_{\parallel}$) and the broken line corresponds to isotropic motion with a distribution of correlation times (distribution parameter p is defined as being larger for a narrower distribution). The dotted line was calculated for a particular model of DNA, a long rigid ellipsoidal molecule. The case of motion with a distribution of correlation times is apparently distinguished from the case with a single correlation time; this has been discussed in detail elsewhere (Shindo, 1980; Scheafer, 1973).

In order to evaluate these models, poly(U) and a DNA fragment (145 bp long) were studied as model compounds (Shindo, 1980). A model having a distribution of correlation times gives the best fits to the relaxation data of poly(U); this is consistent with a common feature of fully flexible macromolecules (Scheafer, 1973). In contrast to poly(U), 145-bp-long DNA is considered to be a fairly stiff molecule, and hence it is not adequate to treat the DNA as a flexible chain. Assuming an ellipsoidal model undergoing anisotropic motion (i.e., $D_{\perp} \neq D_{\parallel}$), a correlation time of 6.5×10^{-7} s and a ratio of diffusion constants $D_{\perp}:D_{\parallel}$ of 91 were obtained for the rotational diffusions of the DNA by using least-squares fits to the data (T_1 and linewidth). The value 6.5×10^{-7} s for the rotation of the long axis is apparently shorter than the correlation time 2.4×10^{-6} s calculated for a rigid

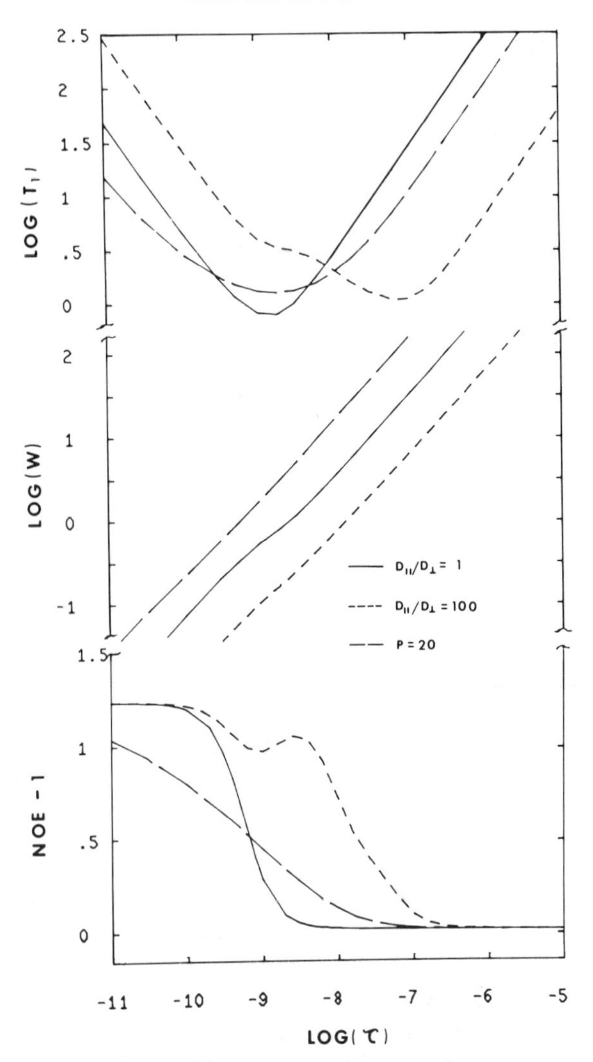

Figure 15. Logarithmic plots of T_1, linewidth $1/\pi T_2$, and NOE vs. τ_c (109.3 MHz) for different models of the DNA motion. The solid line represents isotropic motion with a single correlation time, the dotted line corresponds to anisotropic rotation of an axial symmetrical ellipsoid having the ratio of diffusion constants $D_c : D_{\parallel} = 100$, and the broken line represents the isotropic motion having a broad distribution of correlation times.

rod of this polymer (Broersma, 1960); this discrepancy indicates that a rigid-rotor model for the DNA fragment may not be exactly correct. Even short DNA fragments such as this sample may not be straight, and they may be easily bent to some extent, which may be associated with a twisting motion about the long axis. In other words, the presence of many transient conformers may give rise to a distribution of correlation times within the DNA.

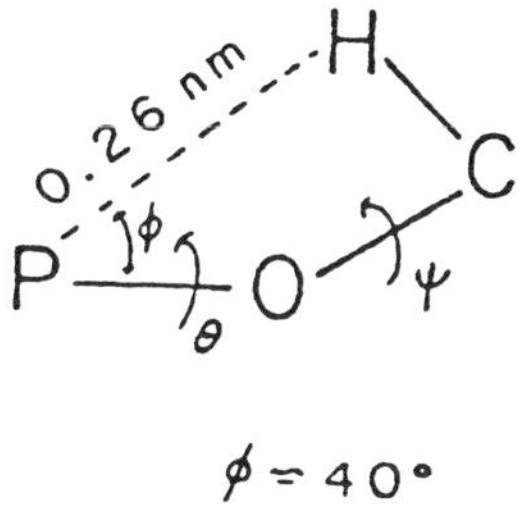

Figure 16. A model for internal motion of the phosphodiester group in DNA. (Bolton and James, 1980; permission of *J. Amer. Chem. Soc.*).

This problem may be simplified by introducing two effective correlation times, one for overall motion and the other for internal motion of the phosphodiester linkages. Klevan *et al.* (1979) proposed fast internal motion of the phosphodiester linkages in DNA based on ^{31}P relaxation data of nucleosome core particles and histone-free DNA in solution. The simple model for the rotation of a $^{31}P—^{1}H$ vector is illustrated in Figure 16, where the rotation axis was taken to be identical with the P—O bond, and the angle ϕ to be the one between the P—H vector and the rotation axis. An effective angle θ for internal rotation of 60° (Klevan *et al.*, 1979) or 40° (Bolton and James, 1980) was used. The overall motion of a large-size DNA was assumed to be more or less isotropic, and a correlation time for the internal motion on the order of 0.5 nsec was deduced from ^{31}P NMR data and also was found to be almost temperature independent (Bolton and James, 1980a).

Based on the above two models, the analysis of ^{31}P NMR relaxation data requires a knowledge of the geometry of the nonbonded P—H vectors with respect to the helical axis or the rotation axis. Furthermore, one model (Shindo, 1980) suggests the necessity to include a distribution of correlation times to fully explain the data. The other model (Klevan *et al.*, 1979; Bolton and James, 1980a) includes the assumption of full rotation of the dipole vectors about the P—O bonds; this is not allowed for in the double helical geometry of DNA. However, it appears that whatever the model for the motions of DNA, the DNA molecule is much more mobile than previously thought.

4.2. Proton NMR Studies

Proton NMR spectra of DNA are usually very broad because of slow tumbling motions of DNA. Figure 17 shows ^{1}H NMR spectra of a DNA fragment as a function of temperature. Apparently, individual signals cannot be well resolved at 25°C, and they are observed only as clusters. Band I in the spectrum is assigned to the aromatic protons H-2, H-6, and H-8, and band II to the pyrimidine proton H-5 and the glycosidic proton H-1′. Band V is assigned to H-2′ protons and band VI to the thymidine methyl group. It has also been shown that the Watson-Crick hydrogen-bonded imino protons

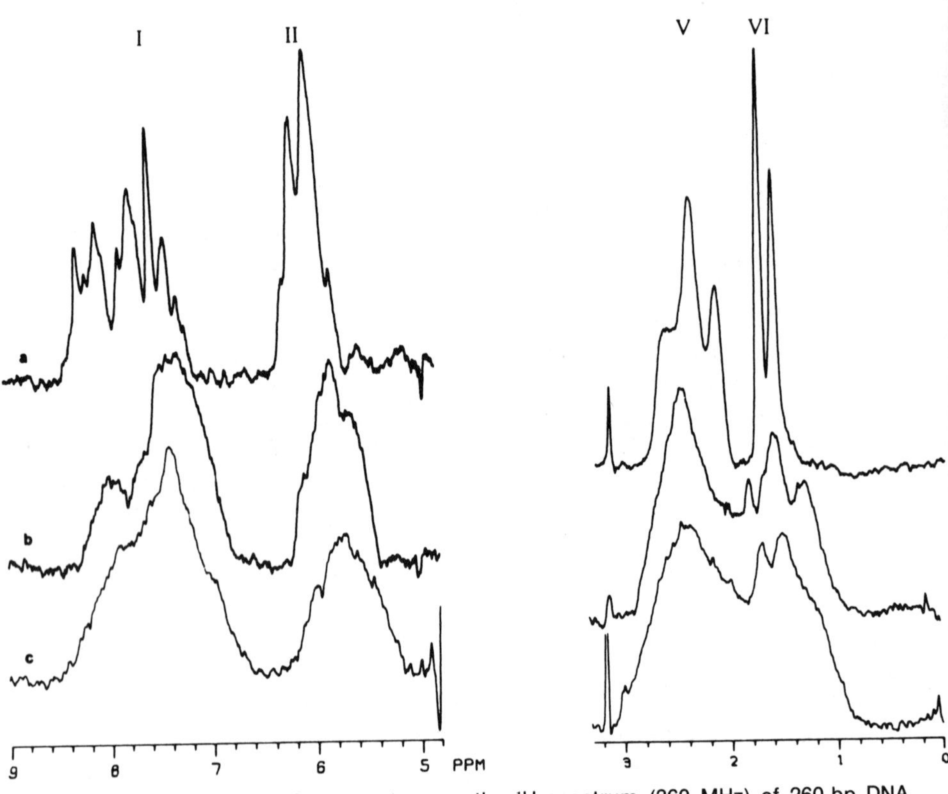

Figure 17. The effect of temperature on the ^1H spectrum (360 MHz) of 260-bp DNA (Hogan and Jardetzky, 1980). (*a*) 75°C, denatured DNA, 30 m*M* ionic strength, pH 7.2; (*b*) 75°C, native DNA, 200 m*M* ionic strength, pH 7.2; and (*c*) 25°C, native DNA, 200 m*M* ionic strength. (Permission of *Biochemistry.*)

appear separately at low fields of 12 to 14 ppm from the TMS peak (Early and Kearns, 1979). It can be stated from the spectra of medium-sized DNA fragments that reliable NMR parameters such as relaxation times T_1 and T_2, and linewidth can only be obtained for the hydrogen-bonded imino protons whose peaks are well separated from the others.

Early and Kearns (1979) studied imino proton resonances of DNA and analyzed their linewidths as a function of the molecular weight. They interpreted the data by employing a model for the motion that occurs in DNA as a flexible double strand. Because of chemical-shift dispersion, the most crucial point in their analysis is how to estimate intrinsic linewidth due to DD interaction. For the hydrogen-bonded protons of the Watson–Crick G—C pairs, they assumed that one out of three imino protons per G—C pair is protonated in a 67% ^2H_2O solvent. Thus a difference in linewidth between 67 and 0% ^2H_2O solvents is taken as a net broadening due to dipolar interactions. The above assumption may tend to overestimate the line broadening because of neglect of the possibility of two protons out of three positions.

For nonexchangeable aromatic protons (band I), the selective "hole burning" and T_2 experiments were employed to estimate intrinsic linewidth because band I contains unresolved peaks. The linewidths of 140 and 30 Hz were estimated for the hydrogen-bonded protons and aromatic protons of a 156-bp-long DNA. Using the atomic coordinates of the DNA structure determined by Arnott and Hukins (1972), the proton–proton vectors were calculated. The ^1H NMR linewidths of imino and nonexchangeable aromatic protons were calculated as a function of DNA size, assuming Woessner's ellipsoidal model for a motional unit of DNA. By comparing the observed linewidth with the theoretical values, the effective hydrodynamic size of DNA was estimated to be 70 and 30 bp or less, respectively, for the hydrogen-bonded protons and aromatic protons. The difference in the motional-unit size at individual sites was attributed to rapid changes in conformation of the sugar ring (changes in pucker) in contrast to relatively rigid bases. This may be taken as a reflection of differences in the degree of internal flexibility. Early and Kearns (1979) also noted that the small motional-unit size estimated from NMR studies compared with that from other hydrodynamic methods is responsible for the fact that the NMR measurements are especially sensitive to a higher frequency in the distribution of local molecular motion and relatively less sensitive to slow overall motion.

There are some uncertainties in the analysis mentioned above. It requires a knowledge of the detailed structure of DNA, such as the B form of DNA, which is still under examination, as described in the previous section. Furthermore, Hogan and Jardetzky (1980) have observed a controversial solvent effect on the imino proton resonances: the linewidth exhibits practically no difference between 70 and 0% of ^2H_2O in aqueous solution.

In recent proton NMR studies of imino protons in DNA restriction fragments (Early *et al.*, 1981a,b), it was shown that the proton exchange becomes important only above 30°C for the imino protons of the A—T base pairs and 55°C for those of the G—C base pairs. From temperature dependence of the relaxation times of imino protons, an activation energy of 15.7 kcal/mol was estimated for the A—T base pair. The result that this activation energy was the same for all three different restriction fragments evidences that the mechanism by which exchange occurs is independent of the length of DNA and the local sequence. This indicates that the opening of base pairs occurs by a single base-pair mechanism, without opening of neighboring base pairs.

Hogan and Jardetzky (1980) proposed a model in which internal motion is a fast two-state motion of dipole vectors relative to the long axis. Overall motion of DNA is treated as being that of a rigid molecule. For long ellipsoids or rods, they described relaxation times T_1 and T_2 as a linear sum of contributions from their internal and overall motions:

$$\frac{1}{T_{2,\text{obs}}} = \frac{\alpha}{T_2(\tau_L)} + \frac{1 - (\alpha + \beta)}{T_2(\tau_S)} \tag{3}$$

$$\frac{1}{T_{1,\mathrm{obs}}} = \frac{1 - (\alpha + \beta)}{T_1(\tau_S)} + \frac{\beta}{T_1(\tau_{\mathrm{int}})} \tag{4}$$

Similarly, a formula was obtained for the NOE. Here $T_{1,2}(\tau)$ is the relaxation time associated with correlation times of the motions of the long axis, τ_L, the short axis, τ_S, and internal motion, τ_{int}, and α and β are related to the average orientation of the dipole vector and the amplitude of the two-state flip. Their original formulas were incorrect and were later corrected by Lipari and Szabo (1981) as follows:

$$\alpha = \frac{1}{16}(1 + 3 \cos 2\bar{\theta} \cos \Delta) \tag{5}$$

$$\beta = \frac{3}{4} \sin^2 \Delta \tag{6}$$

Here, $\bar{\theta} = (\theta_1 + \theta_2)/2$ and $\Delta = \theta_1 - \theta_2$ in the two-state model for internal motions.

Although it remains uncertain how to determine five unknown parameters in Equations (3) and (4) from two observed values of T_1 and T_2, Hogan and Jardetzky assumed a correlation time τ_L for the rotational motion of the long axis calculated from the Broersma's relation (1960), and that of τ_S for the motion of the short axis calculated from another formula (Lamb, 1945; Barkley and Zim, 1979). In the presence of considerable internal motions, the rotational motion of the long axis contributes only little to the relaxation time T_2 (i.e., $\alpha \approx 0$), since a relation $\tau_L \gg \tau_S \gg \tau_{\mathrm{int}}$ may hold for fairly long DNA molecules. The approximate Equations (3) and (4) were applied to the ^{1}H and ^{31}P NMR data, and kinetic parameters for internal motions at individual sites in DNA were evaluated. These are depicted in Figure 18 as an arc directed toward the helical axis. The fluctuation amplitudes of the dipole vectors in phosphodiester, sugar ring protons, and base ring protons are found to be in the range of 40 to 60° relative to the helical axis, and those rates are in the order of a few nanoseconds. Hogan and Jardetzky (1979) tentatively attributed such rapid internal motions to the sugar puckering (C-2′-endo and C-3′-endo) and later (Hogan and Jardetzky, 1980) to a rapid interconversion between the sugar puckered geometries (O-1′-endo and C-2′-endo).

4.3. ^{13}C NMR Studies

Although ^{13}C NMR suffers from low sensitivity because of low natural abundance of the ^{13}C nucleus, the resolution of the spectrum is greater, and analysis of the relaxation data is much simpler than that of ^{1}H and ^{31}P NMR data. For the carbons with protons attached, the observed relaxation data

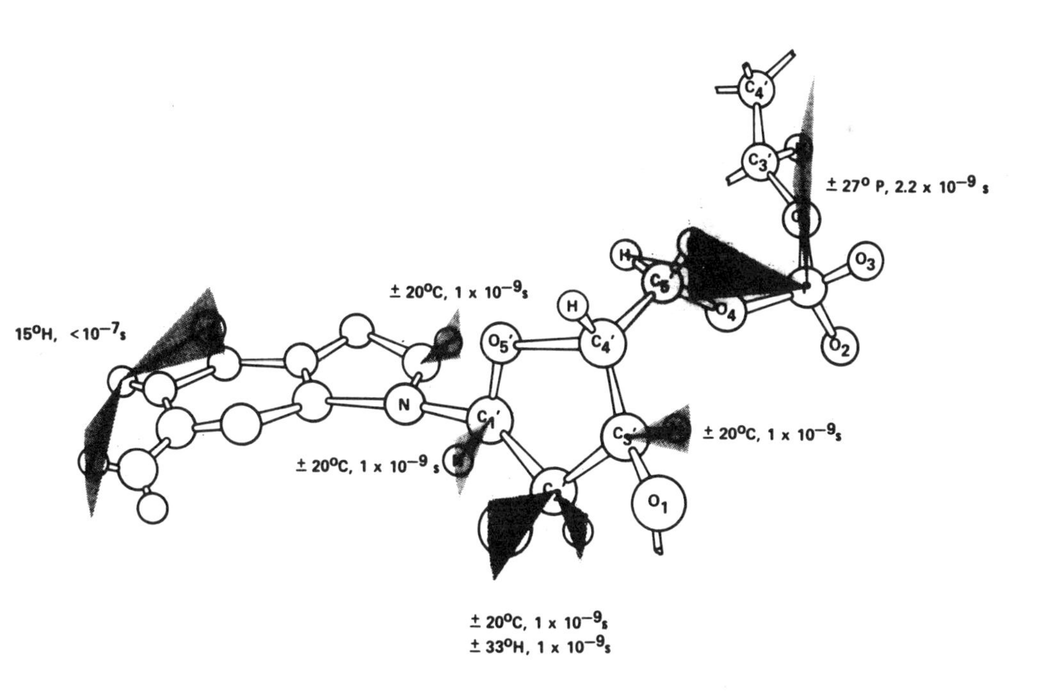

Figure 18. Amplitude and correlation times for the internal motion at individual sites of DNA. (Hogan and Jardetzky, 1980; permission of *Biochemistry*.)

can be interpreted in terms of motions of the C—H vectors alone, since ^{13}C nuclear spins predominantly relax through the protons.

Figure 19 shows the ^{13}C NMR spectrum (at 68.9 MHz) of poly(dAdT)·poly(dAdT) with length of 145 bp at 32°C. On the basis of the spectral linewidth and chemical-shift data for constituent monomers (Dorman and Roberts, 1970; Smith *et al.*, 1973), all possible carbon signals can be assigned. Peaks in the spectrum indicated by *a* and *b* are due to EDTA and a trace of ethanol in the preparation. Table III collects the observed chemical shifts and relaxation parameters for poly(dAdT)·poly(dAdT) at 25 and 32°C. Several points can be noted: (1) each of the deoxyribose carbon resonances, C2′, C3′, and C1′ or C4′ is split into a doublet, (2) low-field peaks are narrower than the upfield peak resonances of each of the doublets, (3) protonated-carbon resonances are broader than nonprotonated-carbon resonances, except for the thymidine methyl carbon resonance, (4) spin-lattice

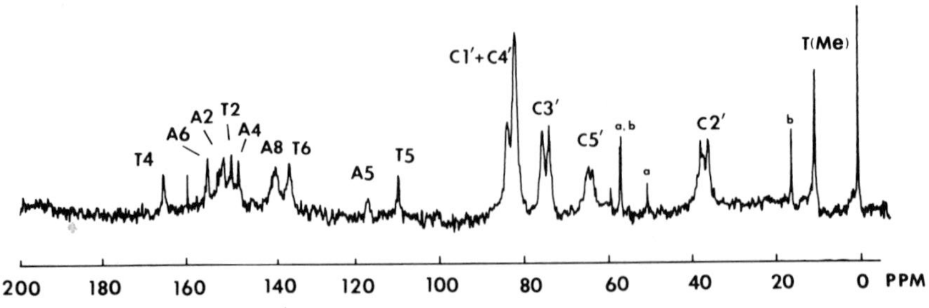

Figure 19. Natural abundance ^{13}C NMR spectrum (67.9 MHz) of 145-bp poly(dAdT)·poly(dAdT) at 32°C.

relaxation times are all the same for the deoxyribose carbons within experimental error, and (5) the NOE values are nonminimum for the sugar carbons except for C5′ but are almost minimum for the aromatic carbons. The fairly large splitting of C2′ and C3′ carbons was considered, and each line in the doublets was attributed to the presence of two distinct puckered conformations of the sugar (Shindo, 1981), probably C2′-endo and C3′-endo in an alternating array, just as observed for the ^{31}P NMR spectrum of the same polymer (Shindo *et al.*, 1979). This implies that an alternating conformation extends into the interior of the double helical structure and that the interconversion between the two forms does not occur in the NMR time scale of the order of tens of milliseconds.

Small DNA fragments such as 145-bp-long poly(dAdT) were shown to be described approximately as rigid ellipsoids of 26 Å in the short axis and 500 Å in the long axis (Broersma, 1960), and hence their motions would be anisotropic. Assuming this model, the predicted linewidth of the carbon with hydrogens attached would be about 500 Hz, which is much broader than the observed linewidth of 90 to 117 Hz at 25°C. The discrepancy between predicted and observed values suggested that a rigid model is not correct, even for such a short DNA piece; significant bending motion or internal motion seems required to account for the data, as indicated for interpretation of the ^{31}P and ^{1}H NMR results in Sections 4.1 and 4.2.

In the first approximation, let us assume a wormlike chain joined with motional units. Then the effective size of the motional unit can be estimated from the observed linewidth, knowing the orientation of the C—H vectors relative to the long axis (helical axis). Figure 20 shows the theoretical linewidth as a function of the length of the motional unit at different angles θ, where θ is the angle between the C—H vector and the long axis. Since the plane of the base rings in B-DNA is approximately perpendicular to the helical axis and therefore $\theta = 90°$ for the C—H vectors of base rings, the length of the motional unit can be estimated from Figure 20 to be about 300 Å (about 90 bp) for the broadest line of the base carbon, and 177 Hz for peak A8. This value is nearly consistent with that obtained from the ^{1}H NMR spectrum of the hydrogen-bonded protons of DNA, within experimental error.

Table III. ^{13}C **NMR parameters observed for poly(dAdT)·poly(dAdT)**

	Chemical Shift[a] (ppm)	Linewidth (Hz)[b]		NT_1 (s) 25°C	NOE 25°C
		32°C	25°C		
Deoxyribose carbon					
C2′	37.8, 36.0	99	127	0.38 ± 0.04	1.28
C3′	75.3, 73.5	79	83	0.29 ± 0.06	1.28
C1′, C4′	83.5, 81.7	83	—	0.35 ± 0.03	1.26
C5′	64	—	—	0.32 ± 0.06	1.00
Base-protonated carbon					
A2	150.5	130	—		
A8	138.3	135	177		
T6	134.9	115	140		
Thymidine methyl	10.4	24	35	2.1 ± 0.3	1.43

[a]Measured from the methyl carbon resonance of ^{13}C-enriched acetonitrile as an internal reference.
[b]The linewidth of the sugar carbons is an average of the two peaks in corresponding doublets.

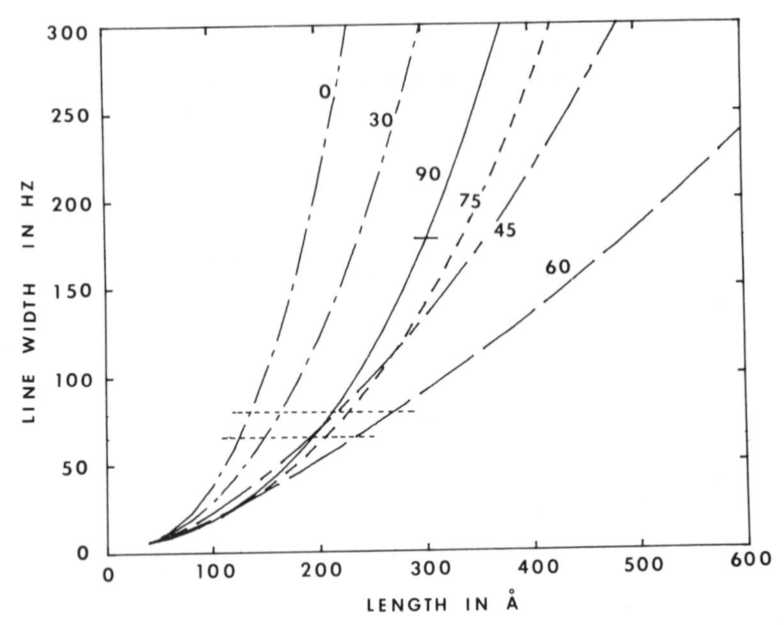

Figure 20. Theoretical linewidth ($1/\pi T_2$) as a function of motional-unit length of DNA, assuming that the motional unit is an ellipsoid. A bar corresponds to the observed linewidth (at 25°C) for the resonance A8, and a boundary of dotted lines indicates maximum and minimum linewidths observed for the sugar ring carbons.

The observation that peak T6 is narrower than peak A8 suggests that relatively large local mobility occurs for thymidine; the motion likely involves fluctuation of the base plane relative to the helical axis. Alternatively, the sharpness of the peak T6 might indicate that a static tilting of the plane of the thymidine base of about 15° relative to the adenosine base plane occurs (estimated from the curves for $\theta = 90$ and 75° at $L = 300$ Å). This latter possibility seems less likely in view of the necessity of maintaining Watson-Crick base pairing, but this remains to be established.

A boundary of the observed linewidth for the sugar carbons is drawn by dashed lines in Figure 20. The length of the motional unit estimated from the linewidths varies from 110 to 270 Å, depending on the values of θ used, although the exact orientations of the C—H vectors are known. Since the magic angle of 54.7° for θ seems unlikely, the upper limit of the motional-unit length (270 Å) would be reduced. Thus it seems unlikely that a wormlike chain model is sufficient to describe the motion of the sugar carbons; it is likely that local motion must be included. In fact, nonminimum NOE values are observed for the sugar carbons (Table III), indicating that the ring carbons are more mobile than the base. It seems likely that the difference in linewidth for the doublets of C-2′ and C-3′ carbons is induced by differences in the local mobility of the sugar carbons of adenosine and thymidine rather than structural difference in the angles θ. It is therefore suggested that the interior of the duplex is relatively immobile in contrast to the relatively more mobile backbone of DNA, and that the local mobilities of the sugar rings and bases are different among nucleotide residues.

Hogan and Jardetzky (1980) reported the natural abundance ^{13}C NMR spectrum at 25 MHz for monodisperse natural DNA with a length of 260 bp. Most of the resonances in the spectrum are very broad, mainly due to chemical-shift dispersion as observed previously by Rill *et al.* (1980). Spin-lattice relaxation time, T_1, and NOE values measured (Bolton and James, 1980b; Hogan and Jardetzky, 1980; Shindo, 1981; and Levy *et al.*, 1981) cannot be directly compared because of the different molecular weight and different conditions used. Hogan and Jardetzky (1980) estimated a correlation time for internal motion of the C—H vectors by fitting Equation (4) and a formula for the NOE to the relaxation data. Although it was not possible to differentiate by ^{13}C NMR the internal motions between the base and sugar carbons, they stated that the base plane at positions C6 and C8 and the sugar carbons at C1′, C3′, and C2′ all experience fluctuations in geometry by $\pm 20°$, and these occur with a time constant near 1×10^{-9} s (see Figure 18).

4.4. Internal Motion and Flexibility of DNA

As described in the previous sections, Tritton and Armitage (1978), Bolton and James (1980a,b), and Hogan and Jardetzky (1979, 1980) proposed that in addition to the overall motion of DNA a rapid internal motion occurs at the rate of nanoseconds. Early and Kearns (1979) and Shindo (1980, 1981) analyzed the relaxation data of individual sites of a DNA molecule in terms of an ellipsoidal model or motional-unit concept. They stated that the size of the motional unit estimated from NMR data is much smaller than that expected from hydrodynamic methods, and that its size is different at different sites in DNA. These results indicate rapid internal motions in DNA, but it is not necessary for the motions to be on the order of nanoseconds. DiVerdi and Opella (1981) and Opella *et al.* (1981) have studied ^{31}P NMR spectra of very large molecular weight DNA and stated that the frequency dependence of the linewidth is reasonably interpreted in terms of isotropic motion of DNA, without including rapid internal motions. Thus it may be possible to say that the fragments of DNA have internal motions that are not present in high-molecular-weight DNA.

Thus two problems arise: (1) how this discrepancy of DNA motion can be interpreted and (2) what the detailed mechanism of rapid internal motion is, if it indeed occurs in DNA. In answer to (1), Opella *et al.* (1981) suggested that rotational motions are significantly damped for a group in the middle of a long polymer compared with those of a short rod DNA, especially since a phosphate group in the middle of a short fragment such as one 145-bp-long is much less than a persistence length from either end. However, strong evidence for rapid internal motion is the fact that a fairly large NOE is observed in ^{31}P and ^{13}C spectra of a relatively small DNA fragment. Therefore, it is desirable to confirm if the NOE can be observed even for high-molecular-weight DNA.

Recently two interesting theoretical papers appeared (Lipari and Szabo, 1981; Keeper and James, 1982). Both treat internal motions as a reorientation of dipole vectors within a cone or jumping between two or more states. Lipari and Szabo (1981) examined two aspects: (1) whether a rapid internal motion at a nanosecond rate can be deduced from relaxation data and (2) whether the relaxation data give a unique solution out of several proposed models. Their conclusions are that there is a rapid internal motion in the nanosecond time scale for all models they studied, but no unique solution exists. On the other hand, the theoretical treatment of Keeper and James (1982) concerns specific conformational changes of DNA, that is, combined motions of six successive bonds in the DNA backbone. Out of two proposed motions, they claimed that sugar ring puckering gives the best fit to the relaxation data. The conclusions of all these investigators are essentially consistent with regard to the rate and fairly large amplitude of the internal motions of DNA. If so, it seems inevitable that the conventional view that the structure of DNA in solution is a rigid B form is incorrect.

It is interesting to explore some relationships between internal motions and somewhat long-range properties such as a cooperative motion of base-sugar-phosphate, telestability (Wells *et al.*, 1977), and acoustic propagation along the DNA chain. However, it will be necessary to accumulate more selective data, especially ^{13}C data, of well-characterized DNA fragments.

5. CONCLUDING REMARKS

Watson and Crick proposed a model of DNA structure in 1953, but it was not until late 1979 that a single crystal of a DNA fragment for X-ray diffraction study, the hexamer, d(CGCGCG), happened to have a left-handed helical structure, the so-called Z-DNA (Wang *et al.*, 1979). Furthermore, the X-ray diffraction study of a dodecamer, $d(CGCGAATTCGCG)_2$ demonstrated that its structure is a B-DNA but with considerable variation along the chain (Drew *et al.*, 1981) and the local motions in the crystal are very interesting (Drew and Dickerson, 1981).

Zimmerman and Pheiffer (1979) demonstrated by X-ray fiber diffraction studies that a highly concentrated solution of DNA gives the same helical parameters as those of B-DNA. Similarly, the analysis of the line shape in ^{31}P NMR spectra of DNAs confirmed a traditional assumption that the helical structure of DNA in solution is very close to that of highly hydrated DNA fiber. However, in contrast to the assumptions about the classical B form of DNA, NMR studies demonstrated strong evidence for the structural versatility and significant fluctuational motions of the bases and sugar-phosphate backbone in DNA. Thus highly hydrated DNA is now considered to have a conformationally flexible structure. It appears that such characteristics may play an important role in the biological function of DNA.

Although the helical structure of DNA seems to be similar along its entire length, there is increasing evidence from enzymologic studies supporting the concept that a specific interaction between a protein and a set of base sequences in the double-stranded DNA occurs on a one-to-one basis. It is tempting to conclude from the above physicochemical results that a protein recognizes primarily a unique helical structure induced by a particular base sequence, rather than the base sequence alone.

REFERENCES

Arnott, S., and D. W. L. Hukins (1972), *Biochem. Biophys. Res. Commun.* **47,** 1504.

Arnott, S., R. Chandrasekaran, P. J. Bond, D. L. Birdsall, A. G. W. Leslie, and L. C. Puigjaner (1981), *Proceedings of the Seventh Aharon Katzir-Katchalsky Conference on Structural Aspects of Recognition and Assembly in Biological Macromolecules*, **2,** 487–499.

Barkley, M. D., and B. H. Zimm (1979), *J. Chem. Phys.* **70,** 2991.

Bolton, P. H., and T. L. James (1979), *J. Phys. Chem.* **83,** 3359.

Bolton, P. H., and T. L. James (1980a), *J. Am. Chem. Soc.* **102,** 25.

Bolton, P. H., and T. L. James (1980b), *Biochemistry* **19,** 1388.

Broersma, S. (1960), *J. Chem. Phys.* **32,** 1626.

Cohen, J. S., J. B. Wooten, and C. L. Chatterjee (1981), *Biochemistry* **20,** 3049.

Davies, D. R., and R. L. Baldwin (1963), *J. Mol. Biol.* **6,** 251.

DiVerdi, J. A., and S. J. Opella (1981), *Biochemistry* **20,** 280.

Dorman, D. E., and J. D. Roberts (1970), *Proc. Natl. Acad. Sci. USA,* **65,** 19.

Drew, H. R., and R. E. Dickerson (1981), *J. Mol. Biol.* **151,** 535.

Drew, H. R., R. M. Wing, T. Takano, C. Broka, S. Tanaka, K. Itakura, and R. E. Dickerson (1981), *Proc. Natl. Acad. Sci. USA* **78,** 2179.

Drew, H. R., T. Takano, S. Tanaka, K. Itakura, and R. E. Dickerson (1980), *Nature* **286,** 567.

Early, A. T., and D. R. Kearns (1979), *Proc. Natl. Acad. Sci. USA* **76,** 4165.

Early, A. T., and D. R. Kearns, W. Hillen, and R. D. Wells (1981a), *Biochemistry* **20,** 3756; (1981b), ibid **20,** 3764.

Fuller, W., M. H. F. Wilkins, H. R. Wilson, and L. D. Hamilton (1965) *J. Mol. Biol.* **12,** 60.

Gorenstein, D. G. (1975), *J. Am. Chem. Soc.* **97,** 898.

Gorenstein, D. G., and D. Kar (1977), *J. Am. Chem. Soc.* **99,** 672.

Gorenstein, D. G., and B. A. Luxon (1979) *Biochemistry* **18,** 3796.

Gueron, S. M., and R. G. Shulman (1975), *Proc. Natl. Acad. Sci. USA* **72,** 3482.

Herzfeld, J., R. G. Griffin, and R. A. Haberkorn (1978), *Biochemistry* **17,** 2711.

Hogan, M. E., and O. Jardetzky (1979), *Proc. Natl. Acad. Sci. USA* **76,** 6341.

Hogan, M. E., and O. Jardetzky (1980), *Biochemistry* **19,** 3460.

Keeper, J. W., and T. L. James (1982), *J. Am. Chem. Soc.* **104,** 929.

Klevan, L., I. M. Armitage, and D. M. Crothers, (1979), *Nucl. Acids Res.* **6,** 1607.

Klug, A., A. Jack, M. A. Viswamitra, C. Kennard, Z. Shakked, and T. A. Steitz (1979), *J. Mol. Biol.* **131,** 669.

Kohler, S. J., and M. P. Klein (1976), *Biochemistry* **15,** 967.

Lamb, H. (1945), *Hydrodynamics*, Dover, New York.

Langridge, R., D. A. Marvin, W. E. Seeds, H. R. Wilson, C. W. Hooper, M. H. F. Wilkins, and L. D. Hamilton (1960a), *J. Mol. Biol.* **2,** 38.

Langridge, R., H. R. Wilson, C. W. Hooper, M. H. F. Wilkins, and L. D. Hamilton (1960b), *J. Mol. Biol.* **2,** 19.

Levy, G. C., P. R. Hilliard, Jr., L. F. Levy, and R. L. Rill (1981), *J. Biol. Chem.* **256,** 9986.

Lipari, G., and A. Szabo (1981), *Biochemistry* **20,** 6250.

London, R. E. (1980), *Magnetic Resonance in Biology*, Vol. 1, J. S. Cohen, Ed., Wiley, New York.

Lyerla, J. R., Jr. (1976), in *Topics in C-13 NMR*, Vol. 2, G. C. Levy, Ed., Wiley, New York, pp. 79–148.

Marvin, D. A., M. Spencer, M. H. F. Wilkins, and L. D. Hamilton (1961), *J. Mol. Biol.* **3,** 547.

McGhee, J. D., and P. H. Von Hippel (1977), *Biochemistry* **16,** 3267.

Nall, B. T., W. P. Rothwell, J. S. Waugh, and A. Rupprecht (1981), *Biochemistry* **20,** 1881.

Olson, W. K. (1980), *ACS Symp. Ser.* **114,** 251.

Opella, S. J., W. B. Wise, and J. A. DiVerdi (1981), *Biochemistry* **20,** 284.

Patel, D. J., L. L. Canuel (1979), *Eur. J. Biochem.* **96,** 267.

Patel, D. J., L. L. Canuel, and F. M. Pohl (1979), *Proc. Natl. Acad. Sci. USA* **76,** 2508.

Patel, D. J., S. A. Kozlowski, L. A. Marky, C. Broka, J. A. Rice, K. Itakura, and K. J. Breslauer (1982), *Biochemistry* **21,** 428.

Pilet, J., and J. Brahms (1972), *Nature New Biol.* **236,** 99.

Pohl, F. M., and T. M. Jovin (1972), *J. Mol. Biol.* **67,** 375.

Rill, R. L., P. R. Hilliard, Jr., J. T. Bailey, and G. C. Levy (1980), *J. Am. Chem. Soc.* **102,** 418.

Scheafer, J. (1973), *Macromolecules* **6,** 509.

Simpson, R. T., and P. Kunzler (1979), *Nucl. Acids Res.* **6,** 1378.

Simpson, R. T., and H. Shindo (1980), *Nucl. Acids Res.* **8,** 2093.

Shindo, H. (1980), *Biopolymers* **19,** 509.

Shindo, H. (1981), *Eur. J. Biochem.* **120,** 309.

Shindo, H., R. T. Simpson, and J. S. Cohen (1979), *J. Biol. Chem.* **254,** 8125.

Shindo, H., J. B. Wooten, B. H. Pheiffer, and S. B. Zimmerman (1980), *Biochemistry* **19,** 518.

Shindo, H., and S. B. Zimmerman (1980), *Nature* **283,** 690.

Shindo, H., J. B. Wooten, and S. B. Zimmerman (1981), *Biochemistry* **20,** 745.

Smith, I. C. P., H. H. Mantsch, R. D. Lapper, R. Deslauriers, and T. Scheich (1973), *The Jerusalem Symposia on Quantum Chemistry and Biochemistry*, Vol. 5, B. Pullman and E. D. Berman, Eds., Israel Academy of Science, Jerusalem, pp. 381–402.

Sussman, J. L., S. R. Holbrook, R. W. Warrant, G. M. Church, and S. H. Kim (1978), *J. Mol. Biol.* **123,** 607.

Tritton, R. T., and I. M. Armitage (1978), *Nucl. Acids Res.* **5,** 3855.

Viswamitra, M. A., O. Kennard, P. G. Jones, G. M. Sheldeick, S. Salisbury, and L. Falvello (1978), *Nature* **273,** 687.

Wahl, Ph., J. Paoletti, and J.-B. LePecq (1970), *Proc. Natl. Acad. Sci. USA* **65,** 417.

Wang, A. H.-J., G. L. Quigley, F. J. Kolpak, J. L. Crawford, J. H. Van Boom, G. van der Mavel, and A. Rich (1979), *Nature* **282,** 680.

Wells, R. D., R. W. Blakesley, S. C. Hardies, G. T. Horn, J. E. Larson, E. Selsing, J. F. Burd,

H. W. Chan, J. B. Dodgson, K. F. Jensen, I. F. Nes, and R. M. Wartell (1977), *CRC Crit. Rev. Biochem.* **4,** 305.

Wells, R. D., T. C. Goodman, W. Hillen, G. T. Horn, R. D. Klein, J. E. Larson, U. R. Mueller, S. T. Neuendorf, N. Panayotatos, and S. M. Stirdivant (1980), *Prog. Nucl. Acids Res. Mol. Biol.*, **24,** 167–267.

Weossner, D. E. (1962), *J. Chem. Phys.* **36,** 647.

Wing, R., H. Drew, T. Takano, C. Broka, S. Tanaka, K. Itakura, and R. E. Dickerson (1980), *Nature* **287,** 755.

Yamaoka, K., and S. Noji (1979), *Chem. Lett.* **1979,** 1123.

Zimmerman, S. B., and B. H. Pheiffer (1979), *Proc. Natl. Acad. Sci. USA* **76,** 2703.

Four

Observations of Amino Acid Side Chains in Proteins by NMR Methods

Jack S. Cohen

Developmental Pharmacology Branch
National Institute of Child Health and Human Development
National Institutes of Health
Bethesda, MD 20205

*Lou J. Hughes**

Department of Chemistry
The American University
Washington, D.C. 20016

and

Jan B. Wooten

Phillip Morris Research Center
Richmond, VA 23261

1. **Introduction** **131**
2. **NMR Methods Applied to Proteins** **133**
   - 2.1. Proton NMR of Aqueous Protein Solutions, 133
   - 2.2. Resolution Enhancement and Spectral Simplification, 135
   - 2.3. Two-dimensional FT NMR, 139
   - 2.4. Solid-State NMR, 144

*On leave from Ithaca College, Ithaca, NY 14850.

3. Effects on NMR Parameters of the Incorporation of Amino Acid Side Chains into Proteins 152
3.1. Protein Microenvironment and Chemical Shift, 152
3.2. Protein Mobility and Relaxation Times, 156
4. NMR Titration Curves 167
5. Studies of Specific Types of Amino Acid Side Chains in Proteins 173
5.1. Histidine, 173
5.1.1. Assignment of Histidine Resonances of RNAse, 175
5.1.2. Properties of Histidine Residues of RNAse, 182
5.1.3. Histidine Residues in Some Other Proteins, 189
5.1.4. Thermodynamic Parameters of Histidine Residues From NMR Titrations, 192
5.2. Aromatic Side Chains, 192
5.2.1. Tyrosine, 196
5.2.2. Phenylalanine, 203
5.2.3. Tryptophan, 203
5.3. Carboxyl Side Chains, 205
5.4. Hydroxyl Side Chains, 214
5.5. Basic Side Chains, 218
5.5.1. Arginine, 218
5.5.2. Lysine, 219
5.5.3. Histones, 223
5.6. Sulfur-Containing Side Chains, 224
5.6.1. Methionine, 224
5.6.2. Cysteine and Cystine, 230
5.7. Aliphatic Side Chains, 230
5.8. Proline, 234
6. Conclusions and Prognosis 235
Acknowledgments 236
Selected Bibliography 236
References 237

1. INTRODUCTION

Nuclear magnetic resonance (NMR) spectroscopy has developed as the method of choice to study amino acid side chains of proteins in solution. This review is intended to provide a convenient source for the use of protein chemists and enzymologists who may be interested in a particular amino acid residue in a given protein system on which they are working and would like to know if NMR methods could assist their understanding of that residue.

As a consequence, this review conforms to certain criteria. First, it is organized on the basis of amino acid types rather than the NMR technique being applied, be it ^{1}H, ^{13}C, ^{15}N, or other observed nucleus. Second, the review emphasizes the side-chain type rather than the particular protein upon which the investigation was carried out. Third, extensive comparisons of data from free amino acids or denatured random-coil polypeptides with the results for the corresponding side chains in native proteins are avoided. Rather, general conclusions about the nature of the effects of incorporating an amino acid side chain into a protein interior are summarized in a separate section.

It is not intended to exhaustively review all NMR studies of proteins that have been reported in the literature. At this time such a venture would in any case be virtually impossible. Instead the focus is upon the information obtained when resonances of individual amino acid side chains have been resolved in NMR spectra of proteins. Explicitly included in this review are those studies in which isotopic enrichment (^{13}C, ^{2}H, ^{19}F, ^{15}N) has been utilized as a direct adjunct of the NMR technique.

Studies of selective effects of perturbants (such as inhibitors of enzyme action, salt, temperature) on the individually resolved signals of amino acid residues of proteins in solution, constitute the most valuable area of applications of NMR to proteins. Efforts to solve the three-dimensional structure of a protein by NMR methods, utilizing paramagnetic ions (Dwek, 1973), networks of coupled nuclear Overhauser enhancements (NOEs) (Wagner *et al.*, 1981), or local shift effects (Dobson *et al.*, 1974), have not met with general success. Indeed the NMR work on a very small, very stable protein, such as the bovine pancreatic trypsin inhibitor (BPTI), has taken considerably more man-years than the solution of an X-ray crystal structure now takes, and the existence of such a structure considerably simplifies the analysis of the NMR spectra. In effect, a few relatively small and stable proteins that give well-resolved signals have facilitated the development of the NMR methods. Such NMR studies provide unique information on local mobility and proximity of residues, and current indications are that NMR is being applied to larger and more biochemically interesting proteins.

Of course, there are significant limitations to NMR methods, such as low sensitivity compared with optical spectroscopy. Among NMR nuclei, ^{13}C and ^{15}N have especially low sensitivities (Cohen, 1978). Also, as the size of a protein increases, there is a loss of sensitivity and resolution because of the increase in the number of resonances and the broadness of their linewidths. However, it should be emphasized that the resolution of any given resonance depends on the local average mobility or correlation time.

Notwithstanding some limitations, NMR methods have been widely used in studies of protein conformation in solution. For comparison, ^{1}H and ^{13}C NMR spectra of the same protein, ribonuclease, are shown in Figure 1, with the main regions of side-chain types indicated. Of particular interest are observations of resonances of side-chain groups that do not contain a UV or fluorescent chromophore, such as Glu, Asp, Lys, Arg, Met, Thr, Leu, and so forth. Such groups when observed by NMR techniques act as intrinsic nondisturbing probes of their local electronic microenvironment in a protein. Enrichment with stable isotopes, such as ^{13}C, ^{15}N, and ^{2}H, provide the advantages of unambiguous assignment and increased sensitivity for the observation of individual amino acid side-chain resonances. It is hoped that this article will draw attention to the value of the various NMR methods in providing unique information about side-chain groups in proteins.

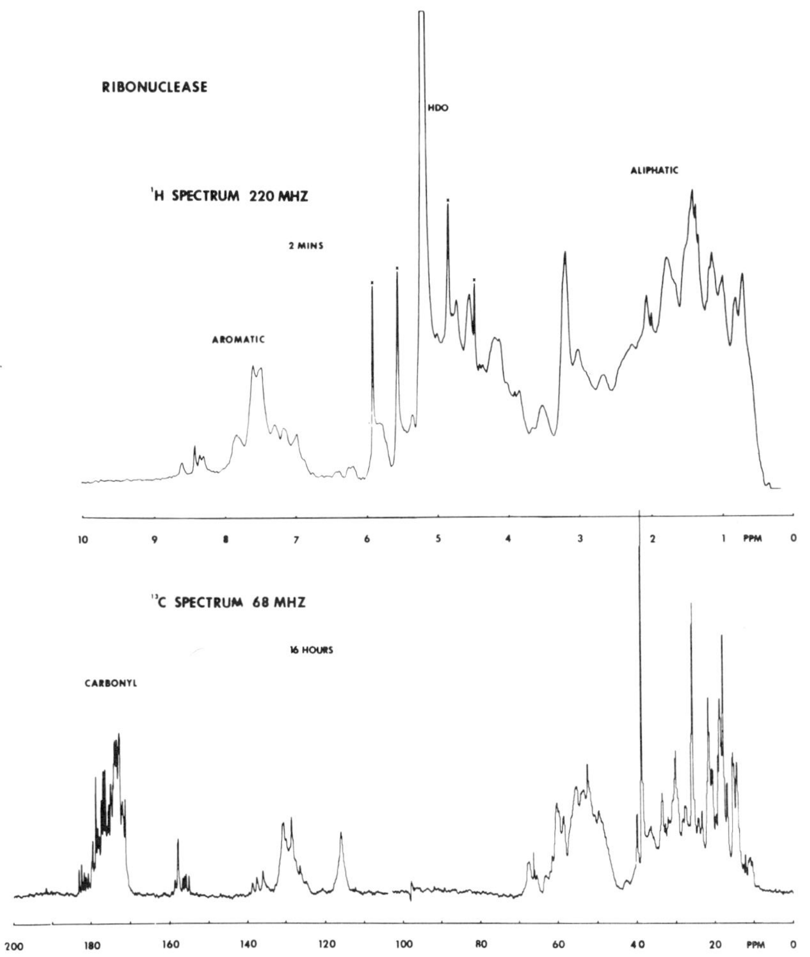

Figure 1. NMR spectra of bovine pancreatic ribonuclease A (Worthington, phosphate-free; 100 mg in 1 ml deuteroacetate, 0.2 *M*, pH 5.5, 22°C). The main regions of the spectra are shown. ^1H spectrum at 220 MHz, in ^2H_2O; ^{13}C natural abundance (1.1%) spectrum at 67.9 MHz, with proton decoupling.

2. NMR METHODS APPLIED TO PROTEINS

Consistent with the purpose of this review, an introduction to the methodology of the NMR techniques used is included here. To obtain a more comprehensive treatment, the reader is referred to the bibliography.

2.1. Proton NMR of Aqueous Protein Solutions

A perennial problem in the observation of protein resonances by proton NMR is the elimination of the exceedingly strong signal due to water. The usual and most obvious strategy to eliminate the water signal is to employ D_2O, but it is necessary to use very highly deuterium enriched water ("100% D_2O") and to lyophilize the protein several times from the D_2O in order to remove the water of hydration within the interior of the protein. In order to remove slowly exchanging resonances, it is sometimes necessary to completely denature the protein in D_2O (Markley and Ibanez, 1978), although this usually leads to some loss of active material. Even so, because of the high sensitivity of today's spectrometers, the desire to observe protein spectra in dilute solution, or to observe weak resonances very close to the water signal, the residual-water-resonance problem is a persistent one. Fortunately, numerous techniques have been developed that work reasonably well in different applications. The methods for suppressing the water signal have been discussed in detail by Redfield (1978) and will be only summarized here.

One of the most common methods for suppressing the water resonance is to apply continuous selective radiofrequency (RF) irradiation to the water signal, which causes the nuclear spin energy levels to be equalized. Since no nuclear transitions are possible under this condition of "saturation," no signal results. The principal disadvantage of solvent saturation is that any signals of any protons that are exchanging with the water protons are also eliminated. Unless the exchangeable resonances are to be observed, in which case D_2O could also not be employed, this presaturation method should work satisfactorily and is generally available on most spectrometers. A second method takes advantage of the fact that the water relaxation time is in general quite long (≥ 2 s) relative to the protein relaxation times and is known as WEFT for "water elimination Fourier transform (FT)" (Patt and Sykes, 1972). It employs the familiar 180°–τ–90° pulse sequence that is commonly used to measure T_1 values. By choosing the value of τ appropriately, the water resonance can be caught at its null point, so the application of the 90° observation pulse results in no net signal. Because the protein proton relaxation times are shorter, there is usually sufficient time for complete recovery during the same interval, resulting in a protein spectrum without the water signal.

When protein spectra must be observed in H_2O solutions, the dynamic range required to observe the protein proton resonances is quite large. The principal problem encountered in effective signal averaging in H_2O solutions is that the analog-to-digital converters available in most NMR spectrometers cannot digitize the weakest signals in the presence of the strong water signal without overflowing. Modern spectrometers have incorporated floating-point averaging, long computer word lengths, and block averaging of transformed spectra in order to represent spectra with huge dynamic ranges, but these methods are to no avail if the digitized signal does not contain the desired information. In these cases it is essential to suppress or greatly reduce the water-resonance intensity. Two methods that have been widely employed to observe proton NMR spectra in aqueous solutions are correlation spectroscopy and long-pulse FT NMR. Correlation spectroscopy is not a Fourier transform method but rather a continuous-wave method quite similar to that of sweeping the radiofrequency employed in early spectrometers. It differs from the continuous-wave technique in that the sweep is extremely rapid and special digital data processing must be employed to recover a nondistorted spectrum (Dadok and Sprecher, 1974; Gupta *et al.*, 1974). Its principal advantage over FT spectroscopy is that only a narrow region need be excited by sweeping through the region of interest. Thus the water resonance can be avoided altogether. It is also possible to design RF pulses that excite only certain regions of a spectrum and permit FT spectroscopy of aqueous solutions. The most common pulse of this type is known as a Redfield 214 pulse and requires RF phase shifts during the pulse. One important advantage of this technique is that FT experiments can be done on aqueous solutions. For example, two-dimensional NMR experiments have been done using the Redfield method for water suppression (Cutnell, 1982). Unfortunately, both correlation and long-pulse spectroscopy are available only on the latest, most modern commercial spectrometers, correlation spectroscopy usually being included only as an option. Much greater detail on these methods has been provided in Redfield's review article (Redfield, 1978).

The simplest, technically least complicated method for removing an interfering residual water resonance is to vary the temperature. The water chemical shift is highly temperature dependent, and a small temperature change, especially on very high field spectrometers, can sometimes move the water resonance away from the region of interest. Wagner *et al.* (1981), for instance, performed two-dimensional experiments on aqueous solutions of BPTI using continuous irradiation to eliminate the solvent resonance. When the sample temperature was raised by 25°C, the water resonance moved sufficiently to allow the resonances previously obscured at the lower temperature to be observed. If the protein is temperature sensitive or if temperature-dependent features are to be observed, this method is obviously not appropriate; otherwise it might be the method of first choice before more exotic procedures are tried.

2.2. Resolution Enhancement and Spectral Simplification

At the highest magnetic field strengths available today, the overlap of protein resonance lines is considerable even for a relatively small protein. This is especially true of proton spectra for which the spectral dispersion is only ~10 ppm and the presence of spin–spin coupling gives rise to many closely spaced multiplets. Aside from recently developed two-dimensional experiments that can effectively improve spectral resolution in some cases, several methods of digital processing and multiple-pulse and double-irradiation techniques have been devised to enhance spectral resolution or reduce spectral crowding. A number of these schemes employ difference spectra to cancel similar features between pairs of spectra obtained under different conditions. For example, one of the simplest, but very effective, methods for spectral simplification is to subtract protein spectra obtained at slightly different pH values (King and Bradbury, 1971; Gettins *et al.*, 1981). Only the resonances that are significantly affected by the pH change are observed. Most modern spectrometers have spectral-subtraction routines and provisions for alternately adding and subtracting the digitized signal from the computer memory during acquisition so that difference spectra can be acquired directly between alternating pulse sequences.

It is customary in FT NMR to apodize the free induction decay (FID) by multiplying the digitized signal by an exponentially decaying weighting function that leaves the initial part of the FID largely unaffected but greatly attenuates the noise in the tail. This procedure reduces the amount of noise in the transformed spectrum but also causes the resonance lines to be broadened by an amount equal to $1/\pi \times T_f$, where T_f is the negative time constant of the exponential function (Martin *et al.*, 1980). If the signal-to-noise ratio is very high, this process can be reversed by applying an exponential with a positive time constant, thereby narrowing the resonance lines by artificially increasing their spin–spin relaxation times. Other more effective schemes of apodization also discriminate against the rapidly decaying components of the FID, which give rise to broad signals, and emphasize the longer-decaying components. These methods, which include sinebell and trapezoidal multiplication, have been compared with pulse methods for line narrowing by Gassner *et al.* (1978). A method quite similar to trapezoidal apodization is the convolution-difference method, which removes broad spectral lines by subtracting an FID multiplied by a steeply decaying exponential from an unaltered or resolution-enhanced FID (Figure 2) (Campbell *et al.*, 1973). Allerhand (1979) gave numerous examples of convolution-difference spectra of ^{13}C NMR spectra and showed that this method is especially helpful for observing nonprotonated aromatic carbons at high field strengths when combined with off-resonance proton decoupling (Figure 3).

Several pulse methods have been found to be effective for the simplification of proton NMR spectra of proteins (Campbell *et al.*, 1975c). One approach is to employ pulse sequences originally conceived to measure relaxation times by selectively observing resonances that have long T_1 or T_2 values. The familiar Carr-Purcell spin–echo pulse sequence (90°– τ–180°–τ–acquisition) and related pulse sequences are especially useful for spectral simplification. For instance, the spin–echo pulse sequence has been used to eliminate the NH resonances in lysozyme that interfere with the aromatic proton resonances yet resist exchanging with D_2O. This is possible because the T_2 for the NH resonances is ~7 ms, whereas the T_2's for most of the aromatic protons are longer than 30 ms. The same method is especially effective in selectively observing the C_δ and C_ε proton resonances of histidines that are exposed to solvent and have particularly long T_2 values. The histidine C_ε proton in lysozyme, for example, has a T_2 of about 150 ms, much longer than the other aromatic proton T_2 values. Recently, Gettins *et al.*

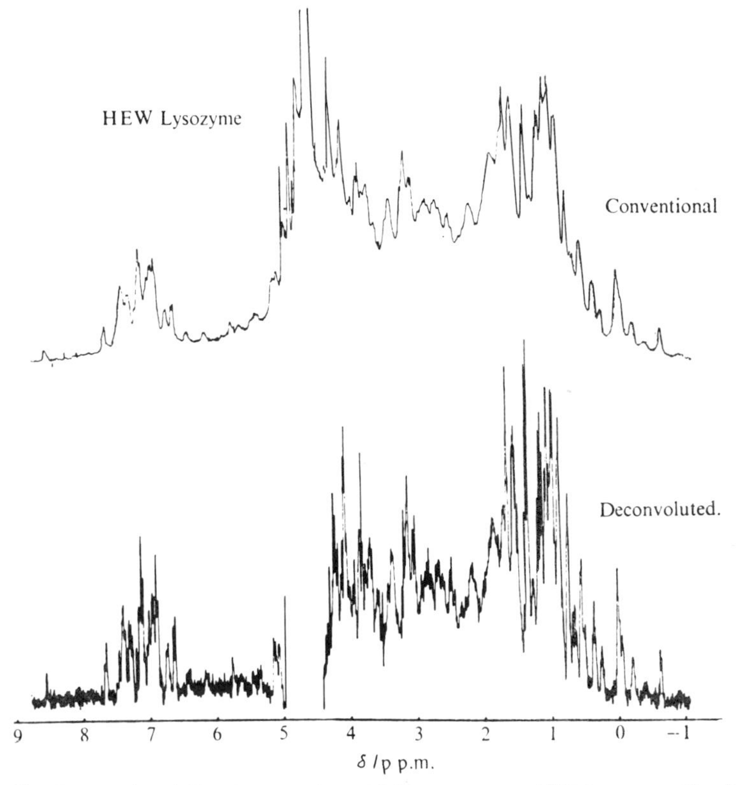

Figure 2. Conventional Fourier transform NMR spectrum of HE lysozyme (5 m*M*) and corresponding deconvoluted spectrum. (With permission from Campbell *et al.*, 1973.)

(1981) employed the Carr-Purcell method to monitor the pH-dependent chemical shifts of histidine C_ε protons of the very crowded aromatic region of the 270-MHz proton spectra of myeloma proteins. By a suitable choice of τ values, spectral simplification can also be accomplished on the basis of multiplet structure. This is possible because the various components of a multiplet do not refocus with the same phase. For instance, in the case of a triplet, the central resonance remains in phase while the two outer components are inverted when $\tau = \frac{1}{4}J$. By adding or subtracting spectra obtained with appropriate τ values, spectra can be obtained that exhibit only doublets, or only singlets and triplets. Bolton (1981) has proposed a related method for observing only singlets in a complex protein spectrum. The signal from proton singlets is not phase modulated by the spin–echo sequence, whereas the phase of most multiplet signals varies cosinusoidally with τ. By adding together a set of spectra obtained with varying τ values, all resonances cancel out except the singlets and the center lines of odd multiplets. This method was very effective in resolving the histidine C_ε and C_δ protons in ribonuclease.

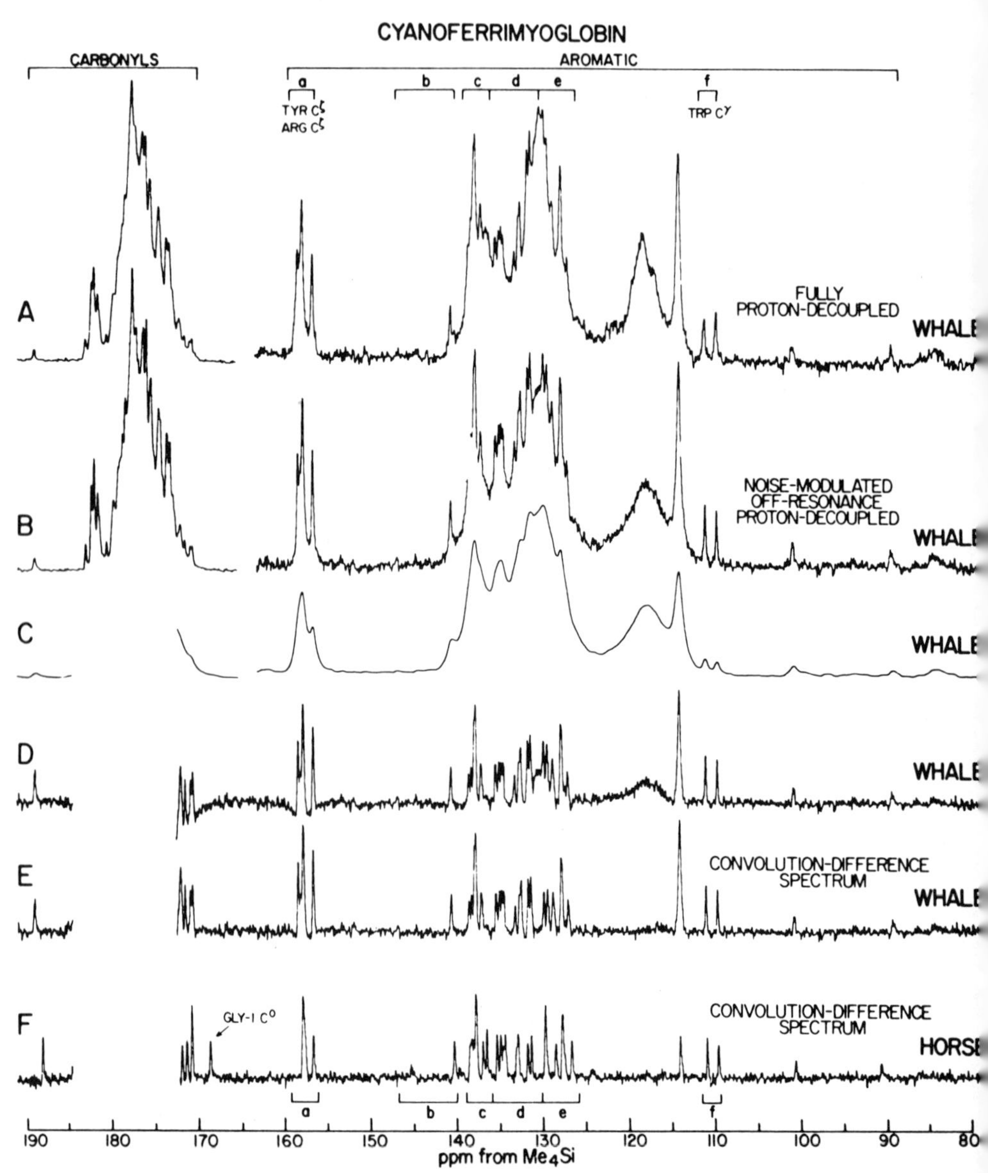

Figure 3. (*A*) Unsaturated carbon regions in the fully proton decoupled natural abundance ^{13}C Fourier transform NMR spectrum of 8.8 m*M* sperm whale cyanoferrimyoglobin in H_2O [0.1 *M* NaCl–0.05 *M* phosphate buffer, pH 6.8, 36°C]. The spectrum was recorded at 15.18 MHz using 8192 time-domain addresses, a 4000-Hz spectral width, 65,536 accumulations, a recycle time of 1.105 s (20 h total time), and a digital broadening of 0.62 Hz (τ_1 = 0.51 s). [The peak at about 114 ppm arises from excess free $H^{13}CN$]. The carbonyl region is presented with 0.3 times the vertical gain of the aromatic region. (*B*) As (*A*) but recorded under conditions of noise-modulated off-resonance proton decoupling. (*C*) As (*B*) but with digital broadening of 9.33 Hz (τ_2 = 0.034 s). (*D*) Convolution-difference spectrum obtained by digitally subtracting 90% (K = 0.9) of spectrum *C* from spectrum *B*. Carbonyl reso-

In some favorable cases, double irradiation can be employed to observe individual multiplets in very crowded spectral regions. Spin decoupling can be combined with difference spectra to observe only resonances coupled to a second selectively irradiated resonance. If the spin-decoupling frequency is not sufficiently selective when applied to overlapping resonances, however, the results can be ambiguous. A much more selective approach is to observe NOE difference spectra instead. A ^{1}H NOE difference spectrum is obtained by subtracting a normal unenhanced spectrum from a spectrum with an NOE resulting from preirradiation (rather than continuous irradiation) of an individual resonance. By presaturation of a group of aromatic ring protons of Tyr-21 and Tyr-35 in BPTI, Richarz and Wüthrich (1978) were able to observe selectively 4 of the 20 methyl resonances in the aliphatic region of the NOE difference spectrum. Similarly, presaturation of a Pro-9 resonance at 0.2 ppm allowed two phenylalanine rings to be distinguished from all other aromatic residues. This method is especially valuable because the NOE enhancements depend on the spatial proximity of protons and thus provide information that can aid in making individual resonance assignments in complex protein spectra.

Various possibilities also exist for simplification of ^{13}C spectra based on recently developed polarization-transfer experiments first reported by Morris and Freeman (1979). The "INEPT" and other pulse sequences for polarization transfer are very specialized pulse sequences in which both the proton and carbon spins are pulsed (Morris and Freeman, 1979; Doddrell and Pegg, 1980; Bendall *et al.*, 1981). Spectral simplification can be accomplished in a somewhat analogous way to the proton spin–echo technique, since the INEPT sequence can selectively adjust the phase of a resonance depending on the $^{13}C—^{1}H$ spin multiplicity. When combined with proton decoupling during the acquisition and by taking difference spectra, subspectra can be obtained that contain only CH and CH_3 resonances or CH_2 and quaternary resonances. These methods have not yet been widely applied to proteins, but the potential for reducing spectral crowding, especially in the aliphatic region, is obvious, and new applications are likely to appear in the literature soon.

2.3. Two-Dimensional FT NMR

During the last four or five years, numerous applications of two-dimensional FT NMR experiments have appeared in the chemical literature. Until recently two-dimensional NMR was a recondite subject that was broached

---

nances have the same vertical gain as the rest of the spectrum. (*E*) Final convolution-difference spectrum, obtained from spectrum *D* by a digital baseline adjustment. (*F*) Same as (*E*) but 8.7 m*M* horse cyanoferrimyoglobin. Sample conditions were the same as for the sperm whale protein, except that KCN of natural isotopic composition was used. (With permission from Oldfield *et al.*, 1975.)

only by those who had the resources and ingenuity to develop the NMR hardware and computer software required to do the experiments. Because of the widespread interest in the experiments and the realization that the two-dimensional methods will have many chemical applications, the necessary spectrometer modifications are now being included in most new commercial instruments. Thus many novitiates in the field are beginning to apply experiments previously accessible only to those who were involved in the development of the methods. Several review articles that discuss the general aspects of two-dimensional NMR (Freeman and Morris, 1979; Freeman, 1980) and reviews that are oriented toward protein studies (Nagayama, 1981; Wüthrich *et al.*, 1979) have appeared.

The two-dimensional NMR experiments represent a dramatic new approach to measuring NMR parameters in which two time domains are acquired in a single experiment. The data matrix obtained in these multipulse experiments is Fourier transformed in one dimension, transposed, and transformed in the second dimension to yield a three-dimensional grid of resonance intensity versus two frequencies f_1 and f_2. Typically these data arrays are as large as 4096×128 data points, which creates special problems in the presentation, storage, and plotting of the two-dimensional spectra. The use of extremely capacious magnetic disk storage and high-speed digital plotters has somewhat overcome the storage and plotting problems, however, and several novel methods of presentation have been employed such as the use of contour plots, projections, and cross-sections of the three-dimensional grid.

There are two classes of two-dimensional NMR experiments that have come into general use. The two-dimensional correlated experiments can determine whether nuclei are interconnected by some interaction such as spin coupling, mutual relaxation, chemical exchange, or chemical shielding. Several of the most useful experiments have been dubbed by their progenitors with the whimsical acronyms COSY, SECSY, and NOESY, which belie their true utility. COSY (correlated spectroscopy) and SECSY (spin–echo-correlated spectroscopy) are slight variations of the same experiment, which establishes that pairs of nuclei or groups of nuclei are spin coupled (Nagayama *et al.*, 1980). And NOESY (nuclear Overhauser enhancement spectroscopy) establishes dipolar-relaxation pathways by providing a complete set of NOEs in a single experiment. Another valuable correlated experiment is the $^{1}H—^{13}C$ chemical-shift correlation, in which the proton shifts are unambiguously associated with the corresponding carbon shifts. A second class of experiments known as two-dimensional resolved spectroscopy can separate chemical shifts and coupling information into separate frequency domains. The most widely applied example of this experiment is the proton *J*-resolved spectroscopy, in which the proton chemical shifts appear along one axis of a two-dimensional plot while the coupling constants appear along an axis orthogonal to the chemical-shift axis. Of course the coupling constants can be obtained in a one-dimensional spectrum, but in a highly complex spectrum such as a protein spectrum in which there is much overlap of spin multiplets, the resolution is effectively improved by the two-dimensional spectrum because the coupling constants are rotated into a second dimension, which reduces the chance that resonances will overlap. Numerous other experiments have been conceived, and new applications to various biochemical systems are constantly being developed.

There have been very few published applications of two-dimensional NMR in which protein amino acid side chains have been observed. Notably, however, Wüthrich and Ernst and their associates have been quick to apply the new two-dimensional methods to BPTI, which has become a paradigm for new NMR experiments on proteins (Kumar *et al.*, 1981; Nagayama and Wüthrich, 1981a; Wagner *et al.*, 1981). Although BPTI has been studied extensively by one-dimensional methods, it provides a cogent example of the efficiency of two-dimensional NMR in the determination of coupling constants and NOEs and in making resonance assignments. In addition, some of the data obtained in the two-dimensional experiments are unique and not readily available from one-dimensional experiments. With regard to observing the side chains, the SECSY and J-resolved experiments appear to be the most valuable. The J-resolved spectrum of BPTI in ^2H_2O at p^2H 7.0 and 68°C is shown in Figure 4 for the region 0.5 to 6.0 ppm. Each proton resonance appears along the chemical-shift axis in the expected fashion, but the multiplets due to spin coupling appear only on the perpendicular axis. The projection of the two-dimensional spectrum along the perpendicular axis is essentially a "proton decoupled" proton spectrum. The coupling constants are obtained by taking cross-sections across the chemical-shift axis at the positions of the peak maxima. The most convenient method of presentation of SECSY spectra is the contour plot, as shown for BPTI in Figure 5 in the same spectral region as that presented in Figure 4. Groups of spin-coupled nuclei give rise to pairs of correlation peaks that appear equally displaced, but on opposite sides of the central spectrum at a distance equal to $\frac{1}{2}$ the chemical-shift difference between the nuclei. A straight line connecting the two correlation peaks forms an angle of 135° with respect to the chemical-shift axis. Some skill must be exercised in matching the correlated peaks since the multiplet structure of each resonance also appears in the correlation peaks. Nagayama and Wüthrich (1981a) published a table of connectivity diagrams for the weakly coupled spin systems of the nonlabile, aliphatic protons of the common amino acid residues, indicating the patterns of correlation peaks of sequentially coupled protons; this should aid in the assignment of side-chain resonances. The characteristic features of strong spin–spin coupling in amino acids that may appear in J-resolved spectra of proteins have also been examined (Wider *et al.*, 1981).

Nagayama and Wüthrich (1981a) employed the following strategy with the two-dimensional experiments for making resonance assignments in BPTI: (1) The individual components of complete spin systems were identified from the pattern of cross-peaks in the SECSY spectra; (2) the proton–proton

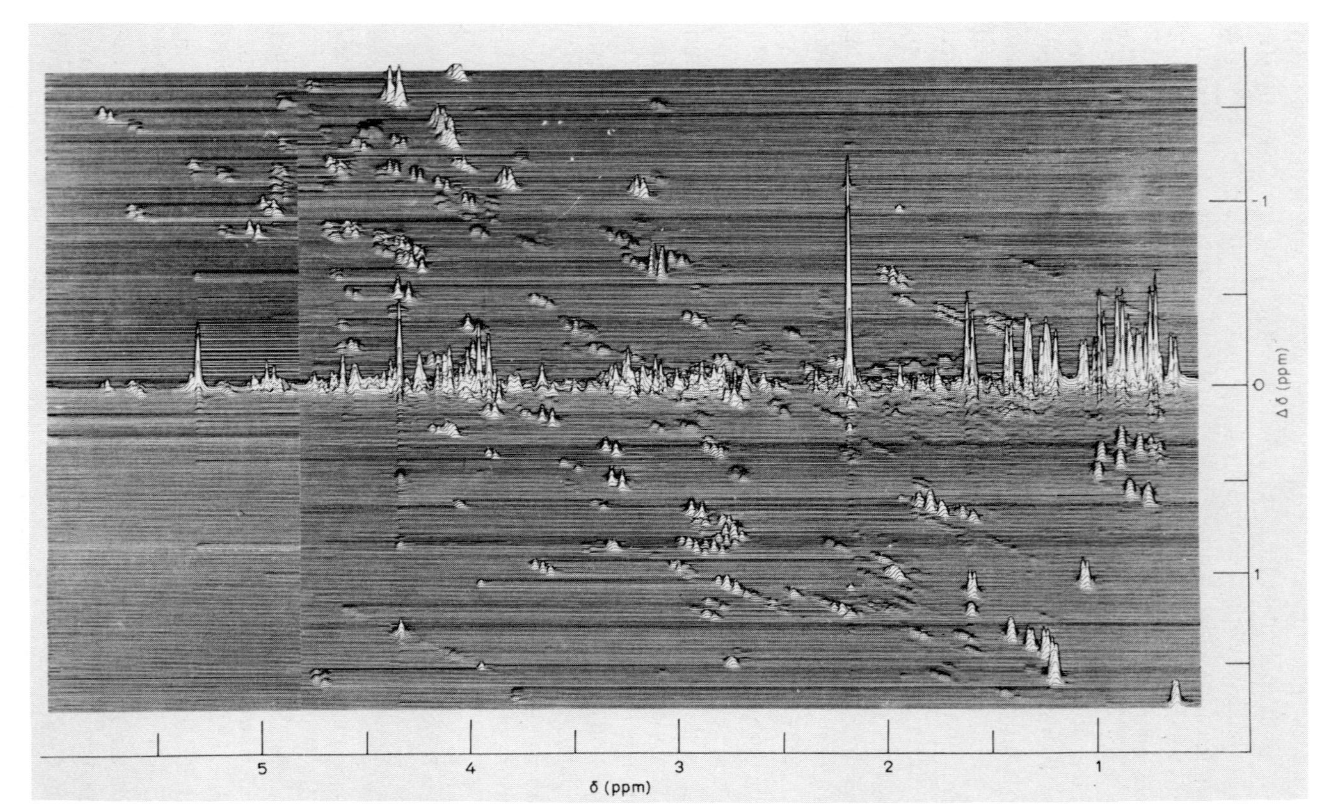

Figure 4. Two-dimensional ^{1}H spin-echo-correlated NMR spectrum at 360 MHz of an 0.01 *M* solution of the BPTI in $^{2}H_2O$, $p^{2}H$ 7.0, and 68°C in the spectral region from 0.5 to 6.0 ppm. The chemical shift on the horizontal axis corresponds to the chemical-shift axis of the one-dimensional spectrum. The $\Delta\delta$ on the vertical axis stands for the difference frequencies between correlated nuclei. Cross-peaks between *J*-coupled protons appear at $\pm 0.5\ \Delta\delta$. (With permission from Nagayama and Wüthrich, 1981a.)

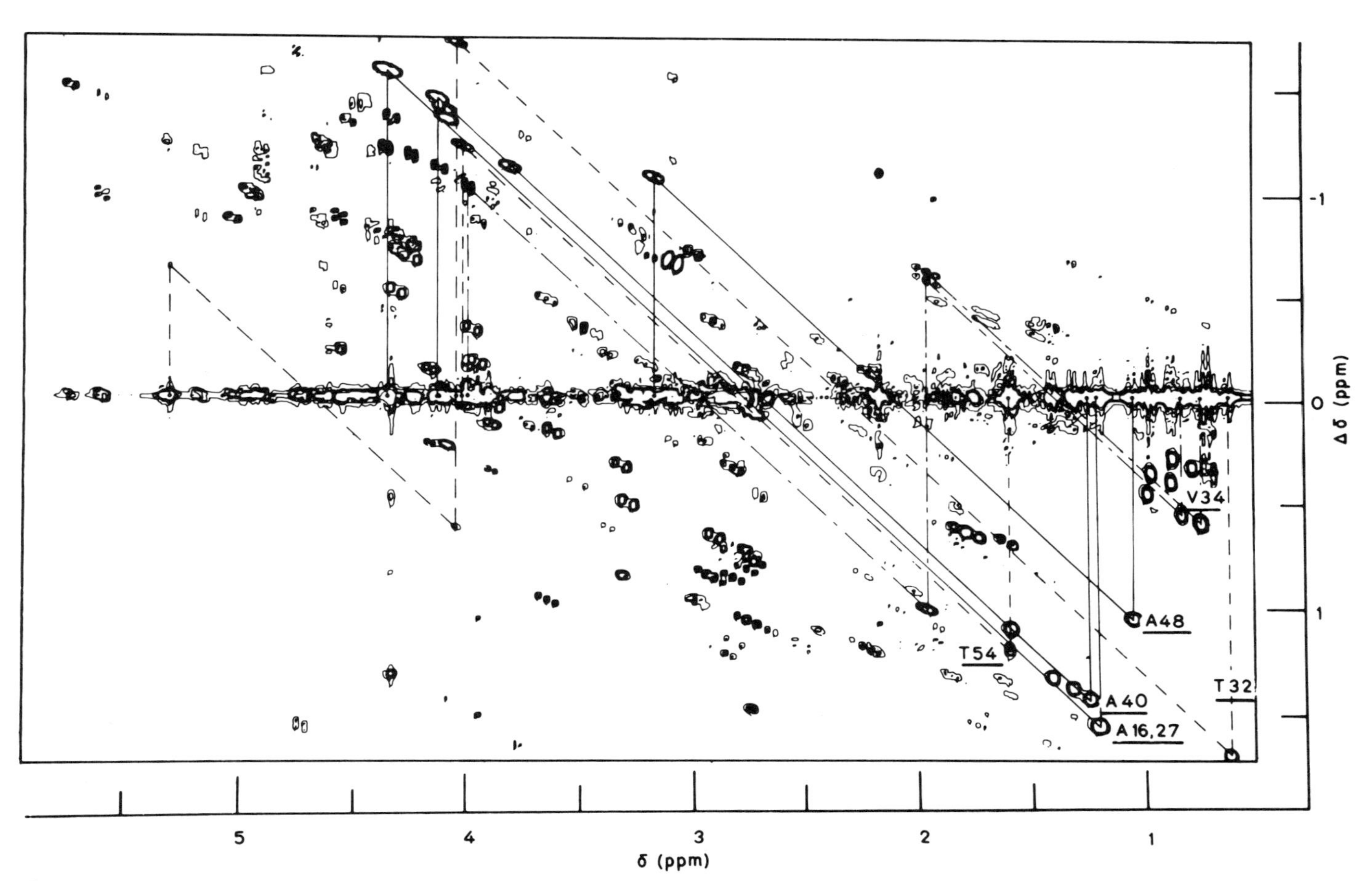

Figure 5. Contour plot of the spin-echo-correlated spectrum of BPTI shown in Figure 4. Connectivities between the components of the spin systems of alanines 16, 27, 40, and 48; threonines 32 and 54; and valine 34 are indicated. (With permission from Nagayama and Wüthrich, 1981a.)

coupling constants were measured by taking cross-sections in the J-resolved spectra; and (3) the SECSY and J-resolved spectra were recorded under varying conditions of pH and temperature to assist in making individual assignments. This process obviously can be tedious, but making resonance assignments in proteins has never been a simple proposition, and the two-dimensional spectra provide a wealth of information in each spectrum. Some amino acids have unique spin-coupling connectivities and can be assigned from the pattern of correlation peaks in the SECSY spectra alone if they are sufficiently resolved. These amino acids are threonine, valine, isoleucine, leucine, lysine, phenylalanine, tyrosine, histidine, and tryptophan. Varying the temperature and pH aids in a fashion similar to one-dimensional experiments by reducing overlap in certain spectral regions and by identifying titrating resonances. When data from the variable pH and temperature experiments, spectra of native and reduced inhibitor, and sequential assignment of C_α backbone resonances were combined, a total of 32 spin systems could be identified. Of these, 25 could be assigned to individual amino acids, and four were new assignments (Asn-24, Gly-36, Asn-43, and Asn-44). The sequential assignment of C_α resonances was accomplished by the two-dimensional NOESY experiment.

The two-dimensional J-resolved experiment allowed the determination of the three bond coupling constants between the α carbons of the protein backbone and the β carbons of many of the side chains. Using this data, Nagayama and Wüthrich (1981b) were able to compare the spatial arrangements of the individual amino acid side chains in solution and in single crystals that are known from X-ray crystallography (Deisenhofer and Steigemann, 1975). The $H—C_\alpha—C_\beta—H$ angles of the side chains of BPTI in solution were determined from a Karplus-type relationship previously established for the coupling constants (DeMarco *et al.*, 1978). In most cases, the coupling constants predicted on the basis of the crystal structure closely matched the measured coupling constants for residues in the interior of the protein, indicating that these side chains are locked into a unique spatial orientation even in solution. Surface residues, however, yielded coupling constants that indicate rapid averaging between two or more orientations.

The value of the two-dimensional experiments appears to be firmly established, but BPTI is a relatively small protein (58 amino acid residues) and had previously been thoroughly studied by more conventional methods. It remains to be seen how well suited the two-dimensional methods are for still more complex systems for which there are fewer well-resolved signals and less background of conventional results. Nonetheless, the combination of very high magnetic fields and two-dimensional methods is likely to extend the range of possible studies.

2.4. Solid-State NMR

The continuing evolution in the development of instrumentation and methods of observation of NMR has created new possibilities for studying material in the solid phase, an area that until recently has been almost exclusively the domain of physicists. With the emergence of new techniques and the availability of spectrometers designed to observe NMR in solid-phase samples, a limited number of scientists have begun to extend solid-state NMR studies to the observation of amino acid side chains in proteins. In general, the proteins studied by solid-state methods are of very high molecular weight and give very broad featureless spectra in solution, if they are soluble at all. Since the resolution of individual resonance lines in the solid phase is much more difficult than in solution, the incorporation of an NMR isotope such as ^{13}C, ^{2}H, ^{15}N, or ^{19}F is essential if a particular residue is to be observed. Usually, but not always, the isotopic label must be incorporated biosynthetically either *in vivo* or in tissue culture. The choice of the isotopic probe depends upon the kind of information desired. The majority of published papers have been studies of molecular motion, although a variety of different motional regimes and types of motion have been observed. The need to enrich the protein with an isotopic label may limit the number of proteins that can be studied, but in those cases in which isotopes have been successfully incorporated, solid-state methods have proved to be uniquely powerful.

A detailed discussion of the techniques required to observe NMR in the solid phase is beyond the scope of this article. A brief description of the principal methods employed, however, is appropriate. The areas of application have included structural proteins, membrane-bound proteins, viral protein coats, and oriented heme-bearing-protein microcrystals. Examples of each of these cases will be discussed.

In the solid phase there are several nuclear interactions that are not manifest in the liquid phase, except that they may influence nuclear relaxation. These interactions are the direct dipole–dipole coupling between nuclei, chemical-shift anisotropy, and the quadrupolar interaction. The dipole–dipole interaction occurs between proximal nuclei, and its magnitude depends upon the orientation of the internuclear vector with respect to the direction of the applied magnetic field. The dipolar coupling can be considerably larger than the familiar scalar (orientation independent) J coupling that is present in solution spectra unless double-resonance decoupling methods are employed. For instance, the scalar coupling between a ^{13}C atom and an attached proton typically falls in the range of 125 to 190 Hz, whereas the dipolar coupling between such nuclei in the solid phase can be many kilohertz. Unlike scalar coupling, which is transmitted between nuclei via the polarization of electrons through the intervening chemical bonds, the dipolar coupling is transmitted directly through space, so any nearby NMR-active nucleus can be involved. In a highly mobile system in which the motions are random and the correlation time for the reorientation of the vector connecting the nuclei is very short ($t \leqslant 10^{-8}$ s), the dipolar coupling averages to zero. Chemical shielding is also orientation dependent, a fact that is usually lost to those who have only run spectra of solutions, since again the rapid molecular motion causes only a mean value to be observed.

In this case the averaged value turns out to be the familiar chemical shift. In a solid the chemical shift of each nucleus is orientation dependent, giving rise to a distribution of shifts known as a *powder pattern*, which can cover a range as large as several hundred parts per million in some cases. The quadrupolar interaction occurs only in nuclei for which the nuclear spin I is greater than $\frac{1}{2}$. The ^2H nucleus ($I = 1$) is of the most interest in biological studies, but samples must be enriched in this isotope because of its low natural abundance (0.015%). For a ^2H nucleus there are two possible transitions between three nuclear spin energy levels, unlike the spin $\frac{1}{2}$ case, in which there are only two energy levels and therefore only one transition. The interaction of the electric-field gradient at a ^2H nucleus with its quadrupole moment causes the spacing between the energy levels to differ, and thus two distinct transitions are observed. Since the spacing depends on the orientation of the field gradient with respect to the magnetic-field directions, rapid molecular motions cause the two transitions to be averaged to the same frequency in solutions. In the slow-motion domain of solids, however, each ^2H nucleus gives rise to two transitions whose frequencies are often very different. For a typical randomly oriented methyl-d_3 group, the separation can be 40 to 80 kHz (typically ~40 kHz for a methyl-d_3 group).

The problem facing the spectroscopist is either to eliminate these second-order interactions so as to obtain a more solutionlike spectrum or to exploit the motional dependence of the interactions to gain information about protein dynamics. In the case of ^{13}C NMR, high-power ^1H coherent irradiation ("dipolar decoupling") is usually employed to remove the ^{13}C-^1H dipolar interactions. Sometimes, however, it is advantageous to use only low-power irradiation ("scalar decoupling") so that only J couplings are removed. This allows carbons in regions of a protein that are immobile and dipolar coupled to be distinguished from carbons in mobile regions in which the dipolar couplings are effectively averaged out by rapid molecular motion. The technique of magic-angle sample spinning has been widely applied in solid-state ^{13}C NMR to remove the line broadening due to chemical-shift anisotropy. This technique involves spinning the sample very rapidly (2 to 3 kHz) about an axis making an angle of 54.7° with respect to the applied field. Although this method has been successfully applied to organic polymers and carbohydrates, protein samples in general are too labile to withstand the extreme shear forces experienced in high-speed spinning. Solid-phase protein samples (and frequently even those in solution) usually exhibit line broadening due to chemical-shift anisotropy. The extent that the full anisotropy is reduced by motional averaging provides information about molecular dynamics. The ^1H—^{13}C dipolar coupling is also exploited to enhance the ^{13}C signal by employing the technique of cross-polarization. This involves transferring spin polarization between the coupled protons and carbons and allows the pulse rate to be determined by the usually shorter proton relaxation rate rather than the carbon relaxation rate. The increased sensitivity and pulse rate result in a substantial time savings in gaining a reasonable signal-to-noise ratio. Since only dipolar-coupled carbons experience the signal enhancement, this method can also aid in distinguishing mobile carbons from immobile ones. In thc case of ^{2}H NMR, information about molecular motion of a protein is obtained by analysis of the quadrupolar powder pattern and relaxation measurements. An extremely broadbanded spectrometer is normally required to observe ^{2}H powder patterns, and this implies the use of a very fast digitizer, very short high-power 90° pulses, and the use of the quadrupolar echo pulse sequence (Hentschel and Spiess, 1979) to observe an undistorted signal.

Applications to Amino Acid Side Chains

Collagen Fibrils. The first study of protein amino acid side chains in which isotopic labeling and solid-state NMR methods were employed was made by Dennis Torchia and his associates on type I collagen. Collagen molecules associate to form elementary fibrils that are ~300 to 3000 Å in diameter and many molecules in length. Each molecule is composed of two α_1 chains and one α_2 chain; these are assembled together in a triple-helical structure. The amino acid sequence repeats in units of Gly-X-Y, which allows the glycyl residues to occupy positions within the core of the helix while the side chains of the X and Y residues are located on the surface and are free to interact with side chains from other molecules. Following initial ^{13}C studies of helical-backbone motion using native collagen at natural ^{13}C abundance and with specifically labeled glycine (Torchia and VanderHart, 1976; Jelinski and Torchia, 1979), methods were developed to incorporate enriched amino acids so that different side-chain motions could be studied. The motivation for these studies was to determine whether the side chains are immobilized by fixed interactions or whether they exhibit fluidlike mobility in the contact region between the helices.

The ^{13}C-enriched collagen was obtained from tissue culture of calvaria parietal bones obtained from chick embryos in which the culture medium was supplemented with either [β-^{13}C]alanine, [ε-^{13}C]methionine, [ε-^{13}C] lysine or [δ-^{13}C]glutamic acid. The ^{13}C NMR experiments included the measurements of linewidths, lineshapes, NOE values, and T_1 values. In each case the side-chain motion either completely averaged or greatly reduced the static chemical-shift anisotropy of the labeled carbon both in fibrils and in solution, as shown in Figure 6 for [δ-^{13}C]glutamic acid. The linewidths in the fibrils, however, were in general always broader than in solution because of the decreased motion about the long molecular axis in the fibrils and, in some instances, intermolecular interactions. In the case of the methionine and lysine residues, the terminal-carbon linewidths were even smaller than the methyl-carbon resonance linewidth of the Ala residues because of the additional motions of the side-chain bonds. In solution, collagen molecular motion is highly anisotropic because the end-over-end motion can occur only very slowly. Consequently, even in solution, the ^{13}C—^{1}H dipolar

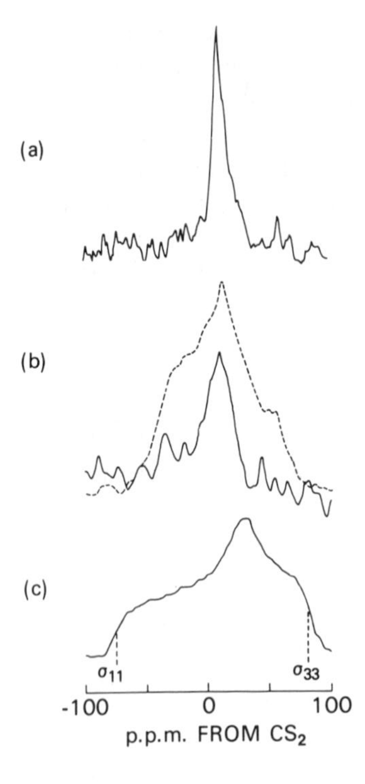

Figure 6. Comparison of the 15.09-MHz ^{13}C NMR spectra of [δ-^{13}C] glutamic acid-labeled collagen (in solution and as fibrils) with the spectrum of [α-^{13}C]glycine-labeled collagen fibrils and the powder pattern for D,L-[δ-^{13}C]glutamic acid. (*a*) Dipolar-decoupled spectrum of [δ-^{13}C]glutamic acid-labeled collagen. The carbonyl peak arising from the solvent has been subtracted from this spectrum. (*b*) (-------) Dipolar-decoupled spectrum of [δ-^{13}C]glutamic acid-labeled collagen fibrils from which the natural abundance background was subtracted. (-----) Spectrum of [α-^{13}C]glycine-labeled collagen fibrils. (*c*) Proton-enhanced spectrum of D,L-[δ-^{13}C]glutamic acid (90 atom % ^{13}C) powder. [With permission from Jelinski and Torchia, *J. Molec. Biol.* **138,** 255 (1980), copyright Academic Press, Inc. (London) Ltd.]

couplings do not always average to zero. Scalar-decoupling experiments of collagen in solution indicate that side-chain motion is sufficient to completely average the dipolar coupling of the carbons in all the methionine residues, only about 40% of the lysine residues, and none of the alanine residues. This variation results because the additional side-chain bond rotations do not exist for alanine and are apparently greatly restricted in some lysine residues, possibly because of electrostatic interactions with negatively charged side chains. The linewidths, T_1, and NOE values of the charged lysine and glutamic acid side chains indicate a high degree of mobility in fibrils as well as in solution, even though their positive and negative charges could in principle result in interactions that would restrict side-chain motion. Taken together, the ^{13}C NMR data indicate that the methionine, lysine, and glutamic acid terminal carbons in fibrils reorient rapidly (correlation times in the range 10^{-9} to 10^{-10} s) and isotropically on a time scale of 10^{-3} s because of side-chain bond rotations. Furthermore, the motions indicate that the contact regions between the molecules in the collagen fibrils are fluid.

The investigation of alanine methyl-group motions in collagen was also the first application of solid-state ^{2}H NMR to study protein dynamics (Jelinski *et al.*, 1980). Subsequent studies have included leucine side-chain motions (Jelinski *et al.*, 1980; Batchelder *et al.*, 1982). The advantage of the ^{2}H

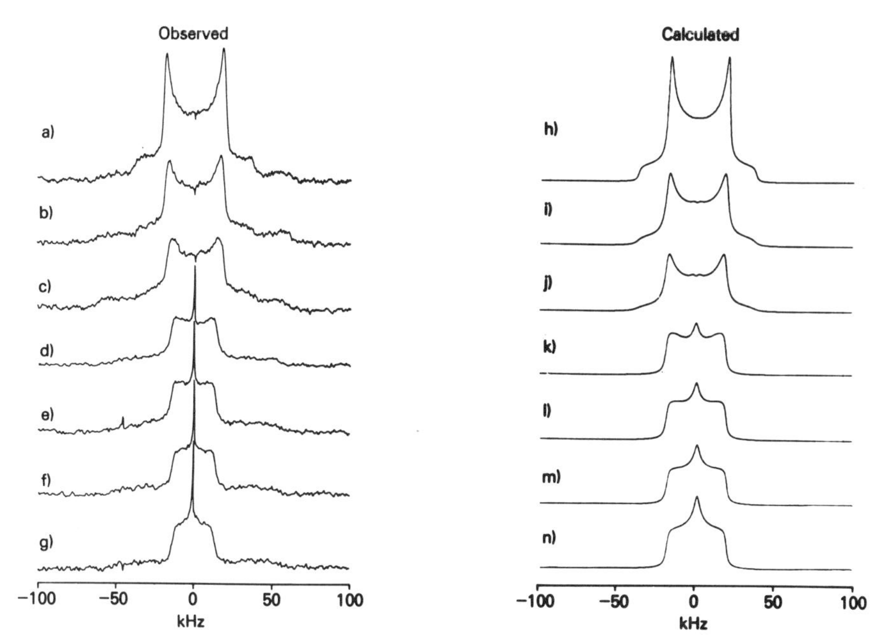

Figure 7. Experimental (*a*) to (*g*) and calculated (*h*) to (*n*) ^2H NMR spectra of [$^2H_{10}$]Leu-labeled collagen in equilibrium with 0.02 *M* Na_2HPO_4 at various temperatures. Spectra were observed at the following temperatures: (*a*) −85°C, (*b*) −43°C, (*c*) −18°C, (*d*) −6°C, (*e*) +1°C, (*f*) +15°C, (*g*) +30°C. Theoretical spectra were calculated using a two-site-hop model in which the CgUCd bond axes are assumed to jump between two sites separated by 108 to 112°. (With permission from Batchelder *et al*., 1982.)

method is that very detailed information can be obtained about side-chain motions by comparing the experimental lineshapes of the quadrupolar powder pattern to calculated lineshapes based on models that include various types of motions. For example, the powder patterns for the alanine methyl-d_3 groups in collagen were compared with powder patterns calculated for a methyl group undergoing a two-site hop between different configurations that might result from backbone motions. The ^2H data indicate that collagen molecules in fibrils reorient about the long helical axis over an angular range of ~30° at 18°C. This result supports earlier conclusions about reorientation of the peptide backbone. A similar study of [$^2H_{10}$]leucine-labeled collagen fibrils showed that the isobutyl side chain of leucine rapidly interconverts between two predominating conformations at a rate proportional to temperature over the range −85° to 30°C (Figure 7) (Batchelder *et al.*, 1982). Recent theoretical considerations suggest that the measurement of T_1 values of quadrupolar powder patterns can aid in distinguishing between simple two- or three-site hops and cases in which motion is essentially diffusional (Torchia and Szabo, 1982).

Membrane-Bound Proteins. Many membranes systems contain protein as an integral and essential component in addition to lipid. The use of NMR methods for the study of protein–lipid interactions is appealing because the crucial question of how lipid influences a protein's structural and motional characteristics is extremely difficult to attack experimentally. One approach is to incorporate a protein into a model lipid system and monitor the lipid with the aid of NMR spin labels (Deese *et al.*, 1981); or conversely, the protein can be labeled and individual or groups of particular residues can be observed. This latter method has been employed by Hagen *et al.* (1978) to study the major coat protein of the nonlytic coliphage M13, which behaves as an integral membrane protein during certain stages of viral infection. Because of its high NMR sensitivity and zero abundance in biological molecules, ^{19}F was chosen as the spin label (Sykes and Hull, 1978). The labeled protein was obtained by inoculating *Escherichia coli* with the M13 phage in a growth medium containing *m*-fluorotyrosine. The isolated fluorotyrosyl coat protein was incorporated into phospholipid vesicles with the aid of sonication. The ^{19}F NMR spectra exhibit features that are intermediate between those expected for the liquid and the solid phase. The two fluorotyrosyl residues give rise to a single broad resonance ($\Delta\nu_{1/2}$ = 300 Hz at 254 MHz) as a result of incomplete motional averaging of the ^{19}F chemical-shift anistropy, as established from the field dependence of the linewidth ($\Delta\nu_{1/2}$ = 60 Hz at 84.7 MHz). The temperature dependence of the linewidth confirmed that the tyrosyl mobility is greatly influenced by the phase state of the phospholipid bilayer. The correlation times for tyrosine bond rotations within the vesicles were determined by relaxation and NOE measurements to be 5×10^{-8} and 2×10^{-9} s for the $\alpha\beta$ and $\beta\gamma$ linkages, respectively. The aromatic residues of fd coat protein solubilized in sodium dodecyl sulfate micelles have also been observed by ^{1}H and ^{13}C NMR, but these spectra more closely resemble solution spectra than in the case of the M13 coat protein (Cross and Opella, 1981; Opella *et al.*, 1980).

Solid-state ^{2}H NMR has been extensively employed to study lipid organization and dynamics (Seelig, 1977). By growing certain bacteria in culture medium supplemented with ^{2}H amino acids, intact membranes can be obtained that contain the integral proteins with labeled amino acid side chains, which can be observed by the ^{2}H method. The first membrane-bound protein to be studied by solid-state ^{2}H NMR was bacteriorhodopsin in the purple membrane of *Halobacterium halobium R1* (Kinsey *et al.*, 1981a,b). Bacterial membranes were isolated that contained high levels of ^{2}H in the methyl groups of the 21 valine residues. Analysis of the quadrupolar powder pattern showed that the rotational motion of the methyl groups is sufficient to reduce the static quadrupolar splitting to ~40 kHz, but no additional motional averaging occurred over a temperature range of −100 to 53°C. Thus in contrast to the isobutyl groups of leucines in collagen fibrils, which undergo two-site hops between different conformations, the isopropyl-group rotations are of much smaller amplitude. The aromatic amino acid side chains in bacteriorhodopsin have also been studied by the ^{2}H method (Kinsey *et al.*, 1981b). The tryptophan side chains were found to be irrotationally bound, whereas the phenylalanine and tyrosine side chains undergo rapid twofold ring flips. The correct analysis and interpretation of ^{2}H powder patterns at this time is still being worked out, and claims by Kinsey *et al.* that some immobile Phe groups may be present appear to be unfounded. Furthermore, ring flip rates of 10^7 s^{-1} are required to match the observed powder patterns rather than the reported 10^5 s^{-1} (Rice *et al.*, 1981).

Viral Protein Coats. Solid-state NMR is especially suited for observing intact bacteriophages. Although sufficiently concentrated solutions of some virus particles can be prepared, the very slow tumbling of these supramolecular structures does not significantly reduce the dipolar interactions or chemical-shift anisotropy; thus the linewidths in solution spectra are prohibitively broad. By employing biosynthetically spin labeled samples, however, much detailed information about the side-chain dynamics can be obtained from solid-state experiments. The aromatic rings of the coat protein in the filamentous bacteriophage fd (MW = 16×10^6) have been studied by ^{2}H, ^{15}N, and ^{13}C solid-state methods (Gall *et al.*, 1981, 1982). High-resolution solid-state spectra were obtained from [γ-^{13}C]Trp-26– and [ε-^{15}N]Trp-26–labeled fd by ^{13}C and ^{15}N NMR, respectively, employing high-power proton decoupling, cross-polarization, and rapid magic-angle sample spinning in both cases. Dynamic information was obtained from an analysis of the motional averaging of the ^{13}C and ^{15}N chemical-shift powder patterns in the ^{13}C and ^{15}N spectra observed in static samples and the ^{2}H quadrupolar powder pattern in the solid-state ^{2}H spectrum of [^{2}H]Trp-26-labeled fd. These data indicate that Trp-26 is very rigid, a result to be expected for the bulky, asymmetrical tryptophan. A small reduction of the ^{2}H quadrupolar powder pattern of the [^{2}H]Trp-labeled fd from that observed for ^{2}H-labeled tryptophan alone, however, indicates the presence of small amplitude motions. Samples of fd were also prepared in which either ^{2}H or ^{13}C was incorporated into the rings of tyrosine (Tyr-21 and Tyr-24) and phenylalanine (Phe-11, Phe-42, and Phe-45). Both the tyrosine- and phenylalanine-labeled samples exhibited ^{2}H quadrupolar and ^{13}C chemical-shift powder patterns characteristic of rapid 180° ring flips. The powder patterns were slightly reduced from patterns calculated on the basis of ring flipping alone, however. Thus these aromatic side chains undergo both large amplitude jump motions and small amplitude librations. One advantage of the multinuclear approach is that the time scales for ^{13}C and ^{2}H NMR differ significantly. Consequently the ring flips and librations have been demonstrated to be rapid relative to both the 10^6 and 10^4 Hz time scales corresponding to the observation frequencies of ^{13}C and ^{2}H, respectively.

Oriented Heme-Protein Microcrystals. In a remarkable new technique, it has been found that microcrystals of heme proteins suspended in saturated solutions of $(NH_4)_2SO_4$ orient spontaneously in the high magnetic fields of superconducting solenoid NMR spectrometers (Oldfield and Rothgeb, 1980). This orienting effect allows NMR spectra to be obtained, in effect, on partially oriented single crystals, since each protein microcrystal aligns in a similar fashion relative to the applied magnetic field. The method is usually limited to heme proteins since the ordering phenomenon results from the torque exerted on the crystals due to the large anistropy in the magnetic susceptibility of the paramagnetic center. In correlation with isotopic-labeling and solid-state NMR techniques, the method has the potential of allowing the observation of individual atom sites in amino acid side chains in the crystalline solid state (as opposed to observing a particular class of side chains), and the information obtained should be directly comparable to X-ray structural data. In the first application of oriented microcrystals, (Oldfield and Rothgeb, 1980; Rothgeb and Oldfield, 1981) observed solid-state ^2H NMR in sperm whale ferrimyoglobin enriched with ^2H in the methyl groups of methionine residues 55 and 131; later ^{13}C-enriched myoglobin was observed by ^{13}C NMR. The ^2H spectrum exhibits a doublet due to the two possible nuclear transitions with a separation of 53.6 kHz corresponding to one of the methionyl methyl groups. A second doublet due to the remaining methionyl methyl groups is obscured by residual deuterium in water. When either the sample is frozen and the sample tube rotated or CN^- ions are added, which bind to the heme iron and change the susceptibility tensor, thereby causing the microcrystals to reorient, a second doublet appears because of the second methionyl group. It is too early to judge how valuable the magnetic-ordering technique will prove to be, but in principle it should provide information about the spatial orientations of labeled residues, by allowing study of the orientation dependence of ^2H quadrupolar splitting, and information about rates and types of motion of individual side chains from relaxation studies.

3. EFFECTS ON NMR PARAMETERS OF THE INCORPORATION OF AMINO ACID SIDE CHAINS INTO PROTEINS

Changes in both chemical shift and line width are observed for amino acid residues as a result of the native-protein-to-random-coil conformational transition. These changes are discussed in general terms in the following sections.

3.1. Protein Microenvironment and Chemical Shift

When an amino acid residue in a protein is converted from a solvated state in a random-coil polypeptide chain to being buried inside the interior of a globular protein, what changes can be expected for the chemical shifts of its

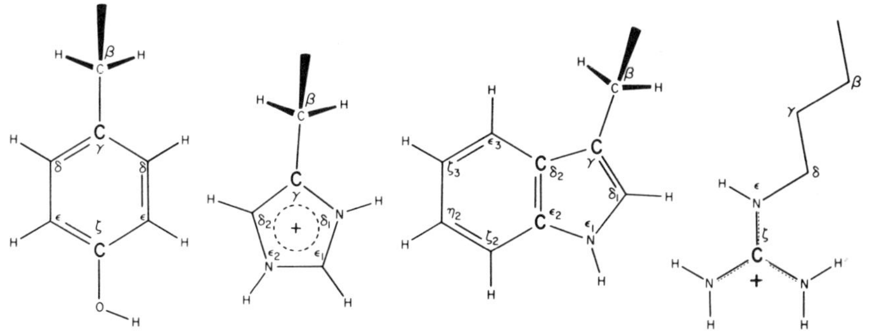

Figure 8. Side chains of some amino acids showing the Greek nomenclature. Skatole and *p*-cresol are the side-chain groups with a hydrogen atom replacing C_α. (Redrawn from Oldfield *et al.*, 1975.)

resonances? Generally one can answer this question by reference to the changes expected in the chemical shift on changing the magnetic susceptibility x. Thus in a series of experiments (Cohen and Glasel, unpublished results) analogs of amino acid side chains, that is, toluene for phenylalanine, *p*-cresol for tyrosine, and skatole for tryptophan (Figure 8), were dissolved in a series of solvents, and their chemical shifts were determined (extrapolated to infinite dilution to avoid problems of solute self-association). The aromatic-side-chain analogs were chosen because most information was available for these types of residues in proteins. In general, the result was as expected on the basis of earlier theories (Glick *et al.*, 1959; Howard *et al.*, 1962), that is, $\delta \propto K$, where K is the volume diamagnetic susceptibility, $K = x/V$, and V is the molar volume of the solvent (Figure 9). There is an upfield shift in going from a solvent of higher K to one of lower K. This is borne out by many results, both on aromatic and aliphatic residues, where the interior of the protein is regarded as being a more diamagnetically shielding environment than the solvent interface. In practice it is impossible to follow a single analog of an amino acid side chain across the whole series of solvents from water to cyclohexane, for example, because of solubility problems. However, the K of chloroform (0.74) is greater than that of water (0.72), so that the solvent series presented (Figure 9) is valid even for protein environments.

This generalization holds only for an isotropic environment, which the interior of a protein certainly is not. Any anisotropy of electron distribution can result in shifts that will be superimposed on the overall diamagnetic upfield shift. Groups such as aromatic rings and carboxyl groups can give regions of anisotropic shifts in either direction (Becker, 1980). In general, a large downfield or upfield shift can be explained by juxtaposition of a side chain to such a group. The presence of a heme prosthetic group, with

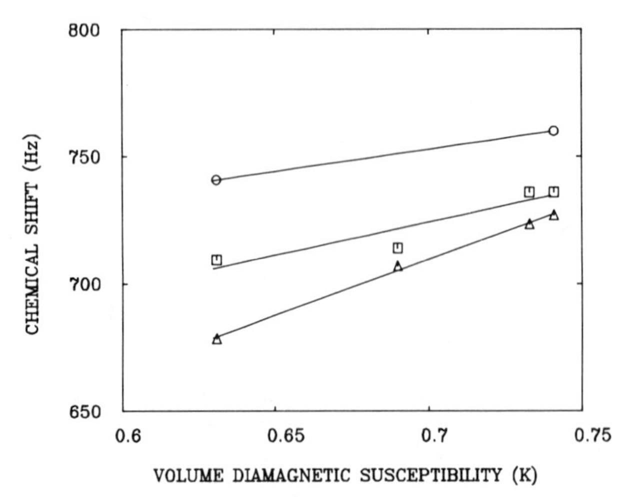

Figure 9. Dependence of chemical shift on volume diamagnetic susceptibility for three solutes, toluene (circles), *p*-cresol (triangles), and skatole (squares) in four solvents, carbon tetrachloride (K = 0.69), chloroform (0.74), methylene dichloride (0.73), and cyclohexane (0.61).

extensive electron delocalization of the porphyrin ring as well as the presence of the iron atom, gives rise to significant shift effects, even in the diamagnetic Fe^{2+} state (Wüthrich, 1970).

Some early examples of specific shift effects due to proximity to aromatic residues were described for lysozyme by Sternlicht and Wilson (1967), and in the case of upfield-shifted methyl groups by McDonald and Phillips (1967). An example illustrating the selectivity of local protein microenvironment is the resolution of all six C_ζ atoms of the six tyrosine residues of RNase A (Figure 10) (Egan *et al.*, 1978). In this series the most buried residues generally give the most downfield resonances, presumably as a result of selective hydrogen-bonded shift effects. Attempts to resolve these resonances by ^{1}H NMR at 250 MHz were not successful (Markley, 1975c), although most were resolved at 360 MHz (Lenstra *et al.*, 1979). It was also found that the ^{13}C chemical shifts of carboxyl groups in several proteins show upfield shifts between the free amino acid and the native protein (Shindo *et al.*, 1978). However, when comparing ionizable groups, one must be careful that the titration of one or more of the groups is not interfering with the appropriate comparison.

A further example of the relationship of chemical shift to protein microenvironment is shown by the ^{19}F chemical shifts of the 11 fluorotyrosine residues of alkaline phosphatase grown on [^{19}F]tyrosine (Sykes *et al.*, 1974). These residues give well-resolved resonances (Figure 11) since the ^{19}F chemical shifts are more sensitive to local environment than ^{1}H shifts, and they exhibit an approximate relationship between chemical shift and

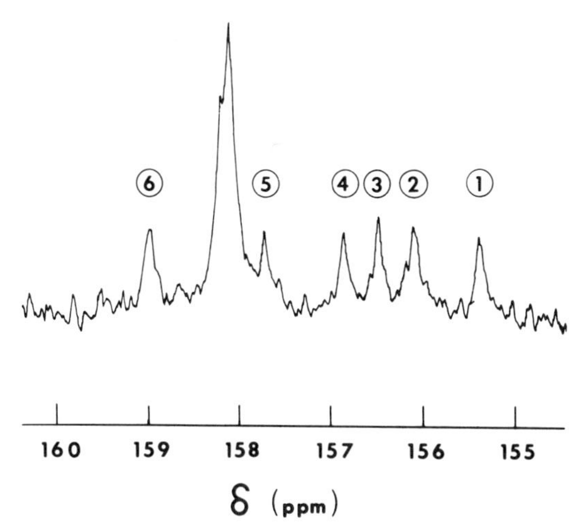

Figure 10. ^{13}C NMR spectrum at 67.9 MHz of the six Tyr C_ζ resonances of bovine pancreatic ribonuclease A (~20 m*M* in 0.1 *M* NaCl, pH 2.25, 20°C). (With permission from Egan *et al.*, 1978.)

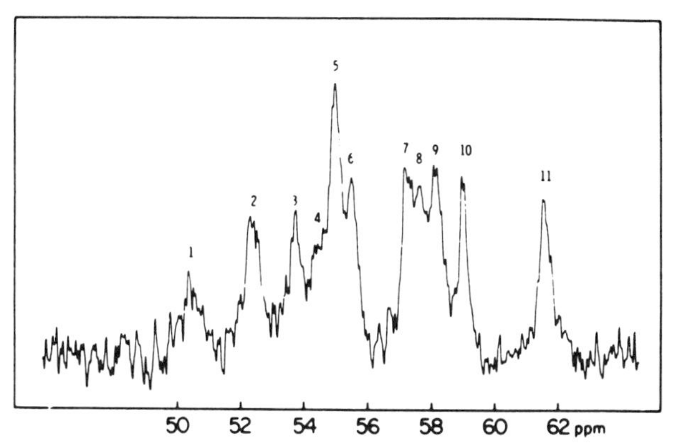

Figure 11. ^{19}F NMR spectrum of native fluorotyrosine alkaline phosphatase at 94.16 MHz. The enzyme concentration was approximately 25 mg/ml in 0.5 m*M* Tris-D_2O buffer (pH 7.85), 25 to 30°C. This represents a concentration of about 0.6 m*M* for each fluorotyrosine. The chemical-shift scale is in parts per million upfield from a capillary of trifluoroacetic acid. This Fourier transform spectrum was obtained with a spectral width of 2500 Hz, acquisition time of 0.2 s, and a pulse delay of 0.2 s. The number of transients per block was 128, and 520 blocks were averaged with a sensitivity-enhancement time constant of 0.1 s. Total time was about 7.5 h. (With permission from Sykes and Weiner, 1980).

linewidth such that the most downfield shifted resonances correspond to the most buried Tyr residues (Hull and Sykes, 1976; Sykes and Hull, 1978).

3.2. Protein Mobility and Relaxation Times

Many NMR studies of proteins have been directed toward understanding protein microdynamics, and in many respects NMR is uniquely suited to the study of both the overall tumbling motion of proteins and the mobility of specific side chains within the protein. This is true in part because many resonances corresponding to individual nuclei or groups of nuclei can be examined by NMR. When necessary, the incorporation of NMR spin isotopes such as ^{2}H, ^{13}C, ^{15}N, or ^{19}F allows the motional features of side chains to be studied with little or no interference from background resonances. The power of NMR studies also derives from the variety of approaches available to the NMR spectroscopist for studying molecular dynamics. These methods include the measurement of hydrogen-exchange rates, the motional averaging of chemical shifts and coupling constants, the measurement of spin–lattice T_1 and spin–spin T_2 relaxation times and nuclear Overhauser enhancements (NOEs), the transfer of saturation between chemically or motionally exchanging groups, and variations in chemical shifts and linewidths due to molecular rearrangements resulting from ion binding, substrate binding, self-association, or variations of pH or temperature. The NMR data have substantially changed the conventional picture of proteins away from the static images presented by x-ray crystallographic results toward images of flexible structures that undergo a variety of different types of motion. Among the types of motions that have been investigated by NMR are the exchange of amide protons, the ring flips of aromatic rings, the librational motion of alkyl side chains, rapid methyl rotations, the overall isotropic tumbling of globular proteins, the unfolding and loss of tertiary structure under the influence of denaturing agents, and various types of segmental motion.

The intense interest and activity in the investigations of protein dynamics by NMR methods has fostered several reviews on the subject. Jardetzky (1981) and Wüthrich and Wagner (1978) have presented brief reviews that outline the utility and range of applications of NMR to the study of protein dynamics. In an excellent comprehensive review, London (1980) discusses the use of various models for molecular reorientation that can be fitted to NMR relaxation data. The internal motions of the side chains of the aliphatic residues methionine and arginine, of the aromatic residues phenylalanine and tyrosine, and of proline are described in detail, and numerous examples are cited. The general topic of motion in proteins has been discussed by Gurd and Rothgeb (1979) and Karplus and McCammon (1981). These reviews are extremely valuable because they are not limited to the NMR method but also discuss results from methods such as solvent exchange, fluorescence, optical absorbance, and protein crystallography. For the novice, several texts on biological NMR have general discussions of NMR relaxation theory and motional effect in protein NMR spectra (Dwek, 1973; Wüthrich, 1976; Jardetzky and Roberts, 1981). Because of these excellent reviews, a detailed discussion of NMR studies of protein dynamics is not given here; rather, a few representative examples are presented in order to demonstrate the usefulness of, as well as some of the problems associated with, NMR methods.

The degree of structural fluctuations can sometimes be assessed indirectly by the measurement of amide proton exchange rates. The underlying assumption of the method is that the intrinsic exchange rate for most amide protons is very rapid and that a slowly exchanging proton results when the amide group is buried within the core of the protein and is thus accessible to exchange with solvent water molecules only as a result of thermal structural fluctuations. Thus the rate of proton exchange is determined by the rate of these fluctuations at any given site. Nuclear magnetic resonance is especially suited to studies of this type, since single proton sites can be monitored, as opposed to the total-NH exchange rate observed by other methods (Hilton *et al.*, 1981). Usually the amide protons observed are those of the backbone amide groups, but in some cases the slowly exchanging amide protons of asparagine or glutamine can also be distinguished, as in BPTI, for example (Wüthrich and Wagner, 1979). The rate constants for exchange of the side-chain protons of lysine and arginine are one or two magnitudes larger at usual pH values relative to those of asparagine and glutamine and thus are usually not observed. Generally the exchange rates are measured by a saturation-transfer experiment, or if the exchange rate is very slow, by simply monitoring the decrease in resonance intensity as a function of time after the addition of ^2H_2O. In ^2H_2O only very slowly exchanging protons can be observed (lifetimes >100 s). In the case of BPTI, proton exchange rates have been measured in ^2H_2O over a p^2H range from 0.1 to 10.9 at temperatures between 10 and 60°C (Figure 12) (Richarz *et al.*, 1979). Hilton *et al.* (1981) have interpreted the complex pH and temperature dependence of the proton exchange in BPTI in terms of two processes for protein unfolding that exhibit widely different temperature dependences. Proton exchange has also been observed for tryptophan NHs in lysozyme (Glickson *et al.*, 1969) and for numerous exchangeable protons on the N-terminal fragment of the *lac* repressor (Wemmer *et al.*, 1981).

Dynamic information can frequently be obtained from a lineshape analysis when an NMR spin-bearing group is interchanging between two sites on a time scale comparable to the reciprocal of the chemical-shift difference between the resonances of nuclei in the two sites. This approach has been most valuable for the study of aromatic ring flips. In an isotropic environment, the ring protons of the phenyl group in tyrosine, for example, exhibit an AA′BB′ spin-coupling pattern in its ^1H NMR spectrum, consisting of two doublets corresponding to two pairs of protons in positions 2 and 6, and 3 and 5 of the aromatic ring. This coupling pattern persists in an anisotropic

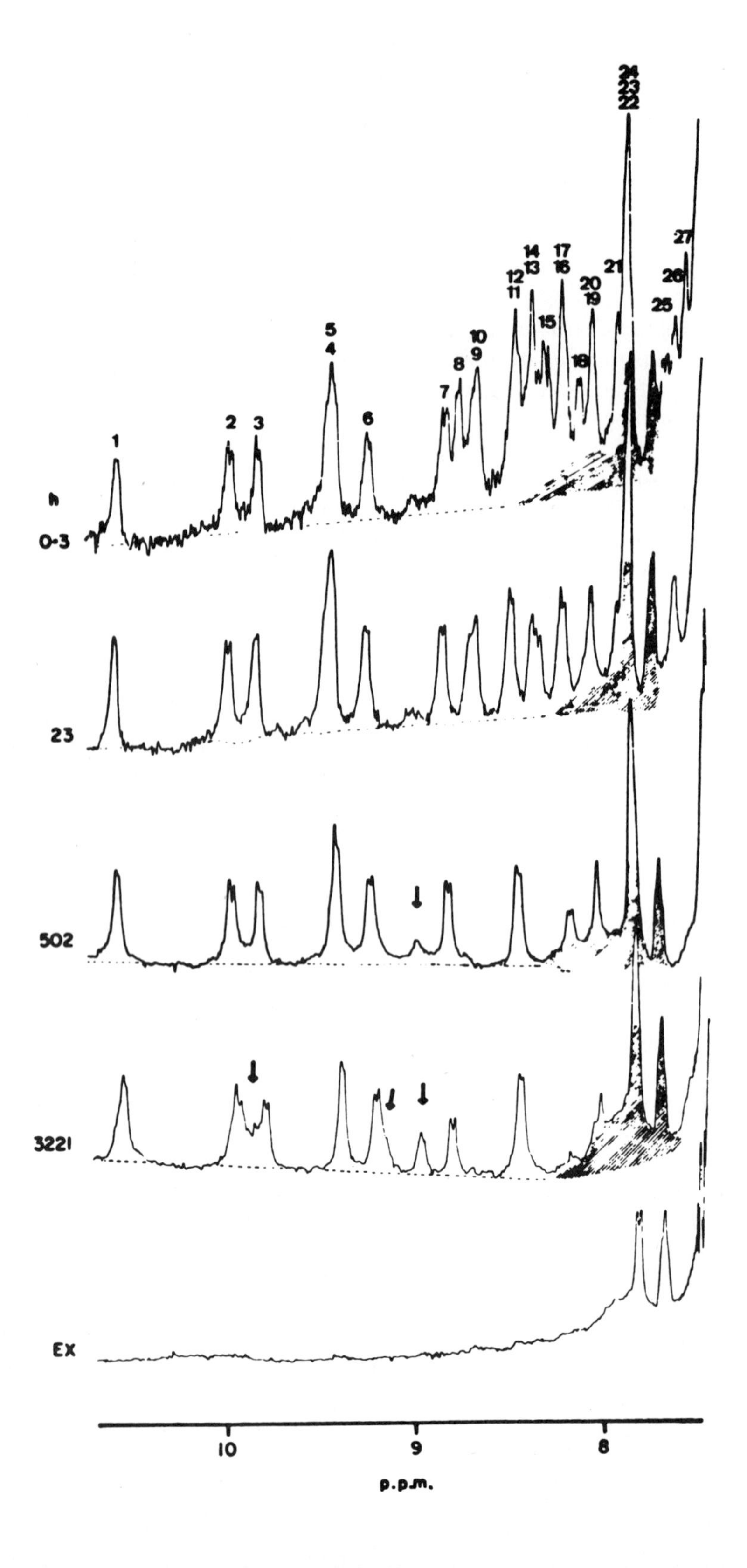

1
2
3
5
4
6
7
8
10
9
12
11
14
13
15
17
16
18
20
19
21
24
23
22
25
26
27
h
0·3
23
502
3221
EX
10
9
8
p.p.m.

environment if the rate of ring flipping is very rapid. Within a protein, however, the ring-flipping motions of phenylalanine and tyrosine side chains are usually restricted, and the chemical shift and coupling constants of each ring proton are in general different, since the C_2 rotational symmetry of the ring is destroyed. In the case of tyrosine, a completely immobilized ring gives rise to an ABCD spin-coupling pattern of four discrete resonances from the four ring protons in distinct chemical environments. In the intermediate motional regime between an immobilized ring and rapid ring flips, the ring protons exhibit a motionally averaged splitting pattern. By simulating the observed lineshapes with a suitable computer program of the kind developed by Kleier and Binsch (1970), a rate constant for the ring-flipping motions can be obtained. Restricted motion of phenylalanine and tyrosine rings was observed in small peptides as early as 1971 (Cohen, 1971). In a frequently cited case, the ring motions of a tyrosine residue in ferrocytochrome *c* were observed over a wide range of motional regimes by varying the temperature from 4 to 37°C (Figure 13) (Campbell *et al.*, 1976). Rate constants were obtained either by NMR spin simulation or by saturation-transfer experiments. The enthalpy of activation at 25°C was determined to be 97 kJ/mol. An obvious requirement for the lineshape method is that the spin system be sufficiently well resolved to compare with a calculated spectrum. Thus the technique is limited to cases in which the phenylalanine or tyrosine protons can be adequately distinguished from other resonances.

The saturation-transfer experiment employed by Campbell *et al.* (1976) is effective for studying exchange processes when the rate of exchange is too slow to study by lineshape analysis. When a resonance corresponding to a nucleus that is chemically or motionally interchanging to another site is selectively saturated by low-power irradiation, a reduction in intensity of the resonance of the nucleus in the second site is observed as a result of a transfer of saturation from the first site. The rate constant for the interchange can be obtained from a combination of T_1 and resonance-intensity measurements (Forsen and Hoffman, 1963; Hoffman and Forsen, 1966). These kinds of experiments have been made much more convenient by modern FT NMR spectrometers that allow double-irradiation experiments to be accomplished with facility.

Much of the information about side-chain motions is derived from the

---

Figure 12. Region from 7.5 to 11.0 ppm of the ^1H NMR spectrum of BPTI at 36°C and p^2H 4.6 recorded at different times after dissolving the protein in ^2H_2O. In the top spectrum the amide proton resonances are numbered in the order of the chemical shifts. The shaded areas represent resonance intensities from aromatic protons. The time (in hours) elapsed between sample preparation and recording of the spectrum is indicated on the left; *EX* is the spectrum obtained after complete exchange of labile protons by heating the solution to 85°C. The accumulation times for the individual spectra were between 5 and 20 min. The arrows indicate three lines that appeared when the solution was kept at 36°C for several weeks. [With permission from Richarz *et al.*, *J. Molec. Biol.* **130**, 19 (1979). Copyright Academic Press, Ltd. (London).]

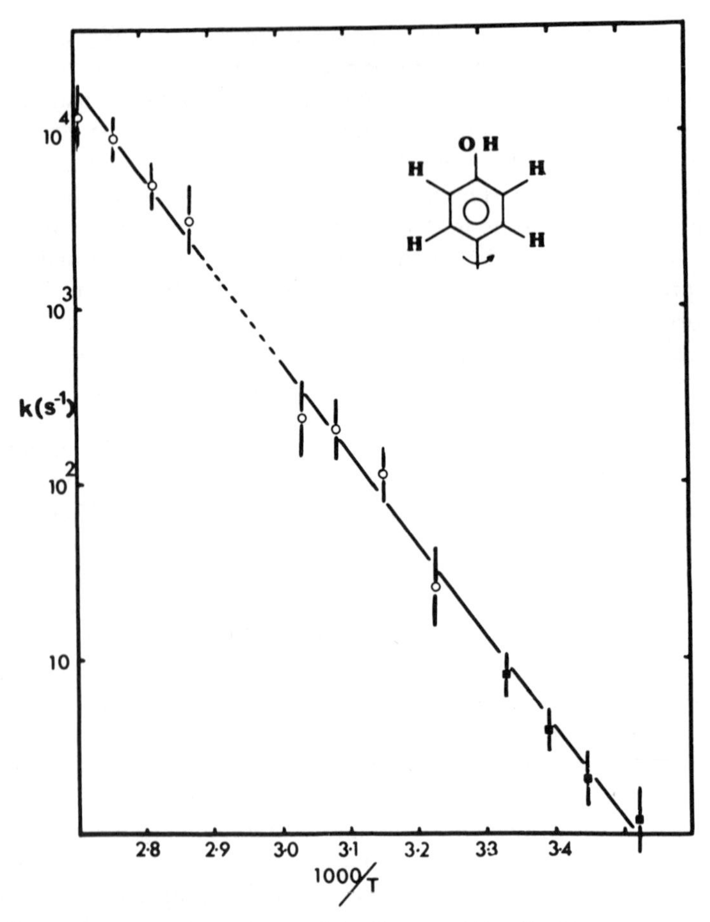

Figure 13. Temperature dependence of the flip rate of Tyr-46 of horse ferrocytochrome *c* as determined by lineshape analysis (circles) and cross-saturation experiments (squares). [With permission from Campbell *et al.*, *FEBS Lett.* **70,** 96 (1976).]

measurement of spin-lattice or spin–spin relaxation times, linewidths, and NOEs. The spin-lattice relaxation time T_1 is the characteristic time constant for return to a Boltzmann distribution of spins among the nuclear-spin energy levels after the system has been perturbed by a radiofrequency field. The spin–spin relaxation time T_2 is a measure of the time required for the nuclear spins to dephase after they have been forced to precess coherently under the influence of a radiofrequency field. The linewidth at half height, which is related to T_2 by the relation $\nu_{1/2} = 1/\pi T_2$, is usually more convenient to measure, but is usually less accurate. The NOE is the fractional change in intensity of one NMR resonance when another resonance is irradiated. Any fluctuating magnetic field at a nucleus can induce relaxation if it has frequency components at the precession frequency of the nucleus being observed. There are several nuclear interactions that give rise to local magnetic fields, and the frequency of oscillation of these fields depends upon the tumbling motions of the molecules bearing the nuclei. The value of T_1 and T_2 measurements lies in the relationship between molecular motion and nuclear spins. The NOE derives from cross-relaxation by mutual spin flips via dipolar interactions and is thus also a sensitive indicator of molecular motion.

A prime consideration in choosing an NMR nucleus for a relaxation study is the mechanism of relaxation. For most dipolar nuclei ($I = \frac{1}{2}$) such as ^{1}H, ^{13}C, ^{19}F, and ^{15}N, the most efficient relaxation mechanism is the weak interaction between proximal nuclear magnetic-dipole moments (dipolar relaxation). In the case of ^{1}H, the nearest magnetic dipole is likely to be another nearby proton; for most ^{13}C nuclei, dipolar relaxation usually results from directly attached protons. Both ^{19}F and ^{15}N, which must be chemically or biosynthetically incorporated into a protein, relax via nearby or attached protons. Another source of dipolar interactions is the unpaired electrons of dissolved molecular oxygen or paramagnetic ions. The magnetic moment of the electron is more than three orders of magnitude greater than that of the proton, and consequently the dipolar relaxation of spin-$\frac{1}{2}$ nuclei by paramagnetic species can be substantial. This is especially true for protons in side chains that can bind to metal ions present at trace levels in aqueous solutions. The effect of paramagnetic impurities on the relaxation of imidazole protons in histidine side chains, for instance, has been demonstrated in lysozyme and ribonuclease (Wasylishen and Cohen, 1974). Usually precautions are taken to remove paramagnetic species by employing deionized water and either degassed or nitrogen-saturated solutions.

A second relaxation mechanism that significantly contributes to both ^{19}F and ^{13}C relaxation in some cases is the fluctuating field due to the anisotropy in the chemical-shielding tensor. The strength of the interaction increases with the square of the applied field; thus T_1 and T_2 can be shorter at higher than at lower field strengths. For ^{13}C nuclei, this chemical-shift anisotropy (CSA) relaxation is usually effective only for nonprotonated carbons such as carbonyls and carboxyls, and nonprotonated aromatics, for all of which the shielding anisotropy is large and there are no nearby protons. Since the linewidth increases as T_2 decreases, the increase in the ^{13}C resolution on going from 14.2 to 63.4 kG or higher field strengths is much greater for aliphatic and methine aromatic carbons than for carbonyl and nonprotonated aromatic carbons, which have significant contribution to relaxation by CSA. One example of the increase in linewidths at higher fields is the broadening of the nonprotonated phenylalanine and tyrosine carbons in lysozyme at 63.4 kG (Norton *et al.*, 1977). Fortunately, only the dipolar mechanism is efficient for methine aromatic carbons at field strengths up to 150 kG.

A final relaxation mechanism of importance for nuclei with $I > \frac{1}{2}$ such as ^{2}H, ^{14}N, ^{17}O, and ^{33}S is the interaction of the nuclear quadrupole moment with the electric-field gradient at the nucleus (quadrupolar relaxation). This interaction is usually so strong that the linewidths of ^{14}N, ^{17}O, and ^{33}S are

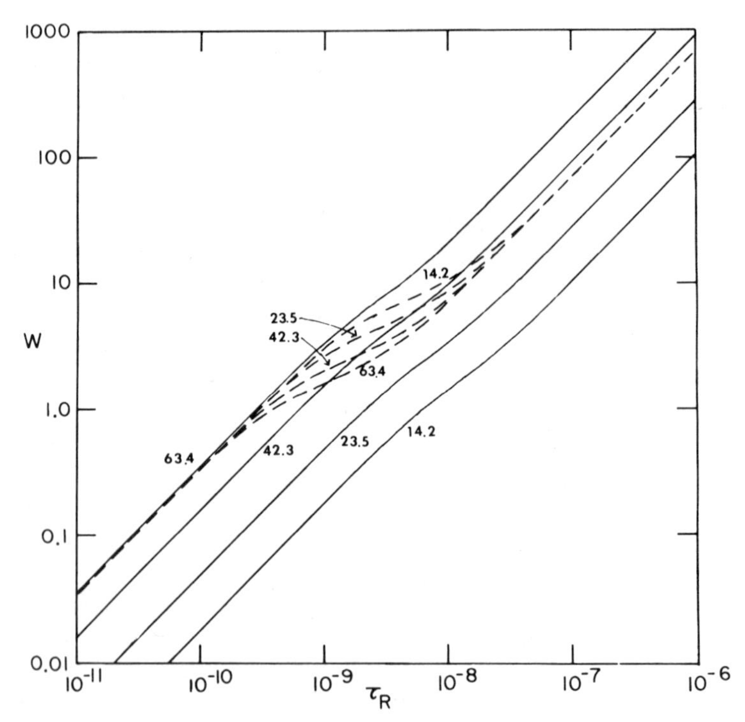

Figure 14. Linewidth (Hz) as a function of correlation time for a ^{13}C nuclear spin relaxing by the dipole–dipole mechanism (dashed lines) and by the CSA mechanism (solid lines) with three protons on a methyl group (2.16 Å away). Isotropic reorientation was assumed. The plots are for four magnetic field strengths (in kG) as indicated. [Reprinted with permission from Norton *et al.* (1977). Copyright 1977, American Chemical Society.]

much too broad to be of value. The quadrupole moment of ^2H, on the other hand, is comparatively small, and ^2H linewidths can be reasonably narrow, especially if the ^2H is incorporated into a mobile side chain. The ^2H magnetic dipole moment is much smaller than that of ^1H; thus quadrupolar relaxation is usually the only mechanism that needs to be considered when interpreting linewidths and relaxation times. Other relaxation mechanisms such as spin rotation and scalar relaxation are usually not significant for most problems of interest.

The effect on the ^{13}C linewidth of dipolar and CSA relaxation is shown in Figure 14. Irrotationally bound or backbone carbons for which the rotational correlation times for overall protein tumbling falls in the range 10^{-9} to 10^{-8} s should exhibit slightly narrower lines at higher field strengths. The linewidths of nonprotonated carbons can have a large CSA contribution to relaxation and frequently are broader than protonated carbons on high-field spectrometers. In the "extreme narrowing region" of rapid motions ($\geqslant 10^{10}\ s^{-1}$), the linewidth is independent of field strength. The functional dependence of ^2H linewidths (Figure 15) is similar to that for the dipolar relaxation of spin-$\frac{1}{2}$

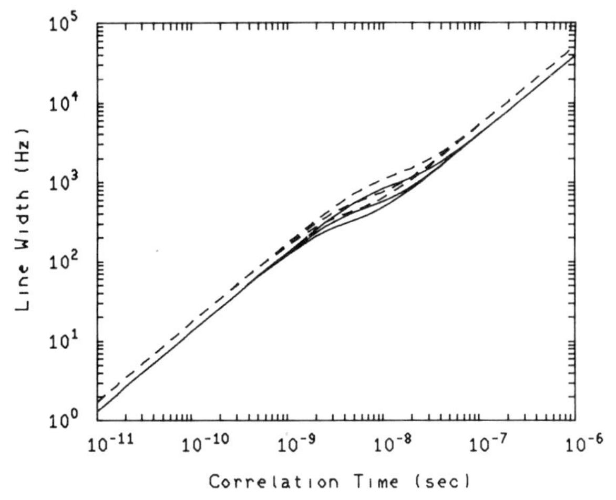

Figure 15. Line width (hertz) as a function of correlation time for isotropic reorientation of a ^2H nuclear spin at 13.8, 27.6, and 41.4 MHz (top to bottom) for Q_D = 168 kHz (solid lines) and 192 kHz (dashed lines). [Reprinted with permission from Wooten and Cohen (1979). Copyright 1979, American Chemical Society.]

nuclei, but the absolute magnitude differs considerably. Comparison of Figures 14 and 15 shows that a ^2H linewidth is about a factor of 10 larger for a group with $\tau_c = 10^{-10}$, whereas for a slowly tumbling group with $\tau_c = 10^{-8}$ s, the ^2H linewidth is about a factor of 100 larger.

Although most studies of protein dynamics have relied on ^1H or ^{13}C, it is sometimes advantageous to incorporate nuclei such as ^{19}F or ^2H. Fluorine has the advantage of very high NMR sensitivity and zero natural abundance background. In addition, the ^{19}F chemical shifts are very sensitive to the local protein environment, and this frequently allows multiple resonances to be observed (Sykes and Hull, 1978; Sykes and Weiner, 1980). One of the most successful examples of the use of ^{19}F to study protein side-chain dynamics has been the use of fluorotyrosine-labeled alkaline phosphatase (Hull and Sykes, 1975). A total of 11 fluorotyrosine resonances were either resolved or partially resolved at 94 MHz (Figure 11). Spectra observed at 235 MHz exhibited only very limited improvement in resolution because of the increase in linewidths at the higher field strength, which is a result of ^{19}F CSA relaxation. The field dependence of the linewidths, however, provides a means for separating the relative contributions of CSA and dipolar relaxation. From an analysis of dipolar T_1 and linewidth data, CSA linewidth data, and the NOE obtained by irradiating the protein protons, it was possible to derive the overall tumbling correlation time τ_c as well as estimates for the rate of side-chain motions for the fluorotyrosine residues. For the overall tumbling motion it was found that $\tau_c = 76 \pm 15$ ns. The rate of rotation about the C_β—C_γ bond was found to be in the range of 10^6 to 10^8 s^{-1} for nine of the Tyr residues and 10^2 to 10^5 s^{-1} for the remaining two.

Except for immobile samples in which ^2H quadrupolar powder patterns can be observed (see Section 2.4), ^2H NMR has been applied to proteins in isotropic solutions only infrequently. There is an inherent problem with ^2H NMR in that the ^2H linewidths may be many parts per million broad, whereas the chemical-shift dispersion of ^2H, as with ^1H, is only about 10 parts per million. In certain cases, however, ^2H can be a valuable aid in studying chain mobility. For an irrotationally bound side chain such as His-15 in [ϵ-^2H]His-15–labeled lysozyme (Wooten and Cohen, 1979), the ^2H linewidth observed was 700 Hz or larger, covering the entire chemical-shift range at 41.4 MHz. The advantage of ^2H NMR, however, is that the linewidths are very sensitive to mobility and can be interpreted solely on the basis of quadrupolar relaxation. For highly mobile groups, the linewidth problem is much less severe. The ϵ-C^2H_3 groups of Met-65 and Met-80 in ^2H-labeled cytochrome *c* (Wooten *et al.*, 1981) exhibited ^2H linewidths in the range of 25 to 65 Hz depending on the mobility at each site (Figure 16). From Figure 15 it can be easily seen that the ^2F linewidth can vary by as much as a factor of 50 between the most mobile groups and groups that reorient only by the overall protein tumbling. Cytochrome *c* is a somewhat special case because the chemical-shift problem of limited resolution is avoided since Met-80 is bound via its sulfur atom to the protein's heme iron. The diamagnetic ring currents in the reduced form and additional paramagnetic interactions in the oxidized form shift the Met-80 C^2H_3 resonance far upfield, thus preventing overlap with the Met-65 resonance. The difference in the linewidths of the two groups (Met-65, 25 Hz; Met-80, 65 Hz) results because motion about the CH_2—S bond in Met-80 is prevented by sulfur binding to the heme iron.

Most detailed information about side-chain motion derives from spin–lattice relaxation measurements. Proton T_1 values, however, are not always useful indicators of molecular motion in large proteins at high magnetic-field strengths. This is because proton spin–lattice relaxation rates tend to be equalized by the effects of spin diffusion. Spin diffusion occurs when mutual spin flips occur by cross-relaxation at a rate faster than the usual spin-lattice relaxation resulting from nearby motionally fluctuating magnetic dipoles (usually protons) (Kalk and Berendsen, 1976). A few mobile groups such as rapidly rotating methyl groups act as relaxation sinks and govern the overall relaxation rate. The effects of spin diffusion are greatest for proteins with molecular weights above 10,000 and at NMR frequencies exceeding 200 MHz. The equalization of proton T_1's by spin diffusion has been demonstrated in a number of proteins (Sykes *et al.*, 1978). The slowest-relaxing protons of Tyr-23 in BPTI, for instance, as a result of cross-relaxation exhibit nonexponential relaxation curves that tend to approach the relaxation curves of rapidly relaxing CH_3 protons. The interproton NOE is also affected by spin diffusion, and the conformational information obtained in small peptides by selective-saturation experiments may be much less specific when applied to proteins. Gordon and Wüthrich (1978) have shown

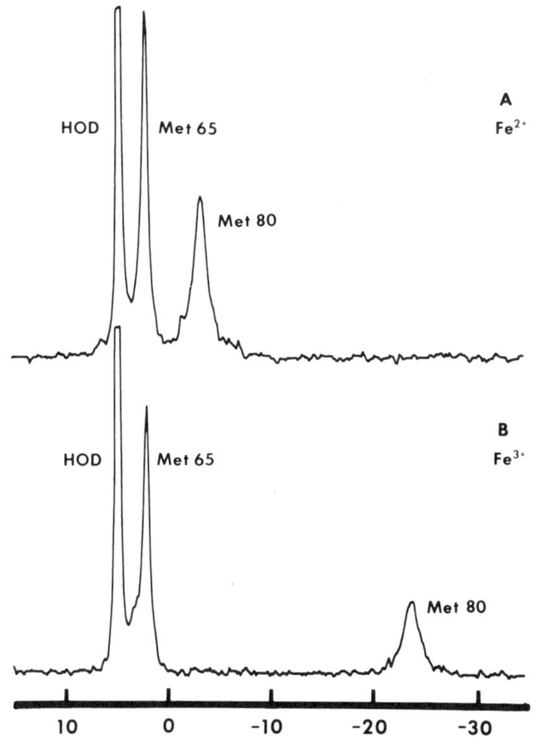

Figure 16. The 41.4-MHz ^2H NMR spectra of (*A*) reduced and (*B*) oxidized cytochrome *c* enriched with 45% ^2H in the methyl groups of Met-65 and Met-80. Each spectrum resulted from a 1-h acquisition on 5 m*M* solutions of cytochrome *c* in 0.1 *M* NaCl at pH 7. The spectrum of the reduced protein was obtained on the same sample as the oxidized protein after addition of sodium dithionite and adjustment of the pH. [Reprinted with permission from Wooten *et al.* (1981). Copyright 1981, American Chemical Society.]

that transient NOEs, in which the rise of Overhauser enhancement is observed following the selective inversion of a particular proton, provides much the same information as the steady-state saturation experiment does on smaller molecules. Recently, the buildup rates of the interproton NOE in BPTI has been measured by two-dimensional nuclear-enhancement spectroscopy (NOESY) (Kumar *et al.*, 1981).

The dipolar relaxation of ^{13}C nuclei via directly attached protons is the most convenient source of dynamic information in proteins because of the preponderance of resolved ^{13}C resonances in protein ^{13}C NMR spectra, the absence of spin-diffusion effects, and the fact that protonated carbons relax predominantly by dipolar relaxation. The details of side-chain motion and protein tumbling motions are obtained from T_1, T_2, and NOE measurements, sometimes at two or more field strengths, usually in terms of a correlation time τ for molecular reorientation. When multiple internal motions are involved, as for aliphatic side chains, for example, in addition to overall protein tumbling, a separate correlation time is in general required for each type of motion. The functional dependences of the relaxation parameters on molecular motion for the case of a completely proton decoupled carbon atom are given by the following relations:

$$\frac{1}{T_1} = \frac{1}{10} \frac{N\gamma_H{}^2\gamma_C{}^2\hbar^2}{r_{CH}^6} [J(\omega_H - \omega_C) + 3J(\omega_c) + 6J(\omega_H + \omega_C)] \tag{1}$$

$$\frac{1}{T_2} = \frac{1}{20} \frac{N\gamma_H{}^2\gamma_C{}^2\hbar^2}{r_{CH}^6} [J(\omega_H - \omega_C) + 3J(\omega_c) + 6J(\omega_H + \omega_c) + 4J(0) + 6J(\omega_H)] \tag{2}$$

$$\text{NOE} = 1 + \frac{\gamma_H}{\gamma_C}\left[\frac{6J(\omega_H + \omega_C) - J(\omega_H - \omega_C)}{J(\omega_H - \omega_C) + 3J(\omega_C) + 6J(\omega_H + \omega_C)}\right] \tag{3}$$

where N is the number of directly attached protons, γ_H and γ_C are the proton and carbon magnetogyric ratios, respectively, $\hbar$ is Planck's constant divided by 2π, r_{CH} is the $^{13}C—^1H$ bond length, and ω_H and ω_C are the Larmor precession frequencies of proton and carbon nuclei, respectively. And J is known as the spectral density function; this represents the available power for relaxation due to molecular motion at the appropriate frequencies.

The exact form of the spectral density functions depends upon the types of motion involved. For a rigid molecule undergoing isotropic reorientation, a single correlation time characterizes the motion, and the spectral density is given by

$$J(\omega) = \frac{\tau_c}{1 + \omega^2\tau_c^2} \tag{4}$$

where τ_c is the isotropic rotational correlation time. When the overall tumbling is anisotropic or internal rotations are present, a more sophisticated model must be employed in order to completely characterize the motion. Various models have been formulated that can be fitted to NMR relaxation data. In general several measurements, such as a T_1 and an NOE measurement, or T_1 measurements at two or more field strengths, must be made in order to obtain an adequate fit to a given model. Craik *et al.* (1982) developed a generalized computer program for the evaluation of molecular-dynamics models, and a convenient tabulation of many of these models is given in their recent paper. An example of the application of molecular-dynamics models to ^{13}C relaxation data in proteins is a study by Richarz *et al.* (1980a) of BPTI. On the basis of a wobbling in a cone model in which the overall protein reorientation is assumed to be isotropic, it was found that the ^{13}C relaxation times of protein methyl groups are affected by the librational motions of the polypeptide backbone and amino acid side chains, as well as the overall protein tumbling and methyl rotations. The results for methyl-group motions are interpreted in terms of three correlation times: τ_R for overall rotational motion, τ_W for the librational "wobbling" motions of the aliphatic side chains, and τ_F for methyl-group rotations about the C—C bond. The value of τ_R was obtained from the average relaxation time of the α-carbon envelope, which is assumed to be dominated by the overall protein tumbling motion. Values for τ_W and τ_F were obtained by fitting T_1's and NOEs at 245.1 MHz and T_1's at 90.5 MHz to the methyl group.

Recently, Lipari and Szabo (1982) have shown that protein relaxation times can be interpreted without reference to a detailed model. In their model-free approach, the spectral-density function is expressed simply in terms of a generalized order parameter S^2, which is a measure of the degree of spatial restriction, and an effective correlation time τ_e, which determines the rate of the motion. The values of S^2 and τ_e are obtained by minimizing the sums of squared differences between experimental and theoretical T_1, T_2, and NOE values while varying S^2 and τ_e. The advantage of the model-free approach is that the complicated spectral-density functions of many models can be avoided. Once numerical values of S^2 and τ_e are obtained, however, they can be interpreted in terms of a particular model. In addition, the order parameter can be used in a diagnostic fashion to determine which models can be applied to a given case; a physically reasonable model must be able to reproduce the fitted value of S^2. The authors have shown how this approach can be applied to existing data for methyl-group rotations in BPTI mentioned above (Richarz *et al.*, 1980a), internal isoleucine motions in myoglobin (Wittebort *et al.*, 1980), and internal motions of methionines in dihydrofolate reductase (Blakley *et al.*, 1978) and myoglobin (Jones *et al.*, 1976).

4. NMR TITRATION CURVES

The position of the resonance of a titrating group shifts between extreme values, reflecting the species on either side of the ionization. This is a general phenomenon, such that the position of the resonance at any given pH depends upon the concentrations of the species present, assuming that the protonation–deprotonation is a fast process relative to the chemical-shift difference between the resonances for the two species. For example, protonation of the imidazole ring (Figure 17) results in a significant change in the electron density experienced by the C_ε ring proton, such that it becomes less shielded. The transition between the characteristic chemical-shift value for the unprotonated form (δ_B) and that of the protonated form (δ_{HB}) results in an NMR titration curve of chemical shift as a function of pH (Figure 18). Assuming a simple proton association–dissociation equilibrium, the chemical shift represents the relative concentrations of the protonated and unprotonated species at any given pH. At the (apparent) p*K* value, 50% of each is present, and an inflection is observed in the experimental curves. If we

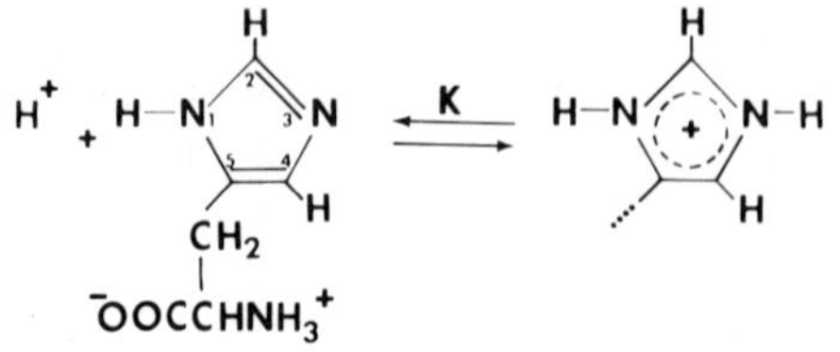

Figure 17. The protonation equilibrium of the imidazole side chain of histidine (C_2 corresponds to C_ε and C_4 to C_δ).

describe the equilibrium by the simple Henderson-Hasselbalch formulation (Cohen *et al.*, 1970a),

$$B + H^+ = HB \tag{5}$$

such that

$$K = \frac{[HB]}{[H^+]\,[B]} \tag{6}$$

then the observed chemical shift at any pH is given by

$$\delta_{cbs} = \frac{\delta_B[B] + \delta_{HB}[HB]}{[B] + [HB]} \tag{7}$$

By substitution of equation (6) into (7), we get

$$\delta_{obs} = \frac{\delta_B + K[H^+]\,\delta_{HB}}{1 + K[H^+]}$$

$$= \delta_B + \frac{\Delta \times 10^{(pK-pH)}}{1 + 10^{(pK-pH)}} \tag{8}$$

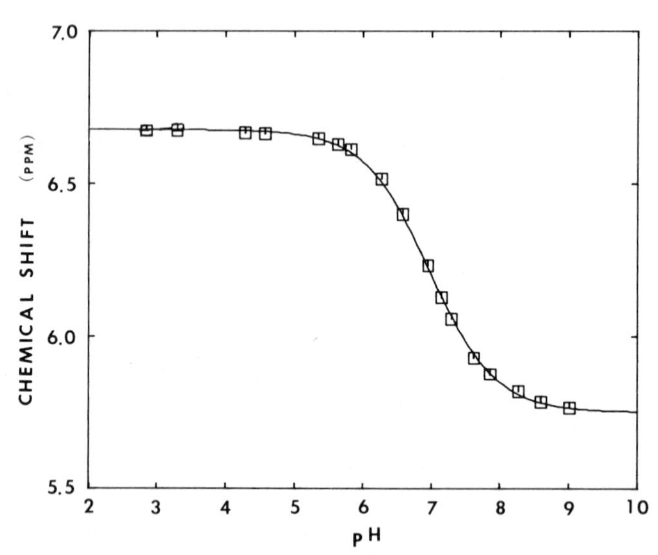

Figure 18. Chemical shift (from CD_3CN) as a function of pH for the ^2H resonance observed by ^2H NMR at 41.4 MHz of [ε-^2H]imidazole (0.5 *M*); the pK_a value is 7.0; the solid line is the best fit with Equation (8). (Redrawn from Wooten and Cohen, 1979.)

where Δ is the chemical shift difference between δ_B and δ_{HB}. This relationship is an excellent representation of the experimental NMR titration curves for the C_ε and $C\delta$ ring protons of imidazole itself and of these protons in proteins. The first quantitative analysis by curve fitting of NMR titration data of the four resolved histidine resonances of staphylococcal nuclease that established this point was published in 1970 (Cohen *et al.*, 1970b).

Markley (1973,1975a) used a variant of this equation that is based on the formalism of Hill (1910):

$$\delta_{obs} = \delta_B + \frac{\Delta \times 10^{n(pK-pH)}}{1 + 10^{n(pK-pH)}} \tag{9}$$

This formalism was developed as an extension of the earlier work of Linderstrom-Lang and others to take account of the electrostatic interactions that occur in proteins due to the presence of many charged groups. The effect of the Hill coefficient in these equations is to be distinguished from the specific effect of an adjacent titrating group, which is discussed in detail below. In general, the use of a Hill coefficient has not been found to be necessary to obtain satisfactory fits of many NMR titration curves of histidine residues in proteins (Cohen *et al.*, 1970b; King and Roberts, 1971).

This is not surprising, since the NMR method provides a monitor of the local microenvironment of the nucleus observed, whereas the Linderstrom-Lang and Hill treatments assume, in effect, a uniform distribution of charge over the surface of the protein. The effect of many adjacent unresolved titrations in the same region of pH is to alter the shape or slope of the titration curve in a way similar to that due to the inclusion of a Hill coefficient. Of course, the introduction of an extra parameter in the fit in general results in a "better" fit, with a smaller root-mean-square error. It is therefore important as a matter of scientific principle to select the simplest model that provides an adequate description of the data in question, without additional assumptions. Indeed, in many cases in which the Hill coefficient has been determined by curve fitting, it has been found to be very close to unity (Markley, 1975a), a value that corresponds to the simple form of the equation (8).

The NMR titration curves of the amino acid histidine were found to exhibit three inflections (Figure 19), including two minor ones at pH values higher and lower than the main inflection that arises from the imidazole transition (Sachs *et al.*, 1971). These were shown to derive from the amino and carboxyl groups on the amino acid by comparison with the blocked derivatives *N*-acetyl histidine and histidine methyl ester (Figure 20). Curve-fitting with an equation consisting of a sum of simple equilibria

$$\delta_{obs} = \delta_B + \Sigma_i \frac{\Delta_i \times 10^{(pK-pH)}}{1 + 10^{(pK-pH)}} \tag{10}$$

is adequate to describe these data if the difference in pK values of the interacting groups are $\gtrsim$1.5 pH units. If not, a more complex treatment is

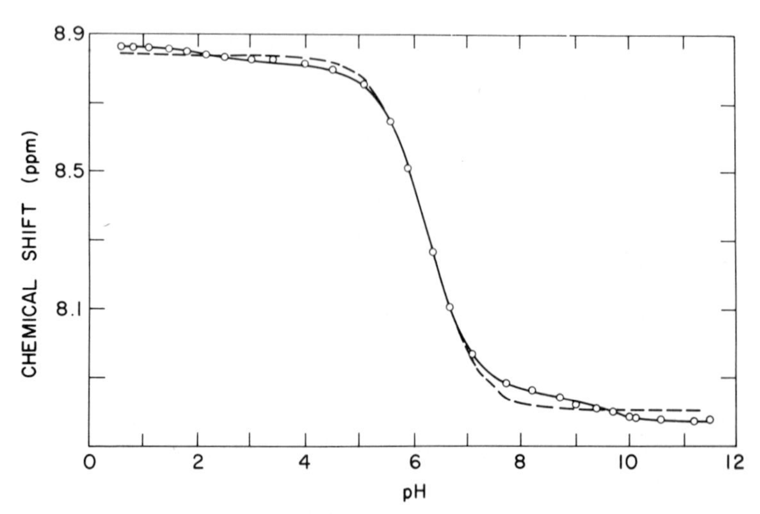

Figure 19. Chemical shift as a function of pH in D_2O for the C_ε proton of L-histidine observed by ^1H NMR at 60 MHz. The dashed line is the fit using Equation (8); the solid line is the fit for three transitions using Equation (10) ($i = 3$). The pK_a values are carboxyl, 2.0; imidazole, 6.2; and amino, 9.2. [Reprinted with permission from Shrager *et al.* (1972). Copyright 1972, American Chemical Society.]

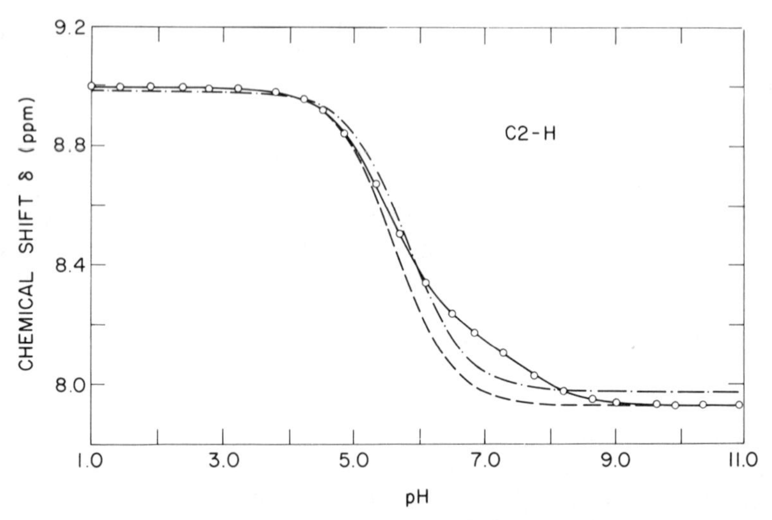

Figure 20. Chemical shift as a function of pH in D_2O for the C_ε proton at 60 MHz of N-acetyl histidine. The best fit obtained by Equation (8) is indicated by the dot-dashed line, and the fit by ignoring the data between pH 3 and pH 9 by the dashed line, which emphasizes the asymmetry in the data. The solid line is the fit obtained assuming imidazole–amino group interaction, and gave pK_a values of 5.5 and 7.5, respectively. [Reprinted with permission from Shrager *et al.* (1972). Copyright 1972, American Chemical Society.]

required, based on the formalism of Edsall *et al.* (1958), which takes into account the competition of the groups for protons in the pH range of the overlap of their p*K* values. An example in which the two pH transitions do in fact overlap is the case of N-acetyl histidine (Figure 20), and the p*K* values derived from analyses of the NMR titration curves agree with the literature values for these model compounds (Shrager *et al.*, 1972).

Studies with histidine peptides support the conclusion that the imidazole ring proton is sensitive to the titration of adjacent groups. This was shown to be the case for the dipeptide His-His; both imidazole C_ε proton titration curves exhibited asymmetry. Fitting with several different models clearly showed that the best fit is obtained assuming each histidine residue interacts with another titrating group, in one case with a lower p*K* than the imidazole p*K*, and in the other case with a higher p*K*. These can be attributed to the carboxyl and amino terminal groups respectively, and thus the two histidine resonances can be distinguished (Shrager *et al.*, 1972). Similarly, the data for the tripeptide His-Gly-His (previously unpublished) clearly show asymmetrical NMR titration curves that can be adequately described by the application of the above theory (Figure 21). One of the two NMR titration curves has a low-pH inflection whereas the other has a high-pH one. Satisfactory fits are obtained by applying a sum of two simple equilibria [Equation (10), $i = 2$] to each curve. Curve 1 may then be assigned to the amino-terminal and curve 2 to the carboxyl-terminal histidine residues, respectively. Although imidazole rings in these peptides can approach each other very closely (although they are not likely to when both positively charged), there is no evidence of mutual interaction between them insofar as the shape of the NMR titration curves are concerned (Shrager *et al.*, 1972; Cohen and Shindo, 1975).

The selective effect of specific adjacent ionizable groups should be clearly distinguished from the generalized effect of many ionizable groups on the protein included in the formulation of Hill described above. This is an important distinction in view of the contention that imidazole groups in proteins can mutually influence each other's p*K* values, even if they are far apart. This distinction takes on added significance in view of the fact that the possible interaction of imidazole residues in bovine pancreatic ribonuclease is relevant to the question of its mechanism of action, as well as being a matter of dispute in terms of the theoretical analysis of the NMR titration curves of the active-site histidine residues 12 and 119. This is discussed in detail in Section 5.1.2.

Extension of the theoretical analysis of NMR titration curves has been made by Rabenstein and co-workers, in that they have elucidated the detailed acid–base chemistry of histidine-containing peptides with biological activity such as Gly-His-Lys (Rabenstein *et al.*, 1977). Others have estimated the effects of charged groups on the ionization constants of imidazole groups in peptides (Tanokura *et al.*, 1976). Calculations of p*K* values for histidine residues in proteins have also been made and compared to actual values in the case of myoglobin (Botelho and Gurd, 1978), although such

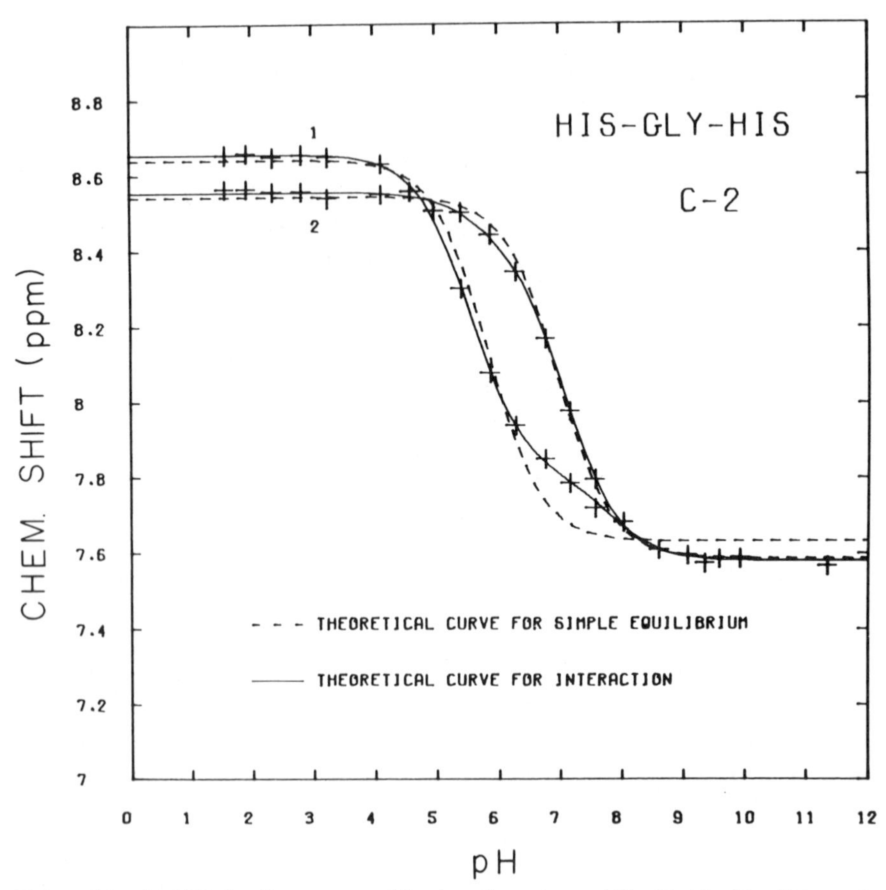

Figure 21. ^1H NMR titration curves of the two C_ε protons of His-Gly-His. The fit to each curve of a single simple equilibrium [Equation (8), dashed curve] is inadequate; the fit with a double transition assuming interaction between each imidazole group and the adjacent terminal titrating group is very good [Equation (10), $i = 2$, solid line). The fit assuming mutual imidazole–imidazole interaction was inadequate (not shown). Curve 1 corresponds to the amino-terminal and curve 2 to the carboxy-terminal His residue, respectively.

calculations are of doubtful validity. Sheinblatt has used the differences in the pH dependences of resonances in small peptides to obtain sequence information (Sheinblatt, 1966).

Most proton NMR titration studies are carried out in D_2O in order to avoid the presence of the large peak of solvent H_2O. This means that one is not strictly measuring pH in these studies. However, the deuterium isotope effect on the glass electrode (which contains H_2O) commonly used to measure pH (Glasoe and Long, 1960) and on the ionization in solution, approximately cancel each other out (both are ~0.4 pH units) (Roberts *et al.*, 1968). For simplicity, therefore, direct meter readings with a glass electrode in a D_2O solution are usually quoted in these NMR studies (see Redfield, 1978).

Nuclear magnetic resonance titration curves have been reported for the carbon atoms of amino acids at natural abundance of ^{13}C (Freedman *et al.*, 1973). For example, it was noted that the six carbon resonances of histidine exhibit the effects of the three ionizable groups present, although of course to a different extent (Figure 22). The direction of chemical-shift change is generally opposite that found in proton NMR titrations on protonation of an ionizable group. However, it was found that in some cases the carbon resonances show opposite directions of shift on protonation. This was taken to indicate the presence of two distinct chemical-shift effects, namely through-bond and through-space effects (Freedman *et al.*, 1973). A more extensive analysis of amino acids allowed the chemical-shift effects due to amino- and carboxy-terminal-group titrations to be generalized. The application of molecular-orbital calculations allowed a rationalization of these effects; for a carbon atom near the site of ionization, a decrease in excitation energy dominates the chemical-shift expression, resulting in deshielding despite an increase in electron density, whereas for more-distant carbon atoms, changes in electron density dominate, yielding shifts in either direction (Quirt *et al.*, 1974). Titration results have been reported for the side-chain carbons of tyrosine at natural abundance ^{13}C (Norton and Bradbury, 1974) and for several ^{13}C-enriched amino acids (TranDinh *et al.*, 1974; London *et al.*, 1978).

Comparison of the ^{13}C NMR titration curves of N-methyl-substituted histidines showed that the N^{τ} tautomer predominates in basic solution (Reynolds *et al.*, 1973). This approach has also been used to study the tautomers of the imidazole side chain in the tripeptide thyrotropin releasing factor (TRF) (Deslauriers *et al.*, 1974).

Similar titration curves have been obtained by ^{15}N NMR of ^{15}N-enriched glycine (Leipert and Noggle, 1975) and histidine (Blomberg *et al.*, 1977) and for arginine at natural abundance of ^{15}N (Kanamori *et al.*, 1978). A recent study by solid-state ^{15}N NMR has shown separate resonances for histidine in different protonation states as a result of slow exchange in the solid state (Munowitz *et al.*, 1982). Also, ^{19}F NMR titration curves of ^{19}F-enriched imidazoles and histidines have been reported; it was found that the chemical-shift changes on protonation cannot be simply explained by electron-density changes (Yeh *et al.*, 1975). The ^{15}N and ^{17}O NMR titration curves of appropriately enriched glycylglycine have been reported (Irving and Lapidot, 1976).

5. STUDIES OF SPECIFIC TYPES OF AMINO ACID SIDE CHAINS IN PROTEINS

5.1. Histidine

In this section we describe studies of the imidazole side chain of histidine residues in proteins. Because of the particular properties of this unique heterocyclic group in proteins and the corresponding unusual NMR prop-

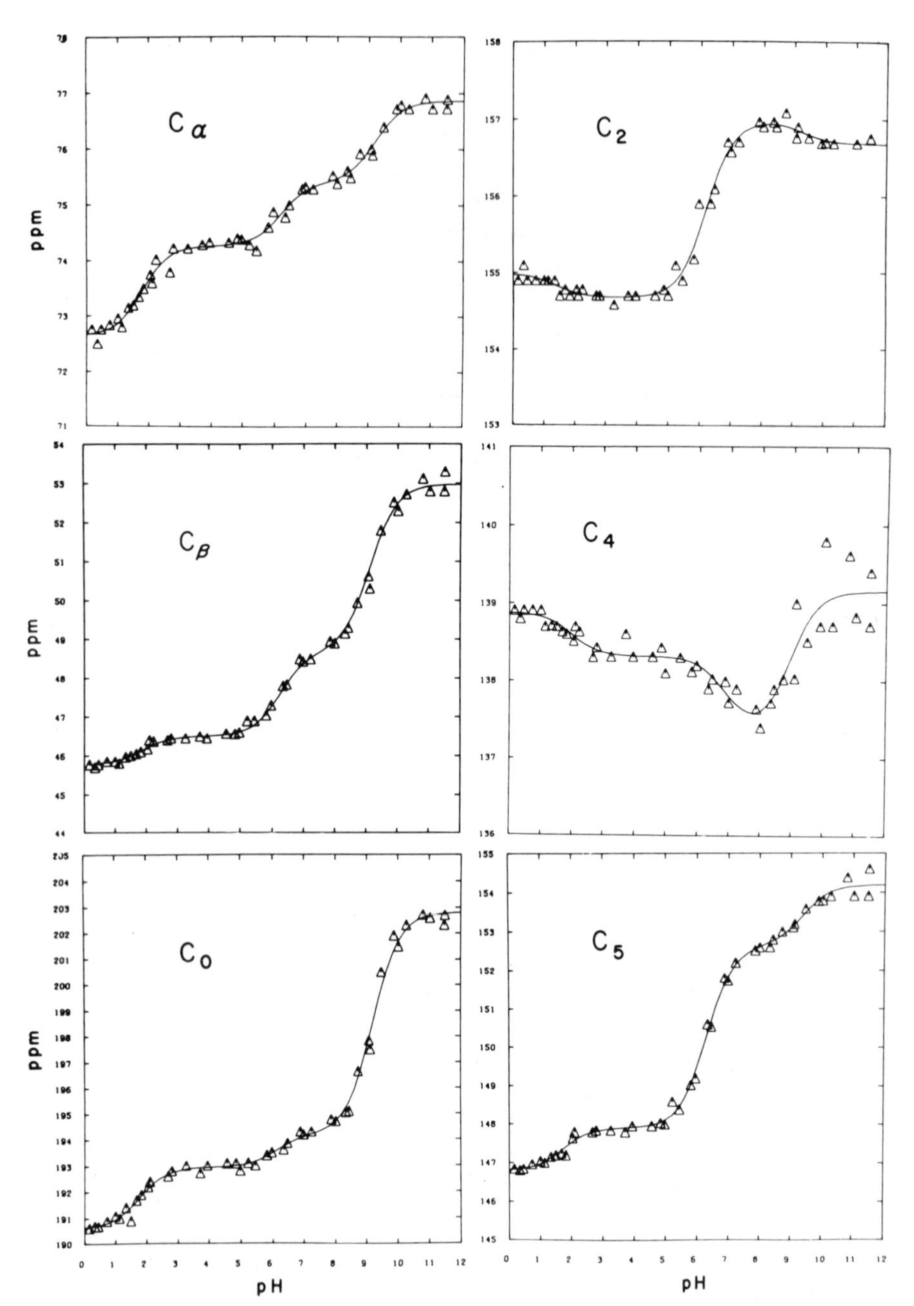

Figure 22. ^{13}C NMR titration curves of the six carbon atoms of L-histidine. All three titrating-group p*K* values are reflected in the data for each carbon atom. (With permission from Freedman *et al.* 1973.)

erties, more studies have been reported on this side-chain group in proteins than on any other. These studies were reviewed by Markley (1975a). The extension of such applications to a very wide range of proteins is listed in Table I.

Because of the importance of the ionization of the imidazole group in protein function, it is necessary to consider whether the theoretical analyses described above for NMR titration curves of side chains in amino acids or peptides are appropriate for application to histidine NMR titration data derived from proteins. The experimental NMR titration data of the four C_ε protons in spectra of staphylococcal nuclease were satisfactorily described by the simple equation (8) (Figure 23) (Cohen *et al.*, 1970b). It was originally reported that the histidine NMR titration curves in this protein were irregular (Markley *et al.*, 1970). However, this may have been due to an overreliance on the accuracy of the data obtainable at that time. The "extra" peak observed in the downfield region of the spectrum of staphylococcal nuclease (Markley *et al.*, 1970), now appears to have been due to an impurity of formic acid (Dobson, 1975). Resonance-area measurements had previously indicated that this peak was not derived from a histidine resonance (Epstein *et al.*, 1971). The fact that these histidine titration curves in staphylococcal nuclease could be adequately represented by a simple equilibrium eliminated the need to invoke a more complex relationship, such as that including a Hill coefficient, which takes into account the charge on the protein (see Section 4). However, the observation of asymmetrical NMR titration curves, although not borne out for staphylococcal nuclease, were reported for ribonuclease (Ruterjans and Witzel, 1969) and subsequently confirmed (Cohen *et al.*, 1970a,1973). In view of the extensive studies reported on the histidine residues of ribonuclease, this protein will be used as an example of the general approach employed.

Ribonuclease (RNase) is a ribophosphodiesterase found in many species (Blackburn and Moore, 1982). Early chemical-modification studies implicated two histidine residues in the mechanism of action of RNase. Since histidine ring proton resonances can be studied individually by proton NMR, unique information on the structure and function of RNase has been derived by the application of this method. Ribonuclease contains four histidine residues (Figure 24), and four histidine C_ε proton resonances were observed in its proton NMR spectrum (Meadows *et al.*, 1967) (Figure 25). Each of these titrates with a specific pK value and exhibits distinct properties. However, before any conclusions can be arrived at regarding the conformation or function of the protein, it is necessary to know the assignment of the four resonances, namely which resonance corresponds to which histidine residue in the amino acid sequence. Consequently, in order to clarify the discussion of the properties of the histidine ring proton resonances it is first necessary to obtain an unequivocal assignment of these resonances.

5.1.1. Assignment of Histidine Resonances of RNase. Derivatives of RNase in which each of the active-site histidine residues 12 and 119 are

Table I. NMR studies of histidine residues in proteins

Protein	Reference	Comment
Aldolase (yeast)	Smith and Mildvan (1981)	6 of 10 His resolved (MW 80 K)
Alkaline phosphatase	Otvos and Browne (1980)	^{13}C NMR study
Carbonic anhydrase	King and Roberts (1971); Cohen *et al.* (1972a); Campbell *et al.* (1974,1975b); Pesando (1975)	Human B and C and bovine B compared
Chymotrypsin A	Markley and Ibanez (1978)	Chymotrysinogen also
Coenzyme A reductase	Veloso *et al.* (1981)	
Colipase (porcine)	Canioni *et al.* (1980)	
Concanavalin A	Carver *et al.* (1977)	Metal-ion binding
Creatine kinase (rabbit muscle)	Rosevear *et al.* (1981)	6 of 16 His resolved (MW 82 K)
Cytochrome *c*	Cohen *et al.* (1974); Cohen and Hayes (1974)	1 of 3 His titrates; several species compared
Dihydrofolate reductase (*L. casei*)	Poe *et al.* (1979); Wyeth *et al.* (1980)	Partial assignments
Elongation factor Tu	Nakano *et al.* (1979)	Effects of GDP and GTP
Glucagon	Rothgeb *et al.* (1978)	S-Me, carbamino adduct
Glutathione hydrolase	Ball and Vander Jagt (1981)	S-2 Hydroxyacyl derivative
Glyceraldehyde-3-phosphate dehydrogenase	Scheek *et al.* (1979)	Photo-CIDNP (MW 145 K)
Hemerythrin	York *et al.* (1980)	
Hemoglobin	Greenfield and Williams (1972); Kilmartin *et al.* (1973); Fung *et al.* (1975); Russu *et al.* (1982); Russu and Ho (1982)	des-His-β146 Hb Comparison of HbA and HbS
Histone H5	Chapman *et al.* (1978)	Zn complex
Insulin (hagfish)	Bradbury *et al.* (1981)	

Lactalbumin	Bradbury and Norton (1975); Berliner and Kaptein (1981)	
Leghemoglobin	Johnson *et al.* (1978)	
Lutropin	Maghuin-Rogister *et al.* (1979); Brown *et al.* (1979)	Subunits also
Lysozyme (hen egg white)	Meadows *et al.* (1967);	Single His, first assignment
	Allerhand *et al.* (1977); Cohen (1969)	Comparison with human
Myeloma proteins	Dwek *et al.* (1975); Gettins *et al.* (1981)	3 His resolved and assigned in Fv
Myoglobin	Cohen *et al.* (1972b); Botelho and Gurd (1978); Bradbury *et al.* (1979); Wilbur and Allerhand (1977a); Hayes *et al.* (1975); Ohms *et al.* (1979)	5 His not observed in Met-Mb ^{13}C several species 7 of 9 His resolved, human oxy-Mb
Neurophysin (bovine I and II)	Cohen *et al.* (1972c)	Plus oxytocin
Nuclease (staphylococcal)	Markley *et al.* (1970); Cohen *et al.* (1970b)	4 His resolved
Papain	Lewis *et al.* (1981)	Cys-His interaction
Phospholipase A2	Aguiar *et al.* (1979)	
Phosphotransferase	Kalbitzer *et al.* (1981);	4 His resolved, *S. aureus*
	Dooijewaard *et al.* (1979)	Phospho-His intermediate, *E. coli*
Protease (*S. aureus*)	Markley *et al.* (1975); Dobson (1975)	Formic acid resonance interferes
Protease, α-lytic	Hunkapillar *et al.* (1973); Bachovchin *et al.* (1978)	^{13}C and ^{15}N NMR study
Prothrombin, fragment 1	Pletcher *et al.* (1981)	
Pyruvate kinase (rabbit muscle)	Meshitsuka *et al.* (1981)	6 of 14 His resolved (MW 237 K)
Ribonuclease (bovine A)	Meadows *et al.* (1968); see text	4 His resolved
	Migchelsen and Benteima (1973); Cohen *et al.* (1973);	Rodent RNases RNases S and A compared
	Wang and Hirs (1979);	No effect of glycosylation in porcine
	Miyamoto *et al.* (1981);	RNase St
	Ruterjans *et al.* (1969); Arata *et al.* (1976)	RNase T1

Table I. (Continued)

Protein	Reference	Comment
Subtilisin inhibitor	Fujii *et al.* (1980)	
Subtilisin	Jordan and Polgar (1981)	Asp-His H-bond
Superoxide dismutase	Stoesz *et al.* (1979); Burger *et al.* (1980)	Exchange properties of NH in H_2O
Tropomyosin (rabbit)	Edwards and Sykes (1978)	2 His resolved
Troponin-C	Levine *et al.* (1977)	Ca^{2+} binding
Trypsin (porcine)	Markley and Porubcan (1976)	
Trypsinogen	Porubcan *et al.* (1978)	Complexes with BPTI
Trysin inhibitor (soybean)	Markley (1973)	

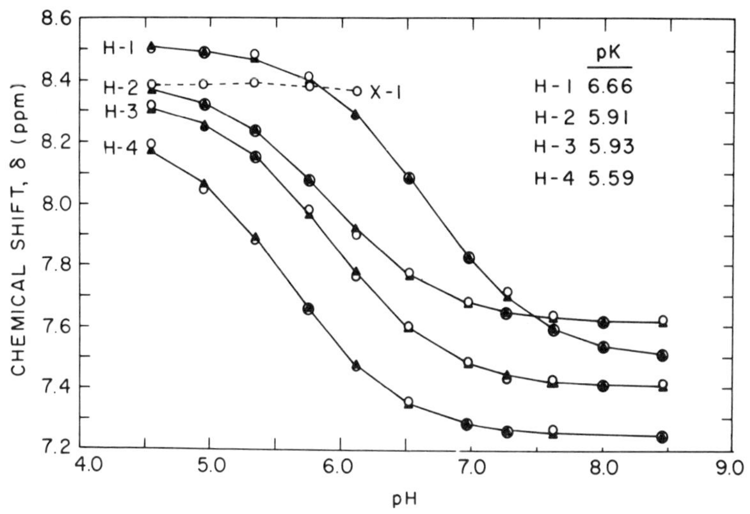

Figure 23. NMR titration curves at 220 MHz of the four C_ε protons of the four histidine residues of staphylococcal nuclease in 0.3 *M* NaCl-D_2O; circles represent observed points; triangles represent calculated points joined by solid lines for the best fit to Equation (8). For the identity of peak *X*, see text. [Reprinted by permission from Cohen *et al.*, *Nature* **228,** 642 (1970). Copyright 1970, Macmillan Journals Limited.]

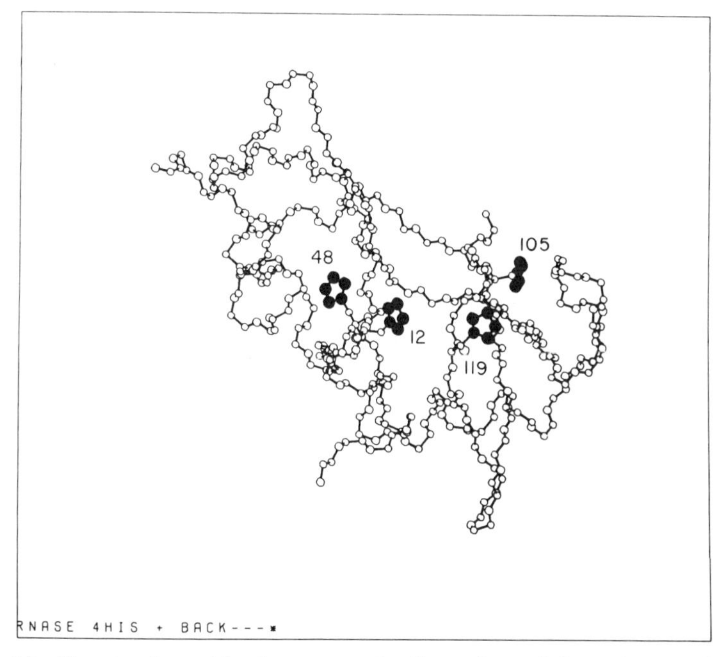

Figure 24. The structure of bovine pancreatic ribonuclease A from the coordinates of Richards and Wyckoff (1973), showing the four histidine residues; His-12 and His-119 are in the active site.

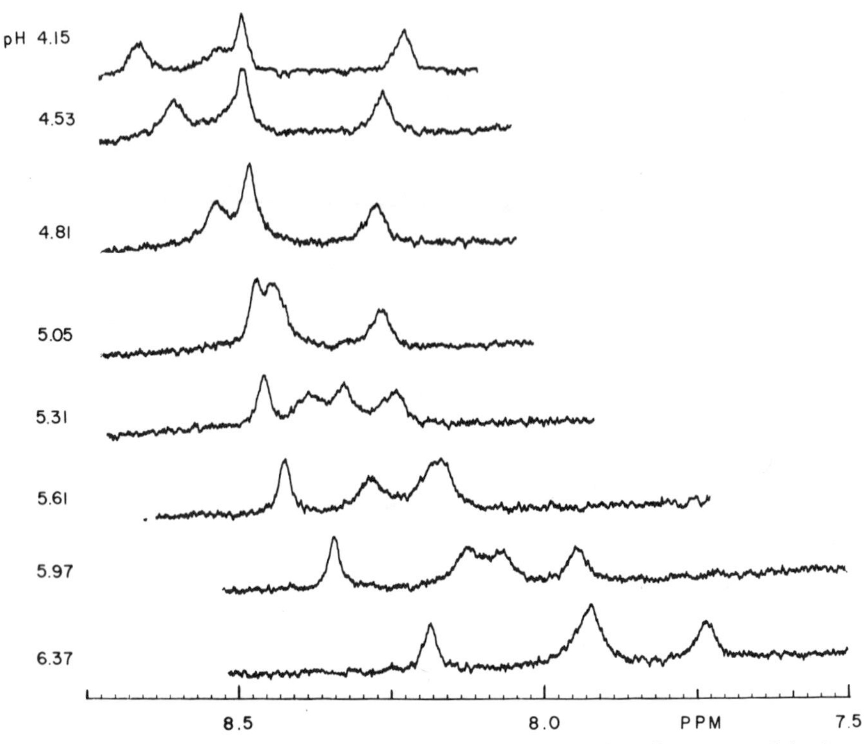

Figure 25. Single-scan 220 MHz proton-NMR spectra of the four C_ε protons of the four histidine residues of RNase S. (With permission from Griffin *et al.*, 1973.)

selectively carboxymethylated have been separated (Crestfield *et al.*, 1963). These derivatives give NMR spectra in which two of the histidine C_ε proton resonances are significantly shifted. This is not due to the carboxymethylation itself, since carboxymethyl histidine derivatives have p*K* values that are quite similar to those of histidine (Shindo *et al.*, 1976; Bradbury *et al.*, 1977). However, since one of the C_ε resonances in each case gives a high p*K* value, this was attributed to the interaction of the carboxyl side chain of the modifying group with the second histidine residue, which must be in sufficiently close proximity to it. It is noteworthy that activity is lost if either of the active-site histidine residues is carboxymethylated (Crestfield *et al.*, 1963). Thus the histidine C_ε proton resonances could be assigned in pairs, that is, the two active-site residues that are affected by carboxymethylation, His-12 or His-119, = H2 or H3, and the two unaffected, His-105 or His-48, = H1 or H4.

A distinction was suggested between the resonances attributed to residues 105 and 48 on the basis of their very different environments within the protein and the very different characteristics of the resonances H1 and H4. Resonance H1 has "normal" titration properties, whereas H4 has a broad resonance and shows an "unusual" titration curve. These were thought to correspond to His-105 and His-48, respectively, since His-105 is exposed on the surface of the protein molecule, whereas His-48 is known to be "buried." Thus the tentative assignments were expanded to H1 = His-105; H4 = His-48.

Two methods have been used to distinguish the active-site histidine resonances in RNase. One method utilizes the fact that the imidazole C_ε protons exchange with deuterium and tritium under alkaline conditions. The other takes advantage of the derivative of RNase known as RNase S, in which the polypeptide chain has been selectively cleaved by subtilisin between amino acids 20 and 21 (Richards and Vithayathil, 1959). This derivative contains two peptide chains that remain noncovalently bound and is fully active. However, the peptide portion 1 to 20, known as the S-peptide, may be separated from the larger portion, residues 21 to 124, known as the S-protein. Since the S-peptide contains a histidine residue at position 12, this allows a selective exchange of this group followed by recombination to give the noncovalent complex with S-protein, designated RNase S′ (Figure 26). This was the initial approach taken by Meadows *et al.* (1968) to the assignment of the resonance of the active-site histidine residue 12. Although their assignment of His-12 to resonance H2 was accepted for several years, it was

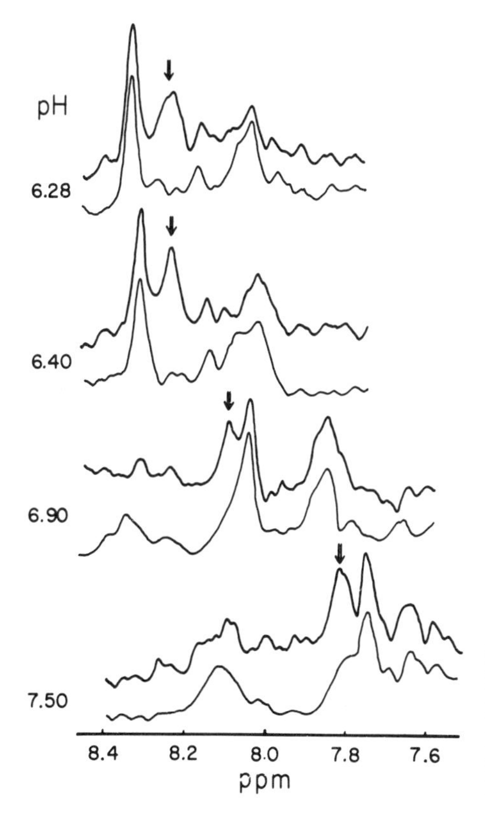

Figure 26. Proton NMR spectra at 220 MHz of RNase S′ (top of each pair) and [ε-^{2}H]His-12 RNase S′ (bottom). The resonance (H3) that is missing in each case is assigned to His-12. (With permission from Shindo *et al.*, 1976.)

subsequently found to be incorrect (Patel *et al.*, 1975; Shindo *et al.*, 1976; Bradbury and Teh, 1975). One major reason for this error was the initial finding that the NMR titration curves of RNase A were different from those of RNase S, implying different conformations of the histidines in the active site (Meadows *et al.*, 1968). However, the NMR titration curves of both active-site histidine residues were found to be essentially identical in both RNases A and S, and the original difference has been attributed to the selective effects of different concentrations of inorganic phosphate P_i in the original samples (Cohen *et al.*, 1973), and/or the selective effect of bound EDTA on one of the titration curves (Brauer and Benz, 1978).

From a comparison of the rates of deuterium exchange of the resonances of RNase with those of tritium exchange, it was concluded that resonance H3 should in fact be assigned to His-12 (Markley, 1975b). However, this comparison of evaluations carried out under different experimental conditions was not entirely satisfactory, particularly since a consideration of the deuterium exchange rates had led others to support the original assignment (Bradbury and Chapman, 1972). A direct comparison of the tritium and deuterium exchange rates of the histidine residues of RNase S under the same experimental conditions (Figure 27) enabled all four resonances to be definitively assigned:

H1 = His-105; H2 = His-119; H3 = His-12; H4 = His-48

The use of RNase S rather than RNase A for the alkaline exchange conditions was advantageous in these experiments since at alkaline pH all four resonances of RNase S are resolved, whereas they are not for RNase A. This allowed the deuterium exchange for RNase S to be carried out in a sealed tube (Shindo *et al.*, 1976), whereas the corresponding experiments for RNase A required an intermittent readjustment of pH in order to resolve the resonance to ascertain the degree of exchange for each resonance as a function of time (Bradbury and Chapman, 1972; Markley, 1975b).

The rate of deuterium and tritium exchange observed for these resonances as assigned also corresponds to the degree of accessibility indicated by the X-ray crystal structure (Richards *et al.*, 1970), namely His-105 > His-119 > His-12 > His-48, and with the ease of alkylation of the active-site histidines, namely His-119 > His-12 (Crestfield *et al.*, 1963). The assignment was confirmed by the proton NMR titration of [ε-^{13}C]His-12 RNase S′ complex, in which only the resonance H3 was not observed, thus proving its identity with His-12 (Niu *et al.*, 1979).

5.1.2. Properties of Histidine Residues of RNase. There have been many descriptions of the NMR titration curves of the histidine residues of RNase (Bradbury and Scheraga, 1966; Meadows *et al.*, 1967; Ruterjans and Witzel, 1969; Meadows *et al.*, 1968; Cohen *et al.*, 1970a; King and Bradbury, 1971; Schechter *et al.*, 1972; Cohen *et al.*, 1973; Haffner and Wang, 1973; Migchelson and Benteima, 1973; Markley and Finkenstadt, 1975; Cohen and Shindo,

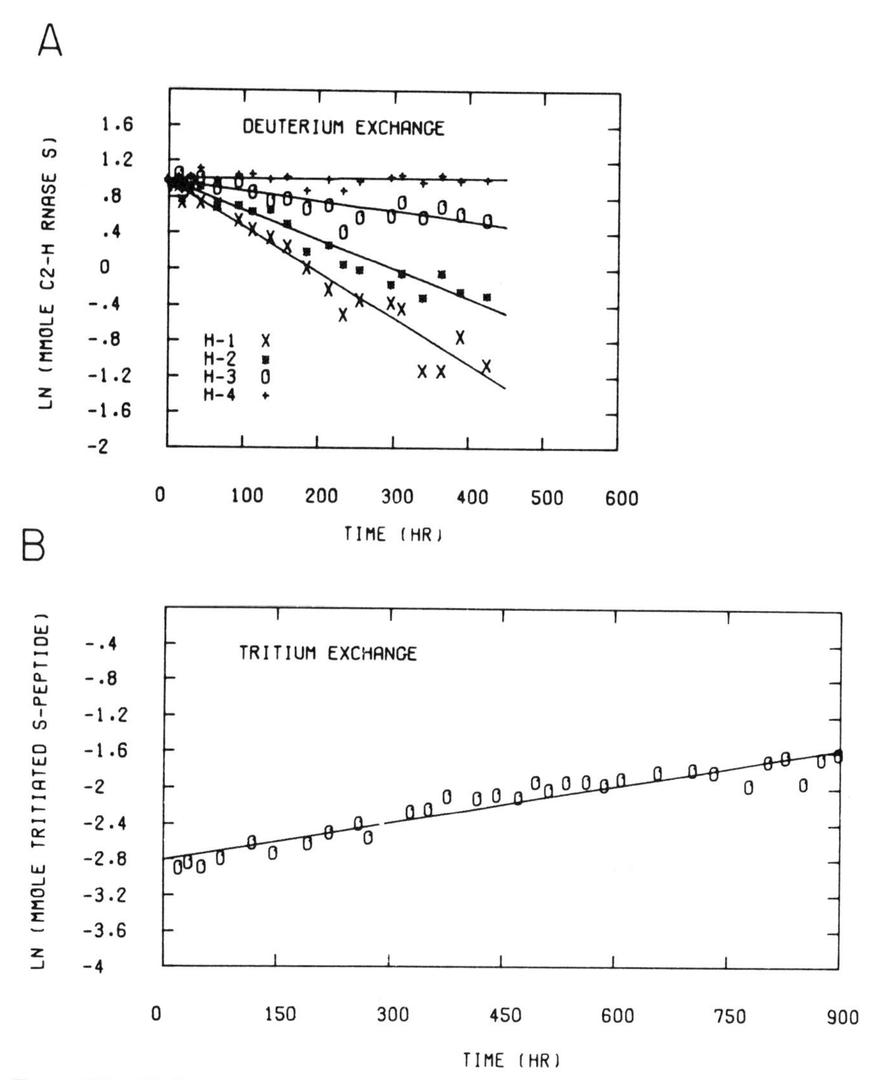

Figure 27. (*A*) Concentration of protons as a function of time in the four C_ε sites of RNase S, derived from intensity measurements of NMR spectra of the four resolved signals at pH 8.8 and 30°C in a sealed sample tube. (*B*) Concentration of tritium incorporated as a function of time into the C_ε site of His-12 of the S-peptide of RNase S under the same conditions, but counted following precipitation of the S-protein. The *slope* corresponds to that of resonance H3 in (*A*). (With permission from Shindo *et al.*, 1976.)

1975). Several reported differences in the properties of these curves lead to confusion, and were partly due to the use of different buffers, salts, and concentrations, as well as to the use of the derivative RNase S. A systematic comparison was made of the titration curves for both RNases A and S with different ions present (Cohen and Shindo, 1975), and the following conclusions were drawn:

1. Histidine residue 105 gives a normal titration curve for both RNase analogs with $pK \approx 6.8$ and $\Delta \approx 1.0$ ppm and is essentially unperturbed by salts, apart from nonspecific bulk effects.
2. Histidine residue 48 gives a broad resonance and cannot be observed for RNase A in chloride solution, but is observed for RNase A in acetate and for RNase S under most conditions. In RNase A in chloride, the broadening of the His-48 resonance may result from slow exchange of His-48 between two environments that are averaged under other conditions. The titration curve of His-48 also shows unusual chemical-shift effects, being generally shifted upfield at low pH and with a downfield titration shift as pH is raised to ~pH 5 (Figure 28).
3. The active-site residues His-12 and His-119 also both show minor inflections at low pH; this is more apparent in the curve of His-12 (Figure 28). These inflections become much more prominent features of the curves on increasing the P_i concentration (Figure 29). Since P_i is a competitive inhibitor, it binds in the active site, and thus the phosphate negative

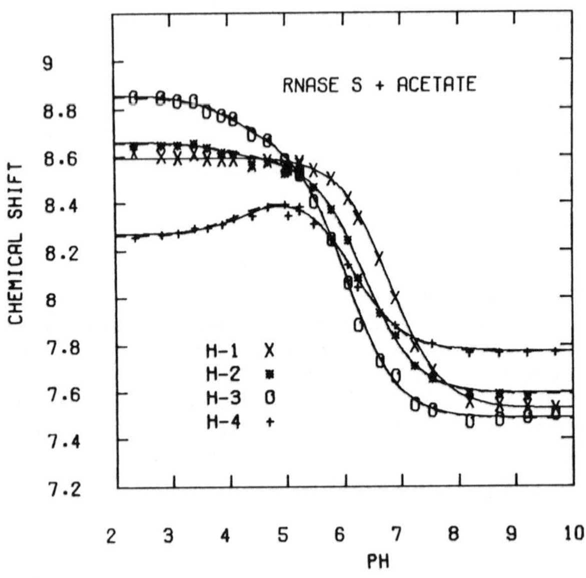

Figure 28. NMR titration curves at 220 MHz of the four C_ε protons of the four His residues of RNase S in deuteroacetate buffer (0.2 *M*). Assignments as in text. (With permission from Shindo *et al.*, 1975.)

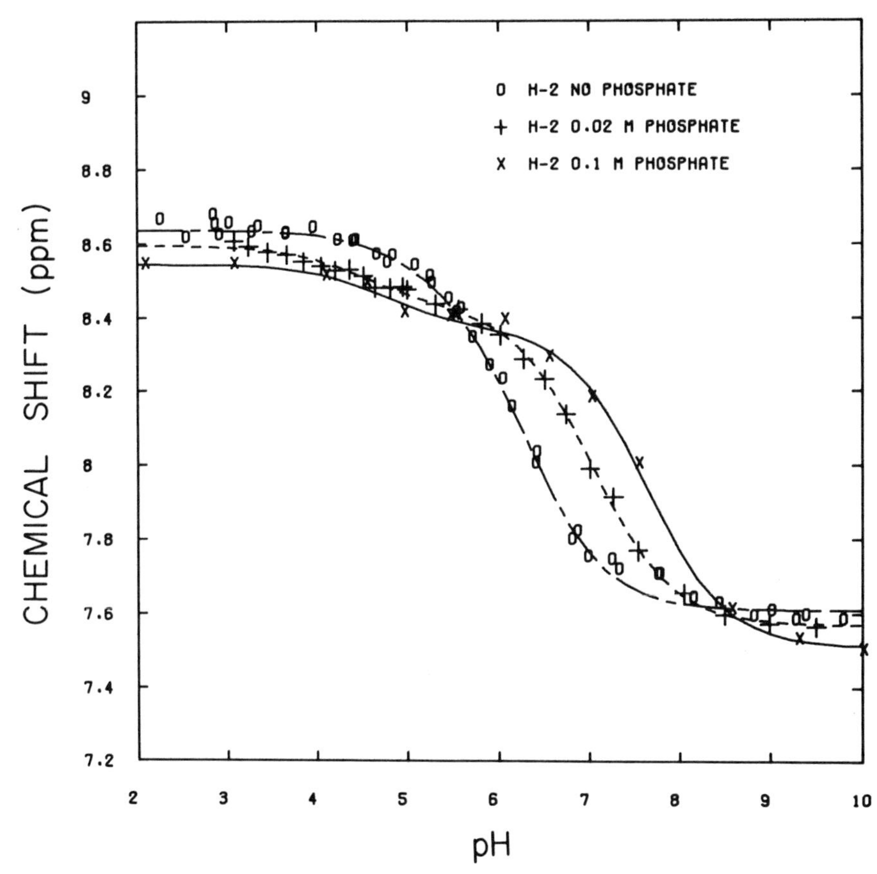

Figure 29. NMR titration curves of the C_ε proton of His-119 of RNase A as a function of increasing P_i concentrations. A similar result is observed for the curve of His-12. The increase in the pK_a of the imidazole group emphasizes the presence of the carboxyl-group inflection in the curves. The lines are fits obtained using Equation (10) (i = 2). (With permission from Cohen *et al.*, 1973.)

charges cause shift effects on the active-site imidazole protons. However, the pH dependence of these shift effects leads one to conclude that the inflections are not due to the titration of the P_i itself, but rather to increases in the pK values of the imidazole groups, whereas the minor low-pH inflection remains essentially constant at p$K \simeq 4.3$ (Cohen and Shindo, 1975).

A detailed understanding of these effects can only be arrived at by fitting these titration curves with various models. Two main types of models were considered as the origin of the asymmetrical titration curves of the active-site histidines: (1) mutual interaction between the two imidazole groups in the active site, and (2) interaction of one or more carboxyl groups with the

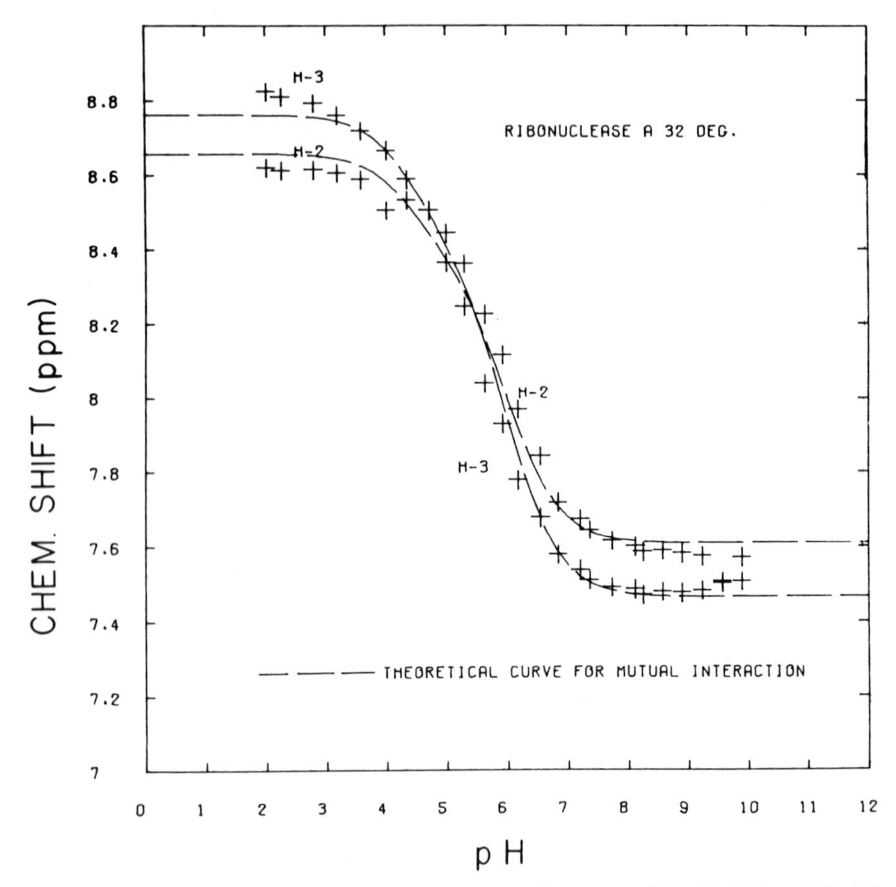

Figure 30. Data for the NMR titration curves of the C_ε proton of His-12 (H3) and His-119 (H2) of RNase at 32°C fitted to a model assuming mutual His–His interaction. The dashed line is the theoretical curve for the best fit.

imidazole groups, giving rise to the low-pH inflections. This analysis has more than academic interest since it provides a means to distinguish between two mechanisms proposed for RNase catalysis, namely that of Witzel (1963), in which the two imidazoles act as a single unit (model 1), and the push-pull type of mechanism of Findlay *et al.* (1962), in which the imidazoles act separately (equivalent to model 2). Application of a mathematical model corresponding to model 1 showed that it was inadequate to represent the data (Schecter *et al.*, 1972). A further example of such a curve fit, not previously published (Figure 30), confirms this conclusion. By contrast, a model corresponding to model 2, in which each curve was fitted assuming interaction with a separate carboxyl group (Figure 31), gave an adequate description of the data (Schecter *et al.* 1972). A more extensive analysis of many data sets of RNases A and S in several different salts indicated that the fit to the data was as good or better assuming a single

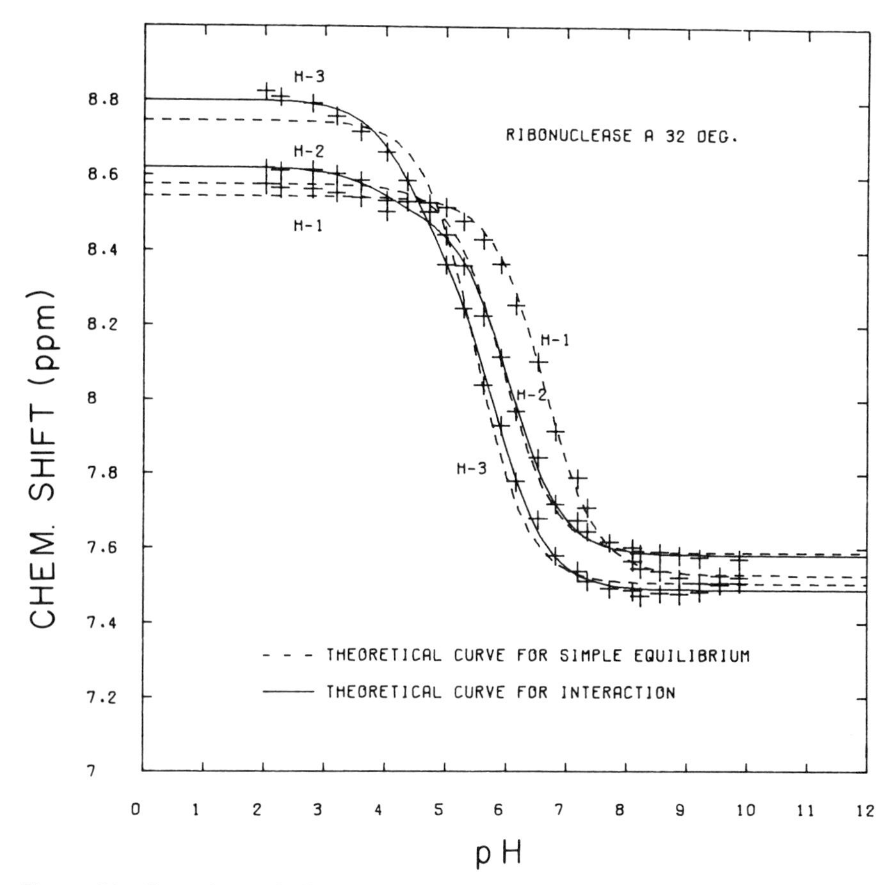

Figure 31. Same data as in Figure 30, including the data for His-105 (H1), but each curve fitted assuming a single transition (short dashed lines), or for His-12 and His-119 assuming two transitions in each case (solid lines). The superior fits in the latter case (compare with Figure 30) are consistent with previously reported results. (Schechter *et al.*, 1972.)

carboxyl group interacted with both (or all three) inflected curves (Cohen and Shindo, 1975). This was taken to indicate a pH-dependent conformational change due to a carboxyl-group titration. Markley and Finkenstadt (1975) attempted to show that the two active-site imidazole residues were mutually interacting *and* that the low-pH inflection arose from an (unidentified) carboxyl group. Of course, the more models one combines, the easier it is to fit the data. Subsequently it was found that bound EDTA affects only the titration of His-12 and not that of His-119 (Brauer and Benz, 1978), tending to confirm the independence of their titrations. A recent reinvestigation of the X-ray crystal structure of RNase (Wlodawer *et al.*, 1982), as well as a neutron diffraction study (Wlodawer and Sjölin, 1981), confirmed that the two active-site His residues are too far apart to mutually interact (Cohen and Wlodawer, 1982).

Although the group giving rise to the carboxyl inflections in the imidazole titration curves was initially presumed to be Asp-14 (Cohen and Shindo, 1975; Santoro *et al.*, 1979; Lenstra *et al.*, 1979), this was subsequently disproved by the use of [γ-^{13}C]Asp-14 (1–15) RNase peptide recombined in RNase S′ (Niu *et al.*, 1979), since the unambiguously assigned ^{13}C NMR titration curve of Asp-14 shows no inflection with a pK value near that expected (4.3), either in the complex or upon addition of inhibitors (Cohen *et al.*, 1980) (Figure 32). Further speculation on the origin of the carboxyl group giving rise to this conformational change has focused on Asp-121 (Cohen *et al.*, 1980). Indeed it has been found that Asp-121 does form a strong hydrogen bond with His-119 (Wlodawer and Sjölin, 1981; Wlodawer *et al.*, 1982).

Studies of binding of mononucleotides, particularly cytidine nucleotides, indicated large and selective effects on the active-site histidine titration curves (Meadows *et al.*, 1969). Subsequent work showed that uridine nucleotides with the phosphate in the same sugar position give very different results (Griffin *et al.*, 1973). In addition, it was shown that a dinucleoside phosphonate analog (UpcA) of the usual RNase substrate gives no significant shift or pK effects on the active-site histidine titration curves (Griffin *et al.*, 1973) or on the ^{13}C titration curve of [ε-^{13}C]His-12 RNase S′ (Cohen *et al.*, 1980). These differences were interpreted to indicate that the effects of the doubly ionized phosphate moiety in the mononucleotides is not an adequate basis on which to make conclusions regarding the mechanism of action of RNase. Indeed the phosphate negative charge in the dinucleoside analog does not appear to directly contact either of the active-site imidazole groups, except in the pentacoordinate transition state or intermediate, in an on-line mechanism of action consistent with the original push-pull mechanism (Griffin *et al.*, 1973) (Figure 33). A possible sideways-attack type of mechanism requiring pseudorotation of the pentacoordinate phosphorus

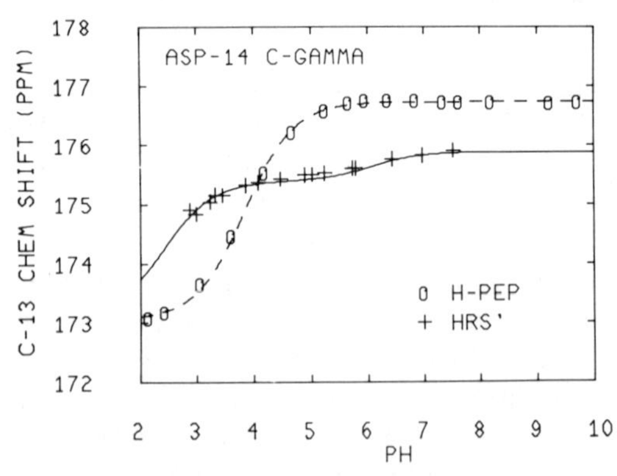

Figure 32. ^{13}C NMR titration curves of [γ-^{13}C]Asp-14 in the RNase (1–15) peptide (○) and the RNase S′ complex (+) formed from it. The pK_a values are free peptide, 3.8; complex, 2.4 and 6.1. (With permission from Niu *et al.*, 1979.)

intermediate (Usher, 1969) was disproved utilizing optically active phosphorothioates (Usher *et al.*, 1970). The recent X-ray diffraction study (Wlodawer *et al.*, 1982) confirmed the interaction between P_i and the two His residues in the active site, and an earlier X-ray study of RNase A in the presence of the inhibitor $U_{2'}p_{5'}A$ showed no strong phosphate–histidine interactions (Wodak *et al.*, 1977). Some attention has been given to the possible role of the acid conformational change in the mechanism of action of RNase. The acid inflection has been found to be present in the NMR titration curves of the active-site histidines in the presence of several competitive inhibitors of RNase action (see above). Thus far no specific role has been identified for this phenomenon (Walker *et al.*, 1976). Recently, a detailed study of the binding of several substances to RNase and a covalently linked 6-chloropurine derivative was reported (Arus *et al.*, 1981). The presence of an acid inflection was confirmed in these studies, and further detailed mapping of the subsites of the RNase active site has resulted.

5.1.3. Histidine Residues in Some Other Proteins. Of the many examples of NMR studies of histidine residues in proteins (Table I) we choose a few illustrative examples.

Human carbonic anhydrase (HCA) is a blood protein (MW $\simeq$ 30,000) that catalyzes the reversible hydration of carbon dioxide. It exists as two main isozymes, HCA-B and HCA-C. The first ^1H NMR study at 100 MHz revealed four titrating imidazole C_ε protons for HCA-B (King and Roberts, 1971). In a study at 220 MHz of both HCA-B and HCA-C, four resonances were resolved for the former and seven for the latter protein (Figure 34). Campbell *et al.* (1974,1975b) confirmed and extended these observations at 270 MHz, substituted Co^{2+} for the Zn^{2+} in the native enzyme, and gave probable assignments of most of the His resonances. Pesando (1975) observed several other resonances at 100 MHz for HCA-B and HCA-C that he attributed to histidine residues; two resonances he assigned to an active-site His with a p*K* similar to that observed for the group controlling enzyme activity. In the other NMR studies cited, it had been concluded that this group was not a His residue on the basis of the resonances observed and the effects of inhibitors. Although this is a more generally held view of carbonic anhydrase mechanism, namely that the pK corresponds to the ionization of a water molecule, Gupta and Pesando (1975) have also reported a titrating NH resonance that they attributed to the active-site histidine.

Several NMR studies of titrating histidine resonances in serine proteases (Hunkapillar *et al.*, 1973; Robillard and Shulman, 1974; Markley and Porubcan, 1976) were interpreted to support the presence of a "charge-relay system" in such enzymes, involving an Asp-His-Ser catalytic triad based on X-ray crystallographic studies (Blow *et al.*, 1969). Subsequently, the validity of this molecular arrangement was questioned in the light of further X-ray (Matthews *et al.*, 1977) and proton NMR studies (Markley and Ibanez, 1978). However, although the biosynthetically ^{13}C-labeled single

A. RNase–UpA Complex

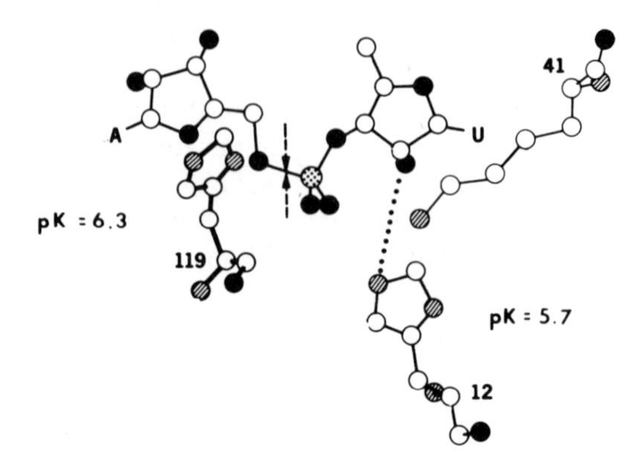

B. Cleavage Step

Penta-coordinate Phosphorus Transition State

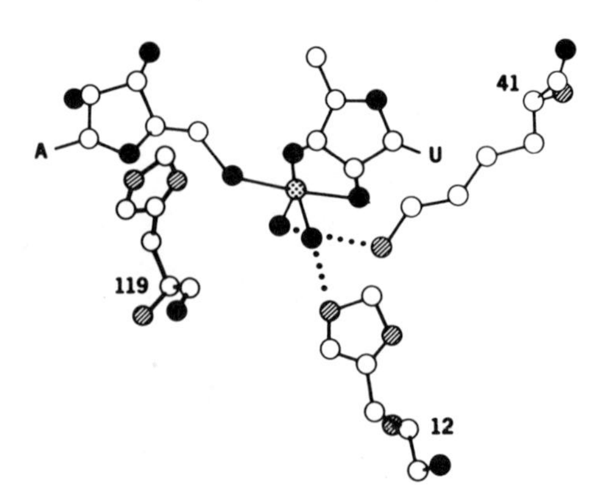

C. Products

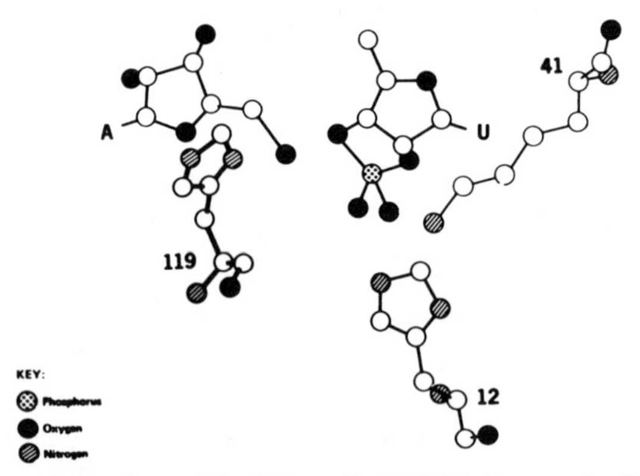

Figure 33. (*A*) Proposed structure of the RNase–UpA Michaelis complex based on the X-ray crystallographic and NMR data from the RNase–UpcA complex. The side chain of

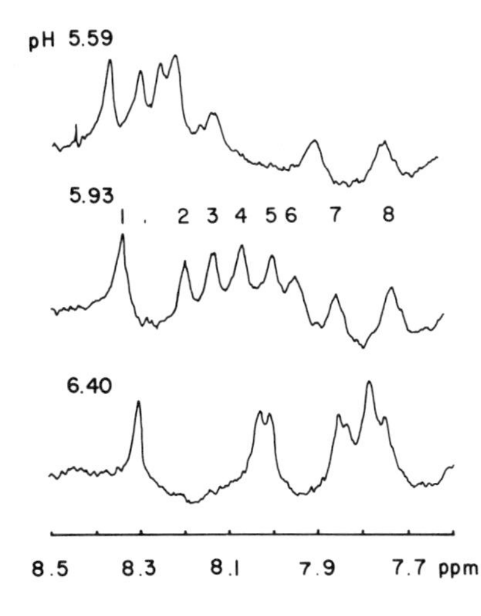

Figure 34. The 220-MHz NMR spectra of HCA-C at several pH values, showing seven titrating His $C_{\varepsilon}H$ resonances. [Reprinted with permission from Cohen *et al.* (1972). Copyright 1972, American Chemical Society.]

histidine residue of α-lytic protease shows a normal titration curve with p$K \simeq 7$ (Hunkapillar *et al.*, 1973), these workers interpreted this pK to arise from the Asp residue in the "charge relay," whereas they considered the His residue to have a much lower pK value, partly on the basis of ^{13}C–^{1}H coupling data that was considered unreliable (Egan *et al.*, 1977). This controversy was resolved by the use of [^{15}N]His (Bachovchin and Roberts, 1978) and a reinvestigation of [^{13}C]His (Bachovchin *et al.*, 1981) in the α-lytic protease system. The results clearly indicated a normal titration of the active-site histidine. Thus at this point, evidence is against the existence of a "charge-relay system" in serine proteases.

Studies of titrating histidine resonances in metmyoglobins from several species led to the conclusion that five His residues in each case do not give observable signals (Hayes *et al.*, 1975; Botelho and Gurd, 1978). One of these resonances could be observed in the deoxymyoglobins, and an additional one was observed in the oxy and azide forms (Ohms *et al.*, 1979). These "extra" titrating resonances were assigned to the near-heme residues His-64 and His-97. The resonance of the proximal His-93, which is bonded to the iron atom, is not observed in the usual region of the spectrum but is

---

His-12 and the ribose portion of UpA are roughly in the plane of the page, whereas Lys-41 is behind the plane of the page. The side chain of His-119 is above the plane of the page with its side chain projecting upward. The dotted line indicates that His-12 may act as a base to abstract a proton from the 2′-hydroxyl group. (See text for further discussion.) The pK_a values of His-12 and His-119 are shown adjacent to them. (*B*) Proposed structure of the transition state for cleavage of UpA by RNase in which the phosphorus atom is pentacoordinate. The location of the phosphorus atom moves so that oxygen atoms are brought into contact with both His-12 and Lys-41 side chains. (*C*) Proposed structure of the RNase–product complex following cleavage of UpA. (Water molecules and hydrogen atoms are omitted from this figure for clarity.) [Reprinted by permission of the publisher from Cohen *et al.* (1980). Liu, Mamiya, and Yasunoba, eds., *Frontiers in Protein Chemistry*, p. 1. Copyright 1980 by Elsevier Science Publishing Co., Inc.]

shifted downfield (La Mar *et al.*, 1977). It is surprising that Wilbur and Allerhand (1977a) in their ^{13}C NMR study of the histidine resonances of myoglobin did not detect the high pK resonances observed in the ^{1}H NMR titration studies.

By comparison of the ^{1}H spectra at 250 MHz of human hemoglobin with its des-His-β146 derivative as a function of pH, it was possible to assign the resonance of this single (out of a total of 19) His residue (Figure 35) and to measure its pK_a value (Kilmartin *et al.*, 1973). Although these results were considered to support the role of this carboxy-terminal His-β146 in the alkaline Bohr effect, subsequent work indicated that this residue is in fact only a minor contributor to the phenomenon (Russu *et al.*, 1980).

Mildvan and his associates recently published studies of the resolved histidine resonances of several high-molecular-weight proteins, namely yeast aldolase (MW $\simeq$ 80,000; Smith and Mildvan, 1981), rabbit muscle creatine kinase (MW 82,000; Rosevear *et al.*, 1981) and rabbit muscle pyruvate kinase (MW $\simeq$ 237,000; Meshitsuka *et al.*, 1981). Although the proton spectra of these proteins at 250 MHz are mainly unresolved envelopes, the local mobility of specific histidine residues is clearly sufficient to allow narrow enough linewidths, and hence resolution of some resonances (Figure 36). Their identification as imidazole C2 protons is based on the characteristic titration properties. These studies are good examples of how the resolution of a few key resonances on the periphery of a largely unresolved NMR spectrum can be used to provide valuable information on the conformation and function of large globular proteins.

5.1.4. Thermodynamic Parameters of Histidine Residues from NMR Titrations. Since it is possible to measure ionization constants for individual histidine residues in proteins, it is possible to obtain enthalpy and entropy of ionization by measuring the pK_a values at different temperatures. Roberts *et al.* (1968) exploited this possibility by measuring the temperature dependence of the pK_a values of the four histidine resonances of ribonuclease A. However, they assumed all the curves were symmetrical, thus obtaining inaccurate pK_a values in several cases, and they used the earlier incorrect assignments (Meadows *et al.*, 1968). Westmoreland *et al.* (1975) reinvestigated this system and were able to determine ΔH and ΔS values. From these values it was possible to conclude that whereas His-105 and His-119 are reasonably exposed to solvent (ΔS = -9 and -6, respectively), His = 12 is in a partially buried environment (ΔS = -18), which is consistent with the X-ray structure and chemical-modification studies (Heinrickson *et al.*, 1965).

5.2. Aromatic Side Chains

Aromatic side chains play significant roles in proteins; research interest in them continues at a high level as a result of (1) their hydrophobic effects, (2) the pronounced influence of their bulkiness on protein dynamics (Section

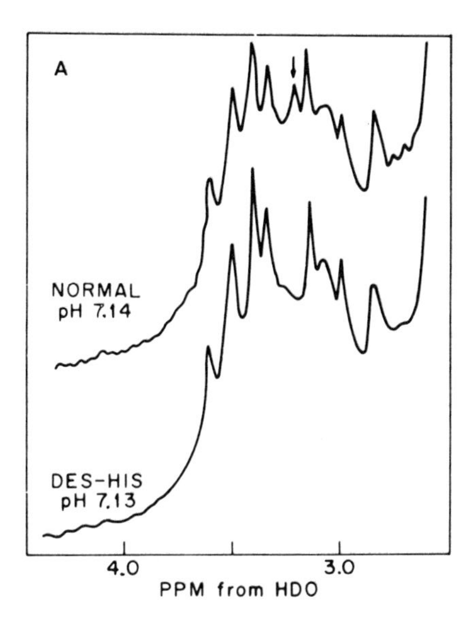

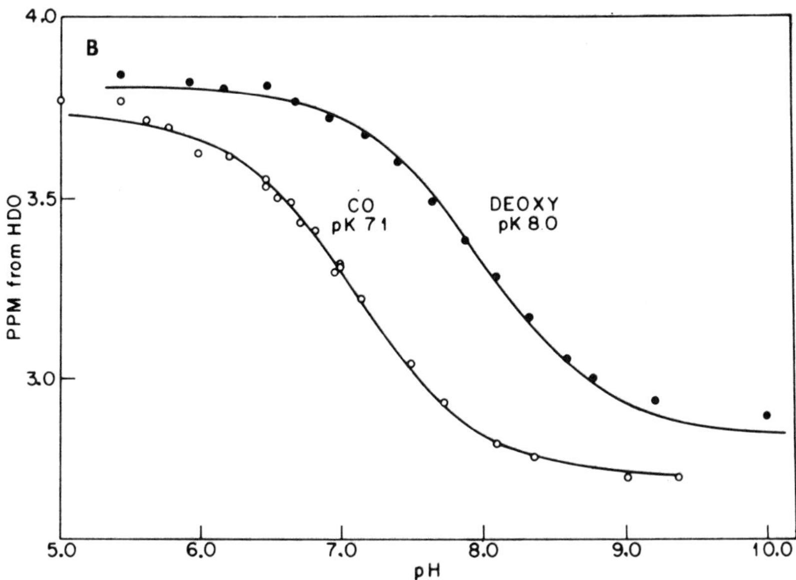

Figure 35. (*A*) A 250-MHz correlation NMR spectrum of carbonmonoxyhemoglobin (upper) and des-His-β146 hemoglobin in P_i buffer. The arrow indicates the position of the C_ε proton resonance of His-β146. (*B*) Proton NMR titration curves of His-β146 C_ε proton for carbonmonoxyhemoglobin (open) and deoxyhemoglobin (closed). (From Kilmartin *et al.*, 1973.)

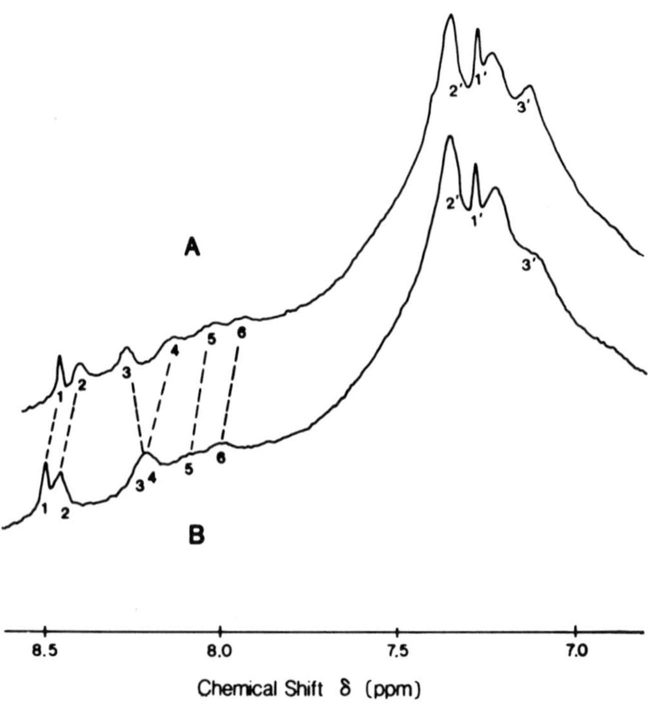

Figure 36. Titration of six C_ε proton resonances in the ^1H NMR spectrum of rabbit muscle pyruvate kinase at 250 MHz. (*A*) Without PEP at pH 5.8, (*B*) With 2 m*M* PEP at pH 5.7. (With permission from Meshitsuka *et al.*, 1981.)

3.2), (3) their ring current interactions, which mediate chemical shifts, and occasionally, (4) their direct participation in enzymatic reactions as active-site residues.

In the earlier period of ^1H NMR studies of proteins, the resolution in the aromatic region was generally insufficient to enable clear-cut resonance assignments (Cohen and Jardetzky, 1968). Consequently attempts were made to simplify the ^1H spectra using ^2H substitution (Putter *et al.*, 1969). When a microorganism requires specific amino acids for growth, this approach is feasible, but it has not been widely used. Initially the use of higher magnetic-field strengths gave superior resolution, particularly when resolution-enhancement methods were applied (Campbell *et al.*, 1973). Also ^{13}C NMR, with much superior resolution in the aromatic region of the spectrum (Figure 1), has become a major direction. The aromatic resonances (and arginine C_ζ, which lies with them) are found in the region of ~105 to 160 ppm downfield from tetramethylsilane (TMS) in ^{13}C NMR spectra (Figure 37). The most upfield peaks correspond to Trp C_γ, and the most downfield resonances originate from Arg C_ζ and Tyr C_ζ (Oldfield *et al.*, 1975). The chemical-shift range for the aromatic region is conveniently and graphically summarized using values obtained from diamagnetic proteins and small peptides by Allerhand (1979).

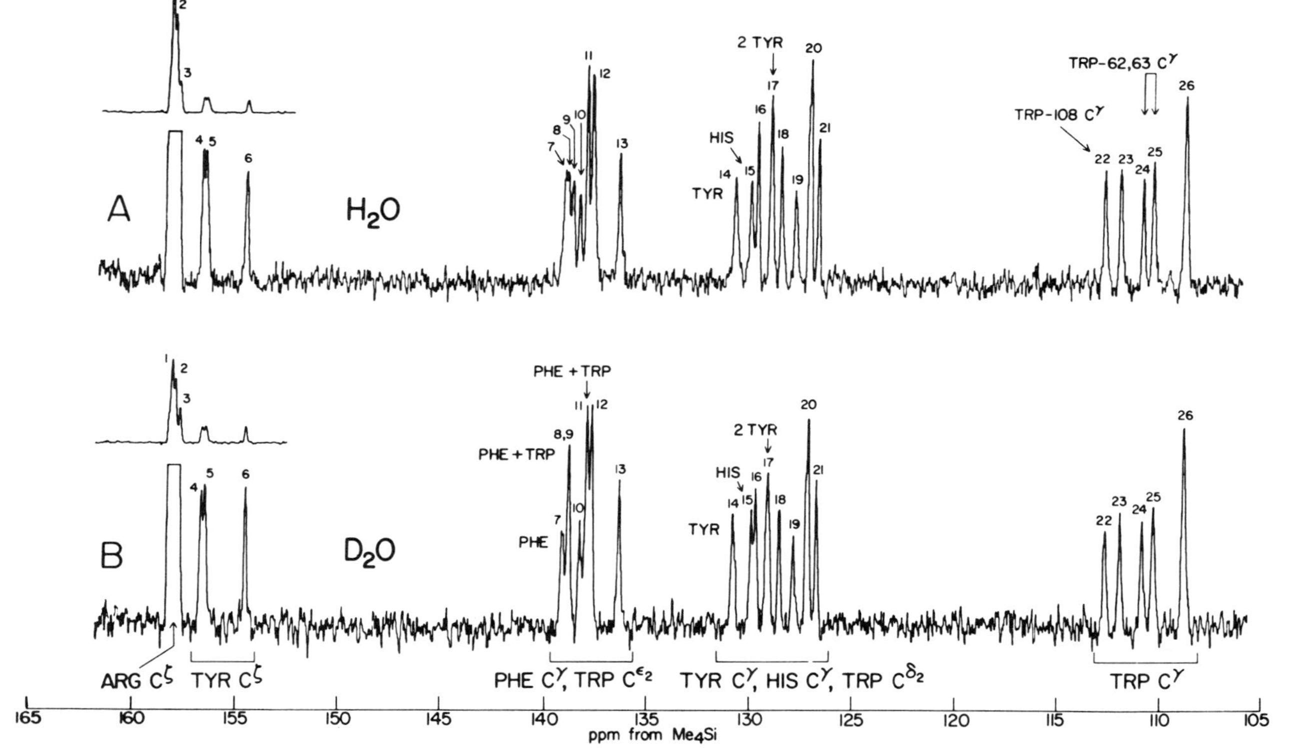

Figure 37. Regions of aromatic carbons and C_ζ of arginine residues in the convolution-difference natural abundance ^{13}C Fourier transform NMR spectra of hen egg white lysozyme. Each spectrum was recorded at 15.18 MHz under conditions of noise-modulated off-resonance proton decoupling, using 8192 time-domain addresses, a spectral width of 3787.9 Hz, 49, 152 accumulations, a recycle time of 2.205 s (30 h total time). The convolution-difference procedure was used with $\tau_1 = 0.72$ s, $\tau_2 = 0.036$ s, and $K = 1.0$. The insets (peaks 1 to 6) are shown with one-eighth the vertical gain of the main spectrum. (*A*) Spectrum of 14.6 m*M* protein in H_2O, pH 3.05, 0.1 *M* NaCl, 44°C. (*B*) Spectrum of 13.8 m*M* protein in D_2O, pH-meter reading 3.08, 0.1 *M* NaCl 42°C. (With permission from Oldfield *et al.,* 1975.)

Attempts have been made to calculate chemical-shift effects due to ring-current phenomena (Sternlicht and Wilson, 1967; Gettins and Dwek, 1977). The chemical-shift perturbations resulting from close proximity to aromatic residues and heme moieties frequently offer a definitive way to discriminate between peak assignments. Many examples are readily found in the literature, and several typical cases will be cited under specific residues. The order of magnitude of such shifts and the direction of the shift are dependent upon ring orientation (see Section 3.1).

Aside from correlation of the individual residues with specific resonances and ring-current shifts, much of the current theoretical and experimental interest in aromatic residues has centered on their dynamics in proteins. Of particular interest is the hindered rotation of aromatic side chains and the 180° ring flips that are observed for phenylalanine and tyrosine rings (Section 3.2). Gurd and Rothgeb (1979) have also recently reviewed the significance and methods of measuring motions in proteins and the dynamics of aromatic residues.

A considerable body of interesting quantitative work has centered around the relatively small protein BPTI (MW 6500, 58 amino acids) since it has sharp resonance signals, high thermal stability, a known crystal structure, and interesting derivatives. It contains eight aromatic residues, four phenylalanines, and four tyrosines; Wüthrich *et al.* (1980) have analyzed the cumulative NMR data for BPTI in terms of hydrophobic domains.

McCammon and Karplus (1980) have addressed the problem of activation of protein processes. Specifically, they have computer simulated the ring rotation of Tyr-35 in BPTI. They conclude that "small, transient packing defects" are involved in the ring rotations, and stress that we do not yet recognize the origin(s) of the free energy for activation of its entropic component. For additional authoritative considerations of motional studies of aromatic residues see London (1980) and Lipari and Szabo (1982).

The NMR studies of aromatic residues are summarized in Table II. In general it is necessary to resort to several stratagems to make peak assignments within a given amino acid type. Procedures that have been helpful include comparative studies of homologous proteins, pH changes, T_1 measurements, decoupling techniques, use of shift reagents, changes of solvent and/or temperature, and, where appropriate, changes in metal oxidation state.

5.2.1. Tyrosine. The hydroxyl group confers upon tyrosine reaction and hydrogen-bonding capabilities not possessed by its aromatic, hydrophobic counterpart phenylalanine. The T_1 measurements also suggest that the phenolic-hydroxyl interactions can restrict ring motions (London, 1980). Undiminished interest and debate have persisted over many years concerning the titratability of the acidic hydroxyl groups, with nontitrating residues or residues with unusual titration curves being considered "buried" or hydrogen bonded to various extents. In this instance, the greater effectiveness of ^{13}C NMR relative to ^{1}H NMR in resolving Tyr resonances was shown by the resolution and identification of the six $^{13}C_{\zeta}$ tyrosine resonances (25, 73, 76, 92, 97, and 115 at 67.9 MHz) of RNase A (Egan *et al.*, 1978) (Figure 10). By contrast, the ^{1}H NMR spectrum at 250 MHz showed insufficient resolution of individual resonances (Markley, 1975c), although four normally titrating Tyr resonances were resolved at 360 MHz (Lenstra *et al.*, 1979). In the ^{13}C NMR study, four of the tyrosines titrated normally ($pK = 9.5$ to 10), a fifth titrated with $pK > 10$ and was presumed to be partially buried, and the sixth did not titrate up to pH 11 (Figure 38) (Egan *et al.*, 1978). The latter was considered to be either partially ionized or hydrogen bonded because of its high chemical shift of ~159 ppm. Following the chemical-shift changes was difficult because of overlapping at some pHs of the four arginine C_{ζ} resonances with the tyrosine peaks. Subsequently, Santoro *et al.* (1979) supported the peak assignment for Tyr-25 by ^{13}C titration studies of RNase A, although they did not go above pH 8. There has been some disagreement regarding the assignment of an aromatic resonance in the ^{1}H NMR spectra of ribonuclease; it has been assigned both to Tyr-25 (Markley, 1975c) and to Phe-120 (Lenstra *et al.*, 1979).

Chemical modification in the case of tyrosine to obtain peak assignments is most appropriately carried out with selective nitration (Cohen *et al.*, 1971; Snyder *et al.*, 1975). Dobson et al. (1978) assigned the ^{1}H aromatic resonances of all three tyrosine residues in the spectra of hen egg (HE) lysozyme, with use of selective chemical modification (Figure 39).

An extensive current survey of the proton resonance assignments for the aromatic amino acids of eucaryotic cytochrome *c* is given by Moore and Williams (1980a), who used many strategies. They discuss in detail their progress on peak assignments of the four tyrosines [48, 67, 74, and 97] in horse cytochrome *c*, although only two assignments have thus far been made for both oxidized and reduced forms of the heme. Activation energies and rates of ring flipping were also calculated for these two residues in horse and tuna cytochromes (Moore and Williams, 1980d).

Single-carbon-atom resonances are most readily resolved in ^{13}C NMR spectra in the spectral region corresponding to the unprotonated carbons of aromatic and arginine side chains. These quaternary-carbon resonances provide the most straightforward approach to initial spectral assignments of tyrosine residues. Thus it was possible to assign specific single-atom resonances (15 MHz) for 22 well-defined peaks originating from 28 aromatic ^{13}C atoms contained in the one histidine, three tyrosine, six tryptophan, and three phenylalanine residues of hen egg white lysozyme (Allerhand, 1979). A useful generalization for tyrosine carbons is that C_{γ} appears upfield from C_{ζ}, and the phenolate and phenolic forms differ appreciably. Only Arg C_{ζ} normally overlaps with Tyr C_{ζ}, and these can be distinguished by proton decoupling and titration.

Titration studies of tyrosine residues in myoglobins by natural abundance ^{13}C NMR (15 MHz) allowed Wilbur and Allerhand (1976) to make specific

Table II. NMR studies of aromatic residues in proteins

Protein	Reference	Comment
Alkaline phosphatase	Hull and Sykes (1975); Sykes *et al.* (1974)	Tyr; [^{19}F]Phe, dynamics, 180° flips
Azurin (*P. aeruginosa*)	Ugurbil *et al.* (1977)	Trp-48; $^{13}C_{\gamma}$ assignment
Bovine pancreatic trypsin inhibitor	Wüthrich and Wagner (1975);	Tyr, Phe; C^{13}, 180° flips
	Maurer *et al.* (1974);	Tyr; ^{13}C, titration
	Snyder *et al.* (1975)	Tyr; ^{1}H, titrations, peak assignments
Collagen (α1-CB2)	Torchia and VanderHart (1976)	Phe; ^{13}C, triple-stranded helix
Cytochrome *c*	Moore and Williams (1980a);	Tyr, Trp; ^{1}H peak assignments
	Campbell *et al.* (1976);	Tyr; chemical-shift nonequivalence
	Oldfield *et al.* (1975);	Tyr, Trp; ^{13}C, ferro, ferri, and cyano forms
	Burns and La Mar (1979);	Phe; ^{13}C, slow dynamic equilibrium
	Boswell *et al.* (1980)	Phe-10 and Phe-18; assignment corrections
Dihydrofolate reductase	Feeney *et al.* (1977);	Tyr; 180° flips
	Kimber *et al.* (1977,1978);	Trp; ^{19}F, 180° flips
	Roberts *et al.* (1977);	Phe; ^{19}F
	London *et al.* (1979) and London (1980)	Trp; ^{13}C, slow exchange for two residues
Ferredoxins	Packer *et al.* (1972,1973);	Tyr-2 and Tyr-30; ^{13}C, distinguish Phe
	Packer *et al.* (1977)	Aromatic ^{1}H
Gene 5 protein (bacteriophage)	Coleman and Armitage (1978); Alma *et al.* (1981b)	Phe and Tyr
Hemoglobin (human fetal)	Oldfield *et al.* (1975)	Tyr-130; ^{13}C, substituted into β chain
IgA Myeloma protein (mouse)	Gettins and Dwek (1977)	Trp-93L; ^{13}C, hapten binding
Lac repressor binding fragment (*E. coli*, 56-residue headpiece)	Arndt *et al.* (1981)	Tyr; ^{13}C, Ser, Glu, Leu, Lys substitutions; protein–DNA interactions

Lysozyme (HE, human)	Campbell *et al.* (1975a);	Tyr, Phe; ^{13}C, 180° flips
	Oldfield *et al.* (1975);	Tyr, Trp, Phe; ^{13}C
	Allerhand *et al.* (1973);	Tyr, Trp; ^{13}C, nonprotonated resonances
	Allerhand *et al.* (1977);	
	Dobson (1977);	Trp-108, ^{13}C
	Blake *et al.* (1981);	Trp-62, ^{13}C, oxindolealanine derivative
	Norton and Allerhand (1977);	
	Glickson *et al.* (1971);	Trp-59; ^{1}H
	Cassels *et al.* (1978)	Trp; ^{1}H, C_{δ}-H, and NH assignments several species
Myoglobin	Wilbur and Allerhand (1976)	Tyr; ^{13}C, peak assignments several species
Nuclease (staphylococcal)	Cohen *et al.* (1971)	Tyr; ^{1}H, [3,5-$^{2}H_2$] and 3-nitrotyrosyl derivatives
Parvalbumin	Cave *et al.* (1976);	Tyr, Phe; ^{13}C, 180° flips
	Opella *et al.* (1974); Nelson *et al.* (1976)	Phe; ^{13}C, internal motions
Ribonuclease	Meadows *et al.* (1969);	Phe; ^{1}H, Phe-120, inhibitor binding
	Freedman *et al.* (1971);	Phe; ^{13}C, semisynthetic [^{13}C]Phe-8 (1–15)
	Markley (1975c);	Tyr; ^{1}H, transitions by pH and inhibitors
	Egan *et al.* (1978);	Tyr; $^{13}C_{\zeta}$, titration and peak assignments
	Santoro *et al.* (1979);	
	Chaiken *et al.* (1974)	Phe; ^{13}C, *p*-fluorophenylalanine T_1's
Rhodopsin (*Halobacterium halobium*)	Kinsey *et al.* (1981a,b);	Trp; $^{2}H_5$, solid state
	Markley and Jardetzky (1970)	Trp; $^{1}H_1$, inhibitor binding
Tropomyosin (rabbit)	Edwards and Sykes (1978)	Tyr; ^{1}H, dynamics

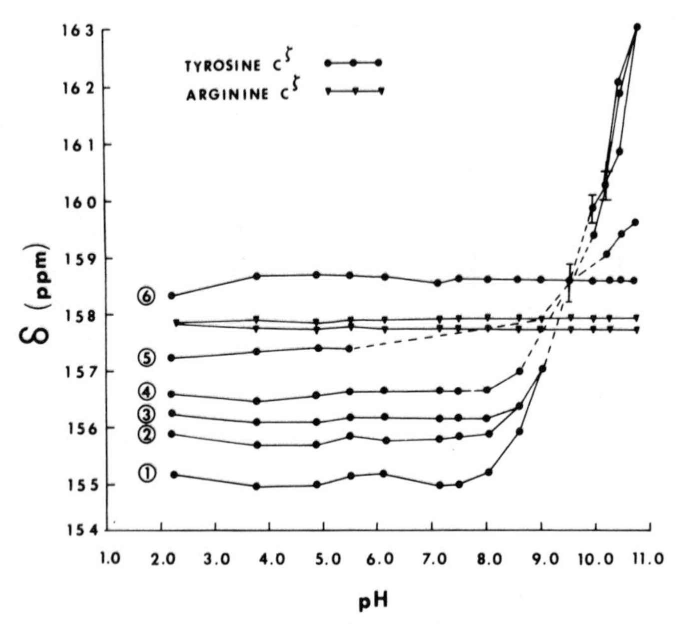

Figure 38. ^{13}C NMR titration curves of the six Tyr C_ζ (circles) and four Arg C_ζ (triangles) resonances of ribonuclease A. (With permission from Egan *et al.*, 1978.)

assignments for C_γ and C_ζ resonances for sperm whale (Tyr-103, Tyr-146, and Tyr-151), horse (Tyr-103 and Tyr-146), and red kangaroo (Tyr-146) and to make deductions on the degree of exposure to solvent. This was an informative demonstration of the value of homologous proteins for peak assignment.

Snyder *et al.* (1975) assigned all four tyrosine proton resonances in BPTI using selective nitration, decoupling, and pH titrations. Their assignments differ in only one pair from those of Wüthrich and Wagner (1975). The ortho and meta protons were examined for information on rate of ring flipping and environmental equivalence. Tyr-10, Tyr-21, and Tyr-23 showed magnetically equivalent environments for corresponding ortho and para pairs, whereas Tyr-35 with four separate resonances was found to be restricted, and the authors calculated an upper ring-rotation limit of 160 s^{-1} (Snyder *et al.*, 1975).

The structural flexibility of the fibrous protein tropomyosin (rabbit), which can form a coiled coil, has been studied by Edwards and Sykes (1978) using ^1H NMR (at 270 MHz, pH 8, high ionic strength). The monomer for the coil has two chains (284 residues each) and six tyrosines, of which five are in interfacial hydrophobic contacts between chains intermeshed in the coil. The NMR dynamic calculations suggest a flexible structure accessible to solvent.

Modifications of tyrosine residues with NMR isotopic markers have been carried out by several groups. Putter *et al.* (1969) used deuteration to

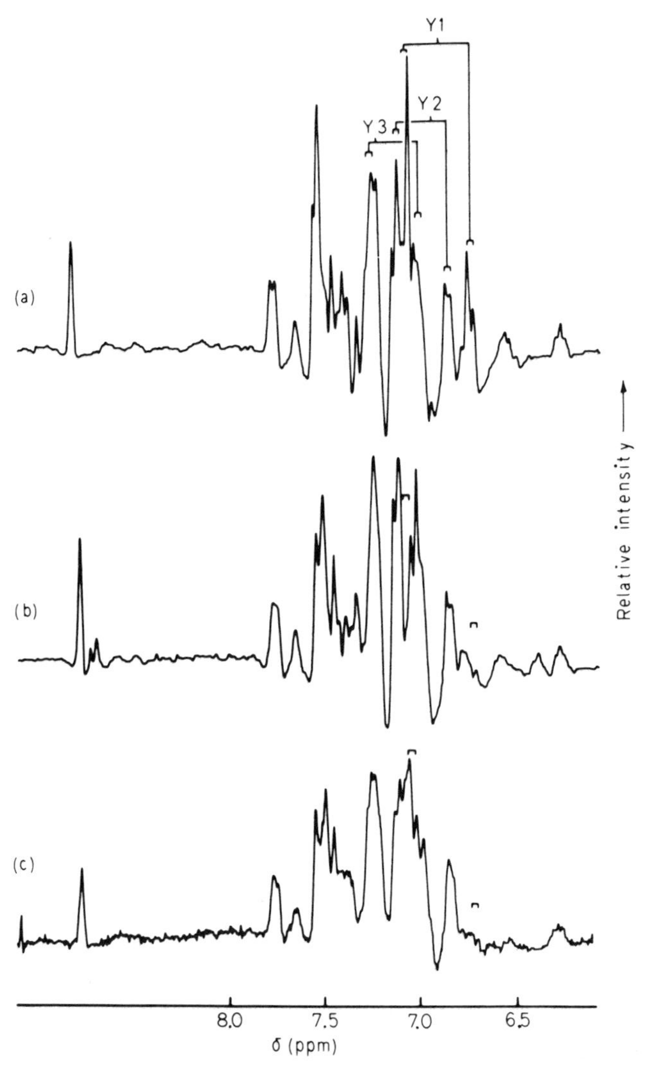

Figure 39. Convolution-difference spectra of lysozyme (5 m*M* in D_2O) at pH 5.3, 54°C. (*a*) Normal lysozyme; (*b*) modified at Tyr-23 by nitrobenzofuran; (*c*) modified at Tyr-23 by iodination. The disappearance of the resonances labeled Y1 on modification assign them to Tyr-23. (From Dobson *et al.*, 1978, with permission.)

simplify the aromatic proton spectrum (Figure 40). Staphylococcal nuclease contains seven tyrosine residues. Cohen *et al.* (1971) prepared [ε-$^2H_2$]tyrosyl nuclease and on reaction with tetranitromethane prepared the 3-nitrotyrosyl derivatives and determined preliminary peak assignments. These and other assignments were utilized by Markley and Jardetzky (1970) in studying inhibitor binding to staphylococcal nuclease.

An intriguing use of genetic substitution is the 360-MHz ^{1}H NMR spectral

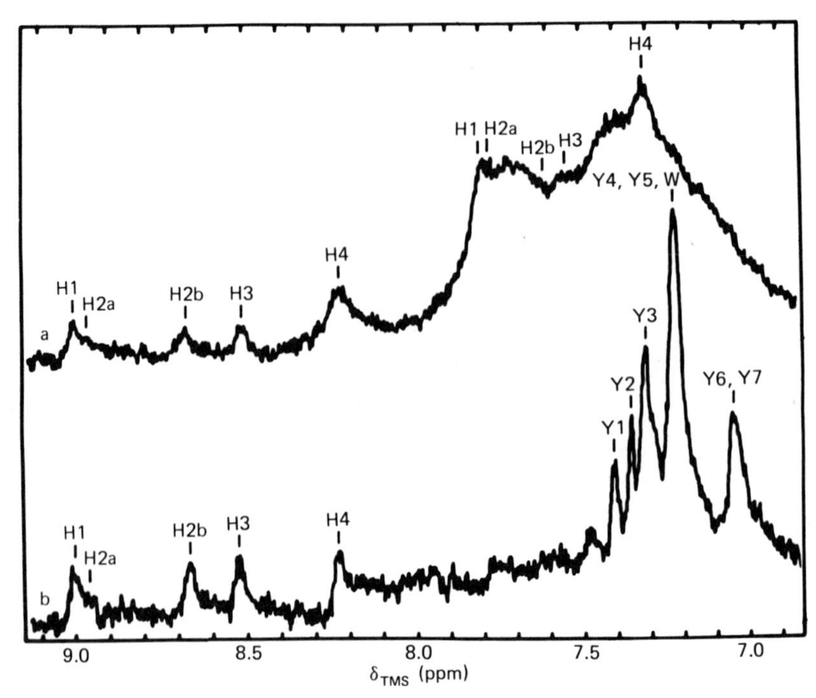

Figure 40. Comparison of the ^{1}H NMR spectra in the aromatic region at 100 MHz of (*a*) staphylococcal nuclease and (*b*) the [δ-^{2}H]Tyr analog grown biosynthetically. Both spectra recorded at pH 6.0 under the same conditions. Assignments are *H*, His resonances; *Y*, [ε-^{1}H]Tyr resonances; *W*, Trp resonance. (From Putter *et al.*, 1969.)

study of the N-terminal, 56–amino acid residue ("headpiece") of the *E. coli lac* repressor binding fragment by Arndt and co-workers (1981). In a continuation of their program to elucidate gene expression by examining protein–nucleic acid interactions, they have carried out individual substitutions of the four tyrosine residues (7, 12, 17, and 47) by various amino acids and used specific leucine-for-tyrosine replacements in conjunction with selective deuteration of the ε or γ tyrosine protons to make the complete aromatic-peak assignments in the NMR spectrum. The protein fragment lacks both phenylalanine and tryptophan and has only one histidine (position 29), so these residues do not complicate the peak assignments. Extensive NOE measurements were found useful in assigning relative distances; for example, His-29 was reported closest to Tyr-12. The significant assignments include correction of some previously assigned peaks (Tyr-12 and Tyr-17 should be reversed from the report by Ribeiro *et al.*, 1980). These two surface residues were deduced to be not only close to each other but possibly stacked. Tyr-47 appeared to be buried. Studies with Tyr-containing peptides implicate these residues in interactions with nucleic acids (Mayer *et al.*, 1979). In other work, Jarema *et al.* (1981) also used [^{19}F]tyrosine-substituted *lac* repressor and the *lac* repressor headpiece obtained from selective genetic mutants to make aromatic spectral assignments.

5.2.2. Phenylalanine. Much of what has been said about tyrosine residues in the previous section applies to phenylalanine, except of course the properties related to the hydroxyl group. Indeed the absence of the selective chemical-shift effects conferred by the hydroxyl group and of its hydrogen-bonding and titration properties makes resolution of multiple Phe resonances exceedingly difficult. A discussion of the 180° ring flipping about the C_β—C_α bonds of phenyl-type side chains was given in Section 3.2. The nonequivalence of ortho or meta protons has been exploited to establish whether or not there is restricted ring rotation (Campbell *et al.*, 1975a). The detection of slow conformational equilibria is summarized by London (1980). Typical $^{13}C_\gamma$ resonances for Phe should appear at ~135 to 140 ppm (TMS reference) with possible overlap with Trp $C_{\varepsilon 2}$. Details of specific examples of NMR spectral assignments for Phe residues are as follows.

Allerhand (1979) summarized and illustrated natural abundance ^{13}C spectra of the aromatic region of hen egg white lysozyme under a variety of experimental conditions, including two field strengths (15 and 68 MHz). All but quaternary-carbon resonances were eliminated by use of convolution-difference spectra. Hen egg lysozyme contains three Phe residues. Peaks corresponding to Phe have been identified, but distinguishing between two specific residues was not possible.

Using semisynthetic incorporation the complex of [^{13}C]Phe-8 RNase (1–15) and S-protein has been studied at 25 MHz (Freedman *et al.*, 1971). Subsequently (Chaiken *et al.*, 1973), this peptide and the *p*-fluorophenylalanine analog were used in binding studies with 2′-CMP; chemical shifts and T_1's were determined. The effects of complexation and incorporation in a helical segment were evident at the [^{13}C]Phe-8 site. The T_1's of the complex and inhibited complex were decreased (in that order) relative to the uncomplexed synthetic (1–15) peptide.

Boswell *et al.* (1980) corrected earlier proton resonance assignments of Phe-10 and Phe-82 in eucaryotic cytochrome *c*; their final assignments are based on a combination of double resonance, chemical modification, ring-current calculations, and NOE techniques.

5.2.3. Tryptophan. It was anticipated that tryptophan with its bulky, aromatic indole ring, which lacks the symmetry (i.e., no twofold rotation axis) found for phenylalanine and tyrosine side groups, would be one of the least mobile residues, and this was confirmed by NMR studies. Recent summaries of the dynamic behavior of Trp are given by London (1980) and Karplus and McCammon (1981). The small number of Trp residues usually found in proteins, sometimes only one Trp residue per protein chain, as in *Staphylococcus aureus* nuclease B, serum albumin, azurin, and horse cytochrome *c*, simplifies its definitive resonance assignments. Other points to be noted about tryptophan are (1) it has substantial ring-current effects on nearby residues, (2) it can hydrogen bond via the indole N—H, and (3) it is located not only as an interior residue of a globular protein but in solvent-exposed positions. It has been observed that a Trp residue forms a hydrogen bond to a heme propionate in most known cytochrome species (Richardson, 1981).

The expected side-chain ^{1}H NMR resonances for Trp consist of five 1-proton C—H peaks appearing as one singlet, two doublets, and two triplets at 100 MHz (Cohen, 1971). Decoupling techniques have been used to make specific assignments. Bradbury and Norton (1973) described ^{13}C NMR spectra of tryptophan and its resonances in peptides and proteins. The development of procedures for single-carbon aromatic-peak assignments has already been mentioned in the phenylalanine and tyrosine sections. Detailed strategy for ^{13}C peak assignments and relaxation data were presented for horse heart ferrocytochrome *c* (one Trp), hen egg white lysozyme (six Trp), horse carbonmonoxymyoglobin (two Trp) and adult carbonmonoxyhemoglobin (six Trp) in the excellent series of papers by Allerhand and coworkers (reviewed in Allerhand, 1979). The use of homologous proteins to make Trp peak assignments was well demonstrated by Oldfield and Allerhand (1975) in the series of hemoglobins. They also utilized selective iodine oxidation of indole to assign the resonances of Trp-108 (Norton and Allerhand, 1976).

Blake *et al.* (1981) examined hen egg white lysozyme by ^{1}H NMR and also converted Trp-62 using N-bromosuccinimide to an oxindolealanine residue. Although crystallographic and ^{1}H NMR studies showed that the native structure of the enzyme is essentially preserved, functional inactivation results because the modified residue does not permit effective substrate (N-acetylglucosamine oligosaccharide) binding. It should be recalled that two diastereomeric forms can result from derivatization and that crystal modifications can occur in several space groups; thus suitable cautions must be exercised in comparing X-ray crystallographic and NMR results so that if mechanistic conclusions are drawn, one is dealing with comparable forms. The six tryptophan residues of lysozyme were studied by observation of the N-bound protons (at 270 MHz), and the procedure for peak assignment was discussed extensively by Dobson (1975). The NH protons were clearly resolved, and rates of proton exchange were discussed in terms of the internal protein mobility and the pH dependence of chemical-shift values. In human and hen lysozymes, Trp-108 shows a downfield shift with increasing pH; a specific interaction between Trp and the carboxyl of Glu-35 was postulated. Bound inhibitors appeared to restrict the mobility of active-site residues. Cassels and co-workers (1978) achieved the complete assignment by ^{1}H NMR (270 MHz) of the C_{δ} and NH protons of the six tryptophan residues of hen egg white lysozyme. The problem of signal overlap with the other aromatic peaks was circumvented by using spin-echo and Carr-Purcell-Meiboom-Gill pulse techniques in conjunction with protein derivatives (e.g., Trp-62 was converted to the oxindole form) and pH and temperature variation. And $^{2}H_2O$–$^{1}H_2O$ exchange studies were used for the NH proton studies. The five low-field Trp NH resonances had previously been assigned by Glickson *et al.* (1971) and Campbell *et al.* (1975c). The sixth resonance was located upfield from the rest, overlapping the region of N—H peptides, by comparison with turkey and bobwhite quail lysozymes.

In their comparison of the solution structures of tuna and horse cytochrome *c*, Moore and Williams (1980d) observed the motional behavior of tuna Trp-33 (which replaces horse His-33). Two of the peaks associated with this residue [TA*5 and TA*7] sharpen as temperature increases, and this was attributed to surface mobility, whereas the persistent narrowness of the signal of deeply buried Trp-59 (tuna and horse) is considered to arise from immobilization of this lonc Trp in horse cytochrome *c*.

An informative example of hindered rotation involving tryptophan is the ^{13}C NMR study of the slow exchange of one of the four [γ-^{13}C]tryptophans in dihydrofolate reductase (*Streptococcus faecium*) between different conformers (London *et al.*, 1979; London, 1980). Two resonances (δ's 109.8 and 110.0 ppm) were considered to arise from a single Trp C_γ in slow exchange between nonidentical environments (Figure 41). Motions of the indole ring were invoked to explain the slow exchange, and temperature-dependence data of linewidths and peak positions supported this view. The exchange was sensitive to the binding of 3′-5′-dichloromethotrexate, which preferred one conformer. Comparison of the native and denatured (urea) spectra for the four quaternary C_γ Trp residues in *S. faecium* dihydrofolate reductase is shown by London (1980) and illustrates the loss in chemical-shift inequivalence upon denaturation.

The five tryptophan residues of dihydrofolate reductase were converted to the 6-fluoro derivative and studied by ^{19}F NMR (94 MHz) by Kimber *et al.* (1977,1978), including effects of inhibitors. Of the five labeled tryptophan residues, two showed ^{19}F–^{19}F spin–spin through-space coupling between nuclei on neighboring tryptophans. Since there are no adjacent Trp residues in the primary sequence, chain folding of the native structure was deduced to have brought the two Trp residues to a proximity of less than 4 Å.

Using deuterium quadrupole-echo Fourier transform techniques, Kinsey and co-workers (1981a,b) did a landmark study supplying the first ^{2}H NMR (55.3 MHz) spectra of the membrane protein bacteriorhodopsin from the photosynthetic purple membrane of *Halobacterium halobium* with the deuterated aromatic residues [$^{2}H_5$]phenylalanine, [$^{2}H_2$]tyrosine, and [$^{2}H_5$]tryptophan (see Section 2.4.1).

5.3. Carboxyl Side Chains

The carboxyl-containing residues, aspartic and glutamic acids, occur frequently in proteins and are well known to participate in important structural and mechanistic functions. These roles include stabilization of conformation by hydrogen bonding and salt linkages, chemical reactivity, often by fast and stereoselective proton transfer as active-site residues, and provision of an appropriate ionic environment for other side-chain groups. Because the carboxyl group is the reactive moiety, it is impossible to study aspartic acid

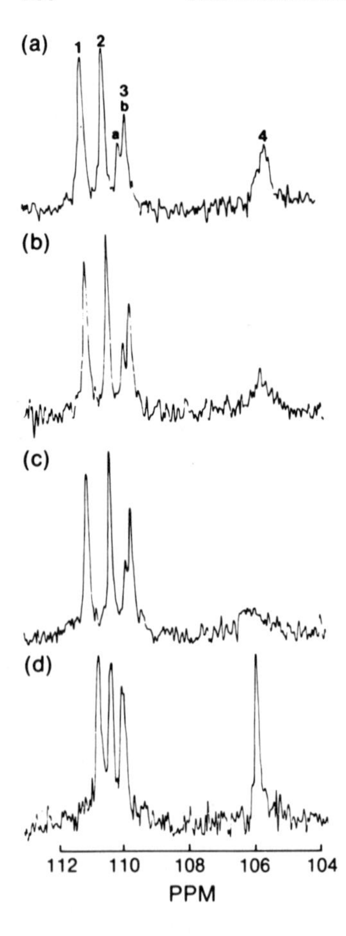

Figure 41. ^{13}C NMR spectra at 25 MHz of [γ-^{13}C]Trp-labeled dihydrofolate reductase (1 m*M*) obtained at (*a*) 5°C, (*b*) 15°C, (*c*) 25°C, (*d*) 15°C plus a saturating concentration of 3′,5′-dichloromethotrexate. (With permission from London *et al.*, 1979.)

and glutamic acid effectively by proton NMR by observing the adjacent methylene protons or the exchanging carboxyl-group proton. The carboxyl side chains are resolvable in natural abundance ^{13}C NMR spectra as several sharp peaks somewhat downfield from the background envelope of carbonyls present in proteins (Shindo and Cohen, 1976). The characteristic chemical-shift values for carboxyl carbons range from ~175 to 182 ppm downfield from TMS as indicated in Figure 42. The Glu carboxyl resonances are generally downfield from those of Asp, and consequently the former residues are ordinarily more readily resolved. In favorable circumstances single-carbon-atom resonances can be resolved and assigned.

Egan *et al.* (1977) reviewed ^{13}C NMR studies on proteins, including carboxyl groups. Instrumental considerations affecting sensitivity and resolution were discussed, and spectra of the carboxyl carbonyl region of HE lysozyme at three field strengths (Figure 43) indicated that a higher field strength is definitely preferable for the observation of resolved carboxyl resonances for medium-sized globular proteins (see also Norton *et al.*, 1977).

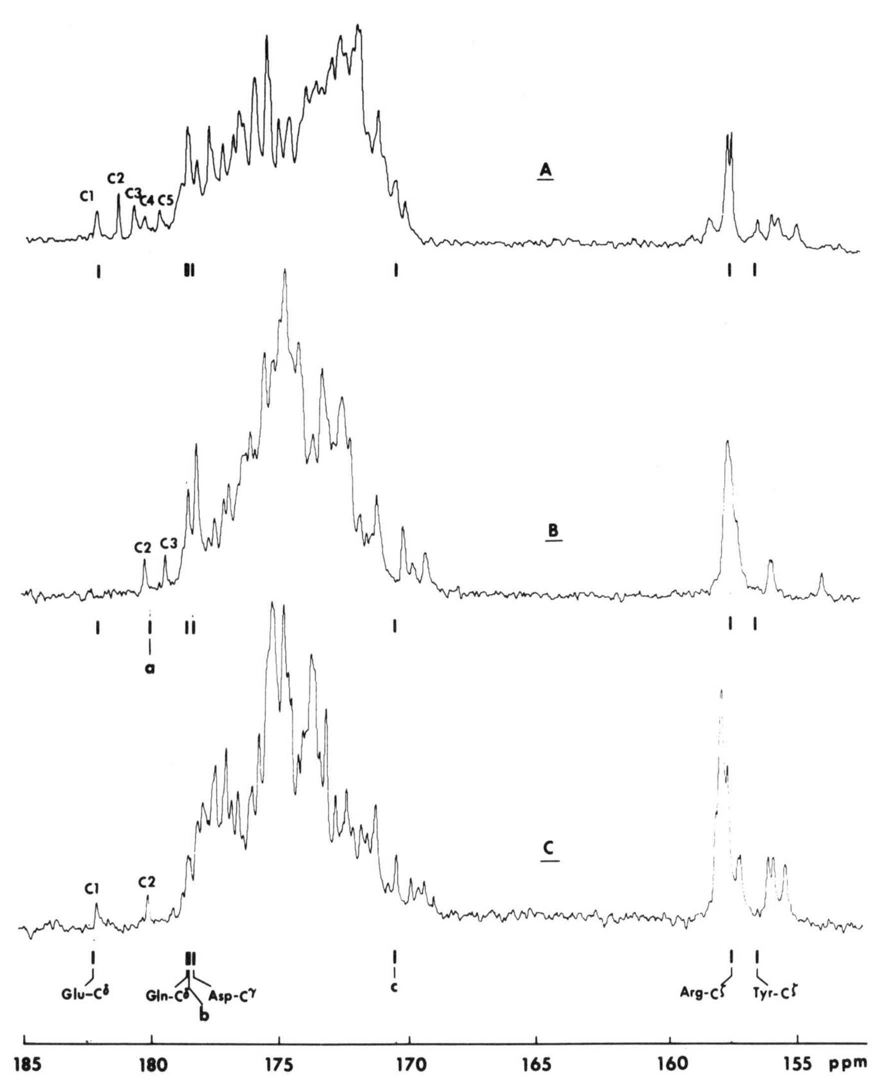

Figure 42. Region of natural abundance ^{13}C NMR spectra recorded at 67.9 MHz containing carbonyl resonances of (*A*) ribonuclease (20 m*M*, pH 8.4), (*B*) HE lysozyme (11.4 m*M*, pH 5.4), and (*C*) HL lysozyme (10 m*M*, pH 7.4). The positions of the respective amino acid resonances are shown by vertical lines below the spectra; in (*A*) and (*C*) these are (from left to right) Glu C_ζ, Gln C_δ, C-terminal Val C_o, Asp C_γ, N-terminal Lys C_o, Arg C_ζ, Tyr C_ζ. In (*B*) the position of the C-terminal Leu C_o replaces the terminal Val C_o in (*A*) and (*C*). (With permission from Shindo *et al.*, 1978.)

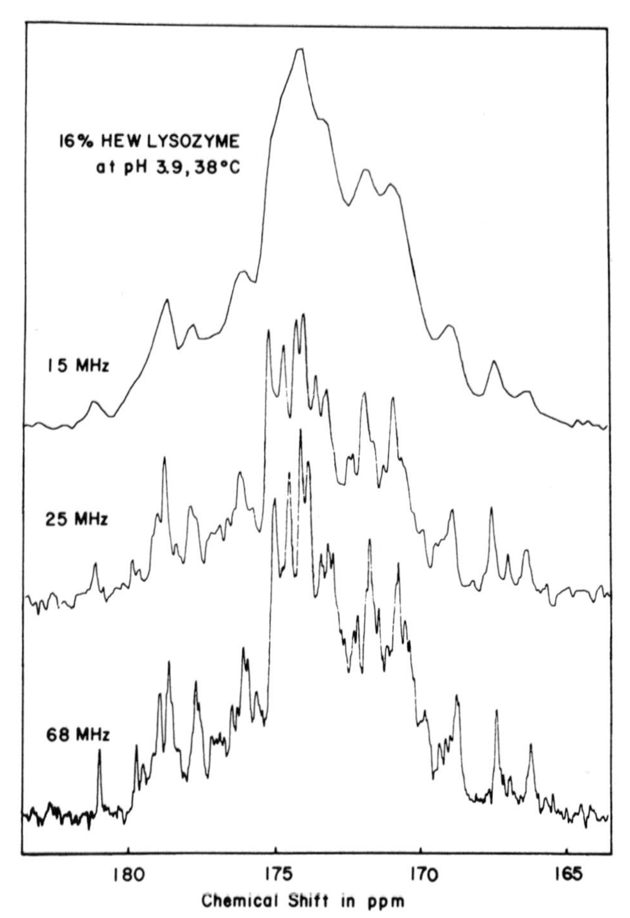

Figure 43. ^{13}C NMR spectra of carboxyl and carbonyl resonances of hen egg white lysozyme (16% solution in 0.1 *M* NaCl) at differing field strengths. The top trace is 15 MHz observing frequency, 4 kHz sweep width, 8K data points, 32K scans with 1.5 s delay time. The middle trace is 25 MHz, 6.25 kHz sweep width, 8K data points with 8K zero filling, 36K scans with 0.2 s delay time. The bottom trace is 68 MHz, 15.2 kHz sweep width, 32K data points, 36K scans with 1.5 s delay time. [Reproduced, with permission, from Egan *et al.*, *Ann. Rev. Biophys. Bioeng.* **6,** 638 (1977). Copyright 1977 by Annual Reviews Inc.]

Effects of paramagnetic impurities on carboxyl carbon relaxation times indicated the need for care in interpreting such data (Egan *et al.*, 1977). Shindo *et al.* (1978) report a general tendency for carboxyl and carbonyl resonances in proteins to exhibit upfield shifts relative to free amino acids and peptides (see Section 3.1). Since isotopic enrichment with ^{13}C may also allow direct examination of the carboxyl carbon, ^{13}C is the preferred nucleus for NMR observation of Glu and Asp residues.

Asp and Glu are normally negatively charged at physiological pH (pK_a's are Glu C_δ, 4.35, and Asp C_γ, 3.90, respectively; Keim *et al.*, 1973) and tend to be solvated surface residues rather than buried ones. However, sometimes a purpose is served by an interior location in which the carboxyl group is hydrogen bonded, such as for Asp-121 of ribonuclease (Wlodawer et al., 1982).

Titration of resolved carboxyl-group resonances permits determination of pK_a values (Figure 44). In favorable cases this may aid resonance assignments and may even provide information on the relative proximity of various other residues. Anomalous titration behavior, such as multiple transitions, is suggestive of hydrogen bonding, ion pairing, or conformational changes that have modified the carboxyl microenvironment. Studies utilizing high-resolution NMR of carboxyl side chains are referenced in Table III.

The lysozyme studies are useful as a model for illustrating carboxyl-group peak assignments in natural abundance spectra, using comparison of homologous proteins, residue modification, enzymatic chain shortening with carboxypeptidase, and Co^{2+} as a specific shift reagent; computer curve fitting also provides individual pK_a values. Resolved natural abundance ^{13}C resonances (68 MHz) of single carboxyl groups were first reported by Shindo and Cohen (1976) for HE lysozyme, which contain two glutamic acid residues, seven aspartic acid residues and one terminal COOH. The paramagnetic Co^{2+} ion, which binds specifically at the active site, causes selective shifts of some resonances. Of the several titrating carboxyl carbon resonances, the most downfield-shifted was assigned to acetic acid strongly bound to the enzyme in the active site and was shifted most by Co^{2+} ions.

This ^{13}C NMR study was subsequently extended (Shindo *et al.*, 1978) to a comparison with human leukemic (HL) lysozyme and to examination of HE lysozyme derivatives by guanidination with O-methylisourea to modify the lysine residues to homoarginyls so salt-bridge formation could be examined. In this way the Glu-7–Lys-1 salt bridge reported in the crystal structure (Browne *et al.*, 1969) was also detected in the magnetic resonance study by the effect on the resolved carboxyl resonance. Cleavage of the terminal leucyl residue with carboxypeptidase to form des-Leu-129 lysozyme allowed assignment of the resonance of this terminal carboxyl group. From these studies three resonances were assigned: Glu-7 (179.98 ppm, pK_a = 2.91); the terminal Leu-129 carboxyl group (178.17 ppm, pK_a = 3.41); and Asp-52 (177.74 ppm, pK_a = 3.90). A key objective of the lysozyme studies had been to examine the behavior of the active-site carboxyl residue of Glu-35, already known to have an abnormally high pK_a. This resonance was not detected, however, presumably because it is shifted under the backbone carbonyl envelope, broadened, or both.

Five carboxyl carbon atom resonances were resolved in the downfield spectral region of RNase A spectra (Shindo *et al.*, 1978). These were given tentative assignments to the five glutamic acid residues of the enzyme in accord with the typical downfield position of Glu C_δ atoms, but the possibility that one of these corresponds to the terminal —COOH of Val-124 was not excluded. Santoro *et al.* (1979) also observed these five titrating ^{13}C

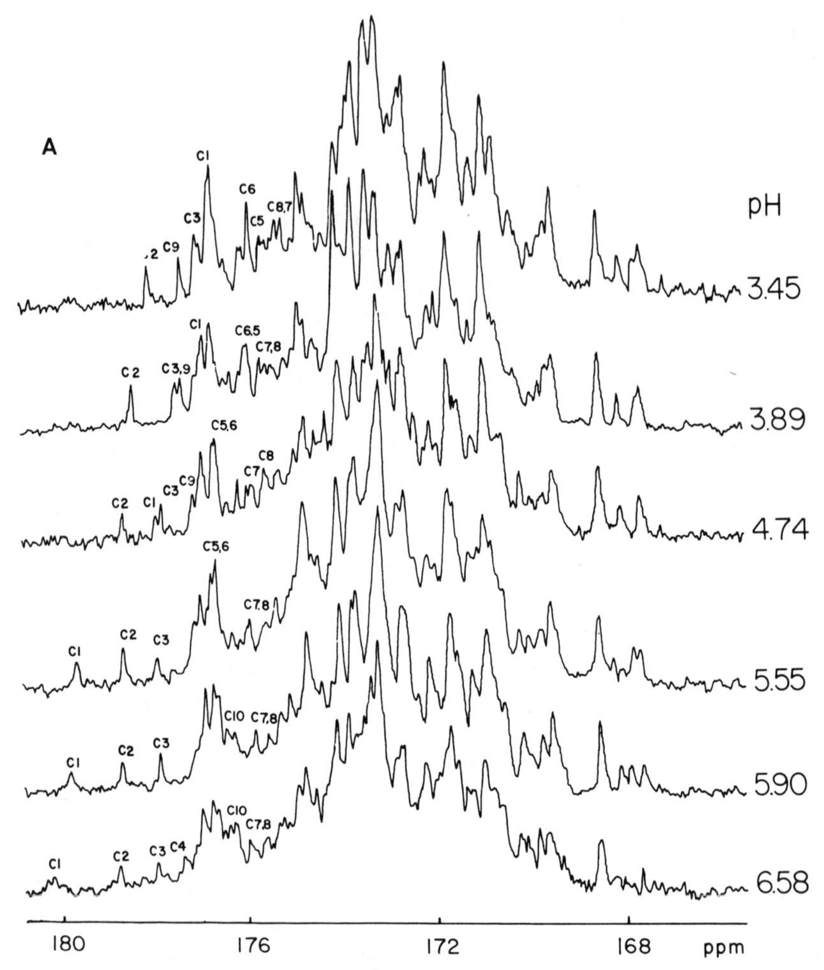

Figure 44. (*A*) Carboxyl and carbonyl carbon resonances of natural abundance ^{13}C Fourier-transformed NMR spectra of HE lysozyme (10 m*M*) in H_2O containing D_2O (4:1), and 0.1 *M* NaCl. Each spectrum was recorded at 67.9 MHz with 8K real points, 15 kHz sweep width, and 30,000 to 40,000 transients (total time 17 to 22 h). The peaks labeled C1 to C10 were observed to titrate on change of pH. (From Shindo and Cohen, 1976.) (*B*) Curve fits of titrating resonances from (*A*), for which reasonable continuities could be determined. (With permission from Shindo and Cohen, 1976.)

carboxyl resonances of RNase A as well as three additional titrating resonances buried in the carbonyl region. Although p*K* values were assigned to these eight resonances, several signals gave very incomplete titration curves. One resonance, from its apparent slow-exchange behavior at ~pH 7 and its p*K* value of ~4.3, was assigned to the carboxyl side chain of Asp-14. As indicated below, this assignment is almost certainly incorrect.

The advantages of selective ^{13}C enrichment were demonstrated in a ^{13}C NMR study of the RNase S′ system. An N-terminal (1–15) RNase peptide

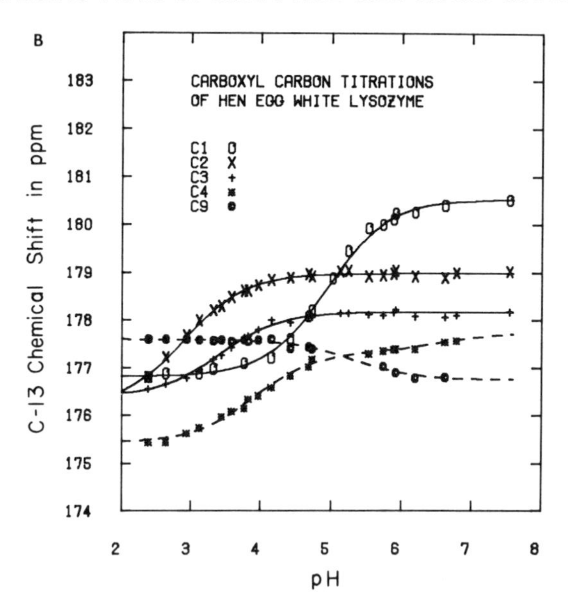

was synthesized with the Asp-14 C_γ 90% ^{13}C enriched. The peptide was then complexed with RNase S-protein to form the semisynthetic RNase S′ complex. Since the complex exhibited a resonance well resolved from the free unbound peptide (Figure 45), it was possible to obtain detailed data for the single ^{13}C-enriched atom in the complex. The unambiguous assignment of this resonance is in sharp contrast to natural abundance ^{13}C studies, and ^{13}C spectra can of course be obtained in a fraction of the time (Niu *et al.*, 1979).

The titration curve of the $^{13}C_\gamma$ resonance of Asp-14 in the RNase S′ complex exhibited a minor inflection with pK 6.1 and a major titration with p$K \simeq 2.4$ (Figure 32). These were assigned to the interaction with His-48 and to the titrations of Asp-14 itself, corresponding to the acid denaturation of the protein. The Asp $^{13}C_\gamma$ resonance is unaffected by the addition of inhibitors (Cohen *et al.*, 1980). Notably in these studies the Asp $^{13}C_\gamma$ resonance showed no inflection in the pH region 4–5, and hence cannot be the group responsible for the pH-dependent acid conformational transition of RNase (Cohen and Shindo, 1975; Santoro *et al.*, 1979; Lenstra *et al.*, 1979). Also, the assignment of peak 4 in the ^{13}C RNase spectra of Santoro *et al.* (1979) cannot be to Asp-14 in view of the entirely different chemical-shift range and titration properties of the resonance observed in the ^{13}C-enrichment studies. Indeed the absence of any significant titration for this resonance above pH 3, indicative of strong hydrogen bonding of Asp-14 to Tyr-25 (Richards *et al.*, 1970), would render it essentially impossible to identify in ^{13}C natural abundance studies.

Gerken *et al.* (1982) used carboxypeptidase to delete the terminal Leu-129 residue from HE lysozyme in their studies on ^{13}C-methylated lysine residues. They found that the terminal carboxylate group is apparently ion-

Table III. NMR studies of carboxylic residues in proteins

Protein	Reference	Comment
Bovine pancreatic trypsin inhibitor	Richarz and Wüthrich (1978)	^{1}H and ^{13}C, terminal Ala-58, pK_a's for Asp and Glu
Collagen (chick embryo)	Torchia and VanderHart (1976)	[5-^{13}C]Glu; fibrils, solid-state
Hemoglobin (human fetal)	Oldfield and Allerhand (1975)	^{13}C, resolved carboxyl
Lysozyme (hen and human)	Shindo and Cohen (1976);	^{13}C, resolved carboxyls, pK_a's
	Egan *et al.* (1977);	^{13}C review
	Norton *et al.* (1977);	^{13}C CSA relaxation
	Shindo *et al.* (1978)	^{13}C individual assignments, pK_a's
Myoglobin	Oldfield *et al.* (1975)	^{13}C
Parvalbumin (carp)	Nelson *et al.* (1976)	Glu-81; ^{13}C, internal H bond
Prothrombin (fragments)	Stenflo *et al.* (1974);	Gla; ^{1}H, Ca^{2+} binding
	Egan *et al.* (1977);	Gla; ^{13}C
	Furie *et al.* (1979)	Gla; ^{13}C, Ca^{2+}, lanthanide shifts
Ribonuclease (bovine)	Egan *et al.* (1977);	^{13}C, resolution of five carboxyls
	Shindo *et al.* (1978);	
	Niu *et al.* (1979); Cohen *et al.* (1980);	Asp-14; ^{13}C-enriched
	Santoro *et al.* (1979)	^{13}C titrations

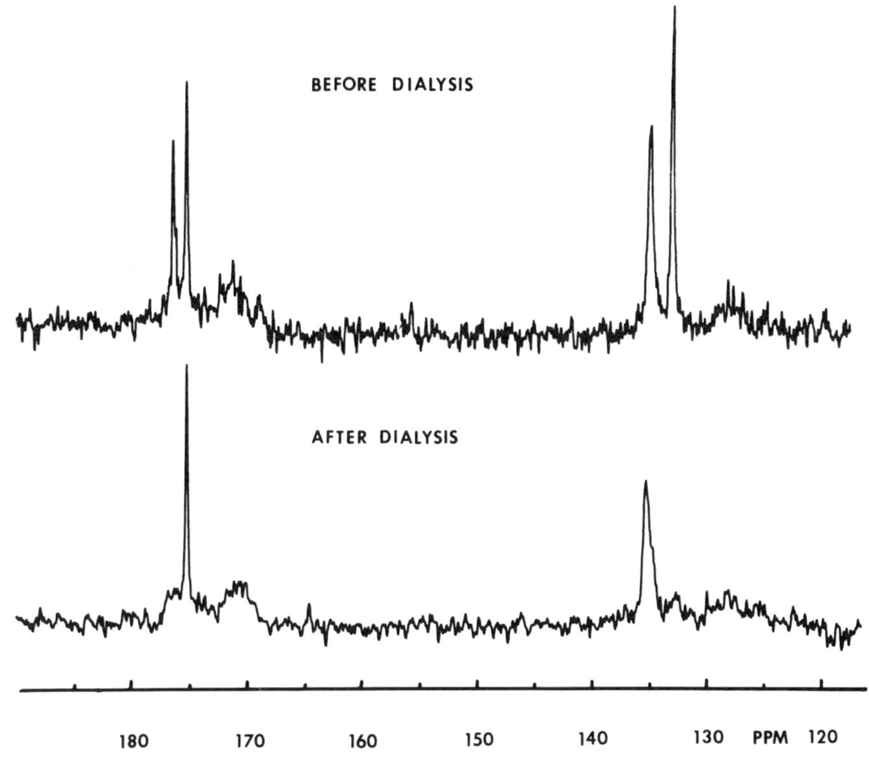

Figure 45. ^{13}C NMR spectra at 67.9 MHz of a mixture of excess ^{13}C-enriched synthetic RNase peptide—[ε-^{13}C]His-12, [γ-^{13}C]Asp-14 RNase (1 to 15)—and RNase S-protein in 0.15 *M* P_i, pH 7.1, before (upper) and after (lower) dialysis. The signals at 175 and 135 ppm correspond respectively to the Asp C_γ and His C_ε resonances of the RNase S′ complex. (With permission from Niu *et al.*, 1979.)

paired with the Lys-13 ε-NH_3^+, as found in the X-ray crystal structure (Browne *et al.*, 1969), although this was not found on observation of the carboxyl residue itself on chemically modifying the lysines (Shindo *et al.*, 1978).

Stenflo *et al.* (1974) identified by proton NMR (100 MHz) and mass spectrometry the unusual amino acid γ-carboxyglutamic acid (Gla), 3-amino-1,1,3-propanetricarboxylic acid (Figure 46). Using a tetrapeptide, they showed that this amino acid participates in the binding of Ca^{2+} ions required for activation of the blood-clotting protein prothrombin. The ^{13}C natural abundance spectrum (67.9 MHz; Egan *et al.*, 1977) confirmed that the molecule has two γ-carboxyl groups at 179.0 and 178.6 ppm (Figure 46). Furie *et al.* (1979) examined in detail by ^{13}C NMR (67.9 MHz) the metal-binding sites of the prothrombin peptide fragment residues 12 to 44 using the paramagnetic trivalent lanthanide ions Gd(III), Pr(III), and Eu(III). This peptide (MW 4100) contains eight of the protein's 10 Gla residues, including two adjacent pairs. Furie *et al.* confirmed that the Gla residues are responsi-

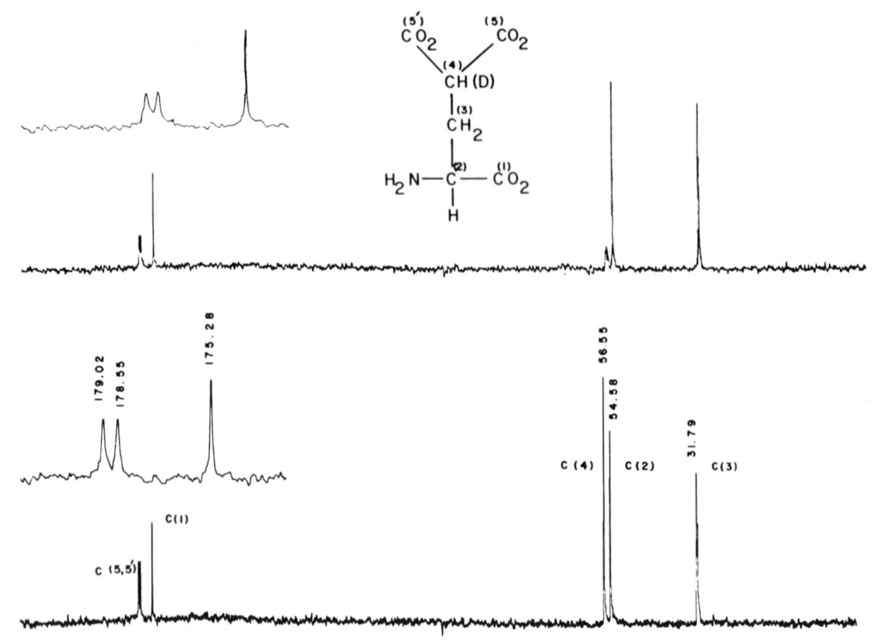

Figure 46. ^{13}C NMR spectra (68 MHz) of γ-carboxyglutamic acid (0.1 *M*, pH 7.4); the region of carboxyl resonances is expanded. The upper spectrum is recorded in $^{2}H_2O$, in which the C_δ position is deuterated. [Reproduced, with permission, from Egan *et al., Ann. Rev. Biophys. Bioeng.* **6,** 638 (1977). Copyright 1977 by Annual Reviews Inc.]

ble for Ca^{2+} binding and concluded that the tertiary and/or quaternary structure of the protein is stabilized by intramolecular or intermolecular metal liganding. Dialysis studies were interpreted as indicating two classes of metal-binding sites. A model was proposed consistent with having one high-affinity site involving two Gla residues per metal ion that can bridge by metal liganding, and four to six low-affinity sites with only a single Gla per metal ion. Titration data with Gd(III) were interpreted as showing that the high-affinity metal-binding site has two nearby arginine residues and is stabilized by a disulfide bond. Estimates of proximity of carbon atoms to bound metal ions were also made using Gd(III). The effect on binding of the disulfide moiety was probed by cleavage and conversion to S-carboxymethyl groups.

Asparagine and glutamine are, of course, the amide forms of the corresponding acids. These nonionic side chains occur in hydrogen bonds, but NMR studies involving these residues have not been frequently reported; ^{13}C T_1 values have been summarized by Howarth and Lilley (1978).

5.4. Hydroxyl Side Chains

The aliphatic residues containing hydroxyl groups in their side chains are serine and threonine (Table IV). The serine side chain is small, polar,

Table IV. NMR studies of hydroxyl residues in proteins

Protein	Reference	Comment
Alkaline phosphatase (*E. coli*)	Otvos *et al.* (1979)	Ser; ^{31}P, phosphorylation at active site
Chymotrypsin	Niu *et al.* (1977)	Ser; [1-^{13}C]acetylchymotrypsin intermediate
Cytochrome *c* (horse)	Moore and Williams (1980b);	Thr-47 and Thr-89; ^{1}H assignment
	Wooten *et al.* (1981)	Thr; ^{13}C, pH conformational transition
Gelatin (calf skin)	Chien and Wise (1973)	Thr; ^{13}C
Lysozyme (hen)	Gerken *et al.* (1982)	Thr-40; [^{13}C]N-methylated Lys-1 interaction
Parvalbumin (carp)	Nelson *et al.* (1976)	Thr; ^{13}C, T_1 and NOE
Proteoglycan core protein	Torchia *et al.* (1981)	[3-^{13}C]Ser and [2-^{13}C]Gly
Ribonuclease A	Allerhand *et al.* (1970,1971);	Thr; ^{13}C natural abundance
	Glushko *et al.* (1972)	Thr; ^{13}C, denaturation and chemical-shift change

hydrogen bonding, and reactive. It has been implicated in the mechanism of action of many enzymes, most particularly the serine proteases, such as trypsin and chymotrypsin. A "charge-relay system" of interacting Asp-His-Ser residues was initially proposed as an activating mechanism for the active-site Ser residue from X-ray studies of chymotrypsin (Blow *et al.*, 1969). However, subsequent X-ray (Matthews *et al.*, 1977) and NMR (Bachovchin *et al.*, 1978,1981) reinvestigations rejected such a catalytic triad in Ser proteases.

Because of the very crowded nature of the proton NMR spectrum of even small globular proteins, there have been no unequivocal observations of Ser β-methylene proton resonances. However, Thr γ-methyl groups are more readily observed, for example, Moore and Williams (1980a) reported the assignments of the methyl-group resonances of Thr-47 and Thr-89 in ^{1}H NMR spectra (270 MHz) of ferricytochrome *c* and ferrocytochrome *c*.

An excellent example of how derivatized serine residues may be studied by ^{31}P NMR (36.4 MHz) is the metal-dependent phosphate-binding study of alkaline phosphatase by Otvos *et al.* (1979). The native dimeric protein (MW 86,000) requires two Zn(II) atoms per subunit for phosphate binding and subsequent phosphorylation at active-site serine. Separate resonances were found for inorganic phosphate, complexed phosphate, and covalently bound phosphorylenzyme–substrate intermediate. Additional metal ions can be bound by the enzyme, and the resulting complex phosphate-binding stoichiometry, including negative cooperativity and the effects on catalysis rates, were analyzed by ^{31}P NMR study; ^{32}P-labeling experiments complemented the NMR study.

Niu *et al.* (1977) prepared the [1-^{13}C]acetylchymotrypsin covalent intermediate, acetylated at the active-site Ser-195, and characterized it by ^{13}C NMR (68 MHz). The ^{13}C-enriched intermediate, which is stable at low pH, showed a resonance for the enriched carbon at 174.0 ppm; T_1's for the carbonyl carbon and the backbone carbonyls were 2.6 s and 2.9 s, respectively, indicating relative immobility of this group in the active site.

Threonine C_{γ}-methyl carbon resonances have been targeted in a number of NMR studies of proteins. For example, Glushko *et al.* (1972) reported the changes in chemical shifts and relaxation times on denaturing ribonuclease, and Richarz and Wüthrich (1977) did a similar study of the ^{13}C methyl resonances of BPTI. A survey of relaxation times and correlation times derived from them is found in the review of Howarth and Lilley (1978).

An example of how Thr can be effectively studied using natural abundance ^{13}C NMR is the pH-induced conformational transition of ferricytochrome *c* (Wooten *et al.*, 1981). This protein has 10 threonine residues (and no serines), giving rise to a sharp singlet and unambiguous C_{β} peak assignment (67.3 ppm at pH 3). The acid conformational transition is characterized by a pK of 3.4 and was monitored both by the loss of intensity of this Thr peak into chemically nonequivalent multiple peaks as the protein folded (Figure 47), and by the exchange broadening of the ^{2}H resonance of deuterium-labeled Met-80 on reducing the pH.

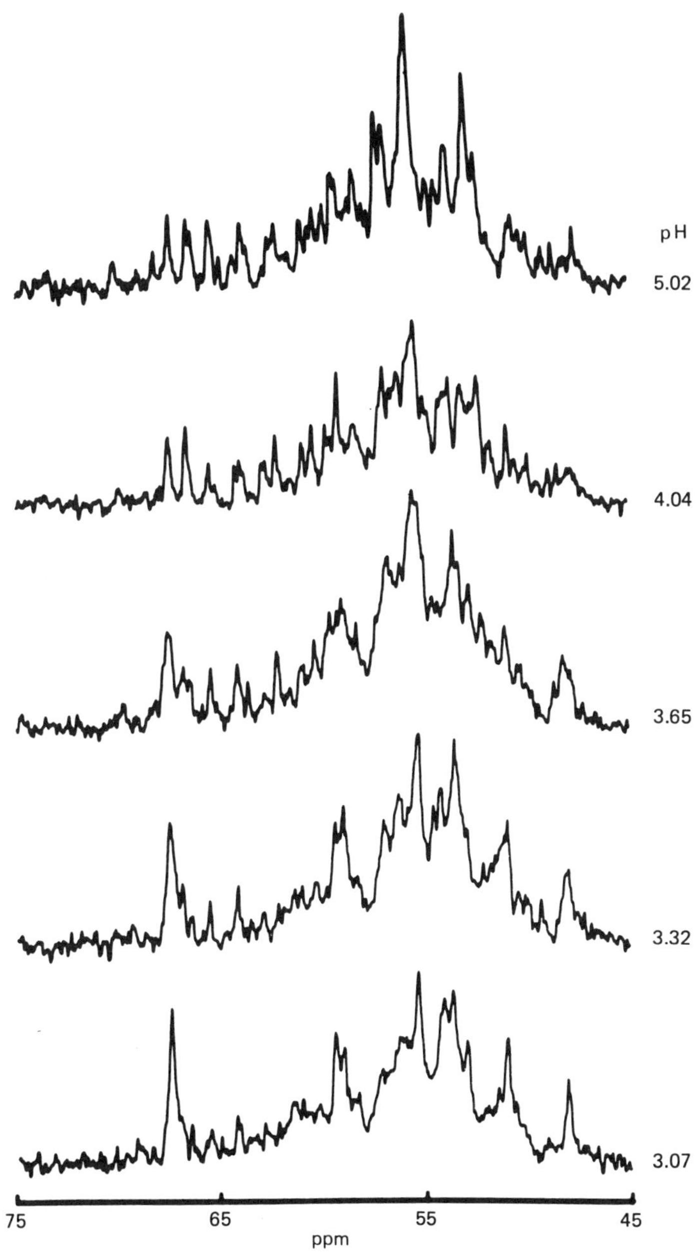

Figure 47. Changes in natural abundance ^{13}C NMR spectra of ferricytochrome *c* due to the acid transition. The sharp resonance at 67.3 ppm at pH 3 results from 10 Thr C_β atoms. [Reprinted with permission from Wooten *et al.* (1981). Copyright 1981, American Chemical Society.]

The N-terminal α-amino group of Lys-1 in HE lysozyme is shown by X-ray diffraction studies to be hydrogen bonded to the hydroxyl group of Thr-40 (Browne *et al.*, 1969). Gerken and co-workers (1982) have studied the ^{13}C-methylated lysine derivative of lysozyme; this study is informative in showing how it is possible to observe such an interaction by examining the other partner in the hydrogen-bonding pair.

5.5. Basic Side Chains

The basic amino acids that are discussed in this section are arginine and lysine (Table V). They are basic because of the side-chain nitrogen atoms, which protonate with characteristic high pK values and are positively charged at physiological pH. Histidine, of course, is closely related, but since its aromaticity markedly influences its role, it was discussed separately (Section 5.1). Basic proteins, that is, those with high content of lysine and/or arginine, appear to play significant roles in protein–nucleic acid interactions, as has been noted in the binding of histones to the negatively charged polyelectrolyte chains of polynucleotides in eucaryotic chromatin. Histones were recently reviewed by Isenberg (1979).

5.5.1. Arginine. Arginine residues are among the most readily located in ^{13}C NMR spectra because of the location of Arg C_ζ among the quaternary-carbon resonances of unprotonated aromatic residues. London (1980) has emphasized, however, that dynamically arginine is more closely related to other aliphatic residues like methionine in the sense that multiple internal rotations with a range of T_1's are indicated. Disparities arise, not unexpectedly, from the differences in volume required for internal motion of methyl groups and motion of the larger branched guanido group. Generally, because of these characteristics and the fact that they are hydrophilic and usually external residues, multiple guanidino C_ζ resonances are not well resolved from each other. Needless to say, proton resonances of the arginine side chain are poorly resolved within the envelope of protein resonances.

Oldfield *et al.* (1975) assigned and interpreted the dynamic behavior of arginine residues from the C_ζ carbon resonances at natural abundance in cytochrome *c*, horse myoglobin, human hemoglobin, and hen egg white lysozyme. The latter has 11 arginines with a range of T_1's, including some with rapid segmental motions. It was pointed out that the arginine C_ζ (as well as Trp $C_{\varepsilon 2}$) T_1 values can show appreciable effects from the ^{13}C–^{14}N dipolar relaxation mechanism. The five guanidino protons are exchangeable in D_2O-containing solvents. Howarth and Lilley (1978) tabulated experimental and theoretical ^{13}C T_1 values for arginine in cytochrome *c*, lysozyme, and hemoglobin in both H_2O and D_2O in their review.

As an example of how arginine may create difficulties in peak assignment for aromatic residues, one can cite the occurrence of a substantial resonance from four arginines in the midst of six $^{13}C_\zeta$ tyrosines of RNase A and its interference with titration studies (Egan *et al.*, 1977).

An informative study of protein–ligand interactions utilizing arginine side chains involves the enzyme dihydrofolate reductase from *S. faecium* (Cocco *et al.*, 1977). Five resonance signals were obtained for the eight arginine residues known to be present; in the urea-denatured protein these collapsed to a single peak with δ = 157.78 ppm. Binding studies with NADPH and methotrexate demonstrated that in some circumstances Arg is a useful probe (with a range of chemical-shift values of 1.2 ppm) in contrast to results with myoglobin, hemoglobin, cytochrome *c*, and lysozyme (Oldfield *et al.*, 1975), which showed only small (~0.4 ppm or less) shifts.

In a subsequent study (Cocco *et al.*, 1978) using D_2O instead of H_2O, improved resolution was obtained, and T_1 data as a function of temperature were determined for a series of complexes including both NADPH and $NADP^+$. Two classes of arginine residues were distinguished by their NMR parameters. One corresponds to solvent-exposed, mobile surface residues with chemical shifts analogous to those of denatured protein, whereas a second buried type is characterized by downfield chemical shifts and restricted motion. The exposed residues show motion within the extreme narrowing limit ($\omega_0\tau_c \ll 1$) in contrast to the other class, which was designated buried on the basis of the T_1 values, small NOEs, and inverse temperature dependence for T_1's. Charged residues such as arginines stabilize their buried, thermodynamically unfavorable positions in hydrophobic environments by forming salt linkages.

A study with Arg at the active site of soybean trypsin inhibitor (STI) enzymatically replaced Arg-63 with ^{13}C-labeled Arg to prepare [$^{13}C^0$]Arg-63 STI and [$^{13}C_\zeta$]Arg-63 STI, and subsequently complexed the labeled proteins with trypsin (Baillargean *et al.*, 1980). The chemical shift for the enriched, uncomplexed [$^{13}C_\zeta$]Arg-63 STI is 157.81 ppm downfield from TMS in H_2O or D_2O at pH 7 and 158.04 for the complex with trypsin. The corresponding values for [$^{13}C^0$]Arg-63 STI are 173.8 and 174.7 ppm (δ's were pH dependent). Analysis of the data led the authors to conclude from the small chemical shifts observed that the guanidinium group does not deprotonate, and the STI–trypsin complex does not involve a covalent, fully tetrahedral intermediate.

5.5.2. Lysine. Sharp, single ^{13}C resonances of C_ε carbons of lysine are well documented to occur in the aliphatic region at ~40 ppm (see Figure 1). These NMR peaks are frequently so narrow that they are highly conspicuous, even in natural abundance spectra, as a result of the high mobility possible for these long, external side chains. Lysine residues have been observed as active-site residues, as chain stabilizers in salt linkages, and in polycationic regions for interaction with polynucleotides. (See Table V for specific examples of lysine residues in proteins that have been studied by NMR techniques.)

The ^{13}C NMR relaxation data for poly-L-lysine (DP 129) have been reported at two frequencies and analyzed by Wittebort *et al.* (1980). They deduced considerable restriction for the C_β and C_α and a trend of increasing

Table V. NMR studies of basic residues in proteins

Protein	Reference	Comment
Azurin (*P. aeruginosa*)	Ugurbil *et al.* (1977)	Arg; ^{13}C
Basic pancreatic trypsin inhibitor	Richarz and Wüthrich (1978);	Lys; ^{13}C, titration
	Richarz *et al.* (1980b);	Lys-15; [1-^{13}C], complexes with trypsin, trypsinogen, and anhydrotrypsin
	Brown *et al.* (1976)	Lys-41; ^{13}C, interaction with tyrosine
Collagen (chick embryo)	Torchia and VanderHart (1976); Jelinsky and Torchia (1979)	Lys; [6-^{13}C], solid state
Cytochrome *c*	Moore and Williams (1980c);	Lys; ^{1}H, temperature, and pH dependence
	Oldfield *et al.* (1975);	Arg; ^{13}C, natural abundance
	Kennelly *et al.* (1981)	Lys; ^{13}C, conversion to homoarginines
Dihydrofolate reductase (*S. faecium*)	Cocco *et al.* (1978)	Arg; ^{13}C, motions, and binding
Gene 5 protein (bacteriophage)	Coleman and Armitage (1978);	Arg, Lys; H^{1}, protein–nucleic acid interactions
	Alma *et al.* (1981a)	Arg; ^{1}H, protein–nucleic acid interactions

Hemoglobin (human)	Oldfield and Allerhand (1975)	Arg; ^{13}C, quaternary protein
Histones	Bradbury *et al.* (1975);	^{1}H, salt, and concentration effects
	Lilley *et al.* (1975);	^{13}C, high Arg and Lys content
	Chapman *et al.* (1976);	^{1}H, high Lys content
	Hartman *et al.* (1977);	
	Howarth and Lilley (1978);	^{13}C, histone tetramer, aliphatic region
	Shindo (1981)	^{13}C, DNA–histone octamer complex
Lysozyme (HE)	Allerhand (1979);	Arg; ^{13}C, summary
	Oldfield *et al.* (1975);	Arg; ^{13}C, T_1's, H_2O and $^{2}H_2O$
	Gerken *et al.* (1982);	Lys; ^{13}C, monomethyl and dimethyl ^{13}C-enriched
	Laterbur (1970)	Lys; ^{13}C, narrow resonances
α-Lytic protease	Bachovchin *et al.* (1981)	Arg; ^{13}C, 67.9 MHz
Myoglobin (horse)	Oldfield *et al.* (1975);	Arg; ^{13}C, convolution difference
	Wilbur and Allerhand (1976)	Arg; ^{13}C, aromatic region
Neurotoxin-II (cobra)	Bystrov *et al.* (1978)	Lys; ^{1}H, trifluoroacetylated
Ribonuclease A	Allerhand *et al.* (1970);	Lys; ^{13}C
	Glushko *et al.* (1972);	Lys; ^{13}C, T_1's
	Jentoft *et al.* (1979)	Lys; [^{13}C]formaldehyde
Soybean trypsin inhibitor	Baillargeon *et al.* (1980)	Arg; [O-^{13}C] and [ζ-^{13}C]

mobility proceeding outward from the backbone; for example, the T_1's at 68 MHz for C_α to C_ε are respectively 170, 206, 264, 443, and 738 ms. The authors noted that protonated lysine (NH_3^+) may have slower motions than the unprotonated form. They compared the polylysine data to T_1 values for C_ε obtained on RNase A by Glushko *et al.* (1972) and concluded that the protein lysyl groups are substantially more restricted. It should be recalled that deprotonation of poly-L-lysine can be accompanied by a coil-to-helix transition, and that both relatively high concentrations and high temperatures promote aggregation and β-structure formation. Thus selection of experimental conditions and data interpretation require suitable allowance for these phenomena.

Richarz *et al.* (1980b) carried out ^{13}C NMR studies on [1-C^{13}]Lys-15 BPTI and its complexes with trypsin, trypsinogen, and anhydrotrypsin. They concluded that the protein–protease inhibitor complex does not involve a covalent, tetrahedral intermediate nor an acylenzyme intermediate at the active site. Instead, the data were more suggestive of a noncovalent complex. The experiments were specifically designed to ascertain the nature of the bonding and the geometry of the complexes.

Several modifications of lysine residues for NMR studies have been reported. Shindo *et al.* (1978) converted lysine amino groups to guanidinium moieties with O-methylurea. More recently, Kennelly *et al.* (1981) also reported ^{13}C NMR studies on eight cytochrome *c*'s in which the lysines were converted to homoarginine residues. The guanidination did not destroy the electron-transport activity. A random polycationic copolymer of lysyl and homoarginyl residues was prepared, and its effects as activator or inhibitor of electron transport examined. This study represents an informative example of probing complexation sites by protein modification and demonstrates both the utility and the versatility of the NMR techniques.

Bradbury and Brown (1973) utilized reductive methylation of lysine residues with formaldehyde and borohydride to provide observable methyl resonances for modified lysines. This method was later applied to RNase A (Brown and Bradbury, 1975), in which resonances corresponding to 10 Lys ε-methyl groups and the terminal α-amino methyl group were observed. However, the sensitive active-site His titration curves show deviations from those of the unmodified protein. Nevertheless it is possible to see perturbations of the Lys-41 ε-methyl groups, for example, in the presence of P_i, which indicates its proximity to the active site.

Jentoft *et al.* (1979,1981) used reductive methylation with [^{13}C]formaldehyde and borohydride to investigate lysine residues in RNase A by ^{13}C NMR (45 and 68 MHz). The study focused on the active-site residue Lys-41. A total of four enriched resonances were observed corresponding to (in decreasing chemical-shift order) two from N_ε-dimethyllysyl residues (with the Lys-41 peak at lower field), N_α-terminal dimethylated Lys-1, and N_ε-monomethylated residues. Both chemical-shift determinations in the presence of 3′-CMP inhibitor and titration studies to determine pK_a's were used to verify the peak assignments; Lys-41 was assigned the unusually low pK_a value of 9.0 and was found to reflect a second pK_a value in the presence of inhibitor (Jentoft *et al.*, 1981), which was attributed to interaction with His-12 (see also Cohen *et al.*, 1973). Gerken *et al.* (1982) have extended this method to prepare ^{13}C-enriched derivatives of HE lysozyme, which contains six lysine amino groups and the terminal amino group of Lys-1. The dimethylated derivative retains 36% activity. The NMR studies included relaxation measurements and titration studies. It was possible to make peak assignments and to establish that the two ion pairs known to exist in the X-ray crystal structure, between Lys-1—Glu-7 and Lys-13—Leu-129 (terminal carboxyl) (Browne *et al.*, 1969) are also present in solution. This result differed from that in the study of the carboxyl-group resonances (Shindo *et al.*, 1978), the other residue in the pair, regarding the terminal-carboxyl-group ion pair in solution.

Wilbur and Allerhand (1977b) also described the observation of a sharp resonance for the naturally occurring N_ε-trimethyllysine group in yeast cytochrome *c*.

5.5.3. Histones. Among the significant proteins of current interest are the basic histones, which as nucleoproteins of eucaryotic DNA are crucial to the structure and superhelical folding of chromatin (Sperling and Wachtel, 1981). The packing affects not only the storing of genetic information but also the dynamics and functioning of the DNA. The nature of the interactions between the histones themselves, for example in the octamer core of the nucleosomes, and with the polyanionic polynucleotide chain (~200 base pairs) in the prevailing bead-structure model is being investigated by a wide range of physicochemical techniques including NMR. Bradbury and coworkers applied NMR to the basic proteins of chromatin (Bradbury and Rattle, 1972; Bradbury *et al.*, 1975); they examined the effect of salt concentration on protein folding and delineated hydrophobic core regions. The effect of ionic strength showed that histone H2A has the greatest aggregation tendency. Increasing either sodium chloride concentration or protein concentration leads to a polar (hydrophobic) aggregation, as judged from the linewidths of aromatic and hydrophobic resonance peaks. From the high content of lysine and arginine in the five types of histones and their observed clustering into polycationic regions, it was first assumed that they can interact electrostatically with the negatively charged sugar-phosphate DNA backbone. Chapman *et al.* (1976) and Hartman et al. (1977) also studied the conformation of the very-lysine-rich histone H1 by ^1H NMR.

In their ^{13}C and ^1H NMR study of several histones from calf thymus, Lilley *et al.* (1975) determined that the isolated chains consist of both structured and unstructured regions and that the disordered N-terminal end of the chain contains a considerable number of highly mobile basic residues. Computer simulations were used to assist their deductions. Howarth and Lilley (1978) showed the NMR spectrum (at 22.6 MHz) for the aliphatic region of the histone tetramer (MW 56,000) and confirmed the deduction of mobile chain termini from the many sharp, narrow resonances—particularly in the C_α region. Shindo (1981) in a ^{13}C NMR study of both DNA deoxyribose carbons and those of the lysine and arginine side chains in the histone octomer complex showed that a large proportion of the lysine and arginine side chains are very mobile, in accord with other results indicating that histone–DNA complexation is not purely an ionic interaction.

Alma and co-workers (1981a,b) examined bacteriophage gene-5 protein (dimeric MW 9690) complexes with sodium salts of oligoadenylic acids $(dA)_8$ and $(dA)_{25-30}$ using 500-MHz ^1H NMR. These authors concluded from downfield chemical shifts, line broadening, and decoupling that charged arginines and lysines were participating in binding. They focused on the aliphatic region of 2.9 to 3.5 ppm (relative to DSS) where Lys-ε and Arg-δ proton triplets are located. The implicated residues would appear to be those shown by X-ray diffraction: three Arg (residues 21, 80, and 82) and two Lys (residues 24 and 26) in the DNA binding groove. Gene-5 is involved in phage replication by virtue of being a DNA-helix destabilizing protein.

5.6. Sulfur-Containing Amino Acids

5.6.1. Methionine. Methionine has been used increasingly as an NMR probe (Table VI) both because of the ease of specifically labeling it with stable isotopes and because of its low frequency of occurrence in many proteins. The chemical reactivity and roles of this residue were recently summarized (Torchinsky, 1981).

In accord with the well-known hydrophobicity of methionine, crystallographic studies have frequently located it in the interiors of proteins (e.g., residues 13 and 30 in RNase and 180 in bovine α-chymotrypsin), where it is considered to stabilize the three-dimensional conformation by hydrophobic interactions. Aside from this typical structural role, it is found in cytochrome *c*'s coordinated to the heme iron. In other cases substrate binding and even catalytic functions have been suggested (Torchinsky, 1981).

As for chemical reactivity, the thioether side chain is a polarizable nucleophile. It is readily alkylated to a sulfonium cation, $—(CH_2)_2S^+(CH_3)_2$, or oxidized to a sulfoxide, $—(CH_2)_2S(O)CH_3$. The buried residues can be made more accessible by carrying out the reactions at acid pH, which promotes protein unfolding. The relatively small number of methionine residues present in proteins simplifies the isotopic labeling, which can be accomplished either *in vivo* by biosynthesis or by specific enrichment by S-methylation with $^{13}CH_3I$ (Figure 48). The resulting sulfonium ion $^{12/13}CH_3—S^+—R$ is subsequently demethylated with (for example) dithiothreitol (DTT) to form the ~50% C-enriched methylmethionine (Jones *et al.*, 1976).

More recently this procedure was extended to the formation of deuteromethionine by use of C^2H_3I (Wooten *et al.*, 1981; Figures 48, 16), although it was found that the usual conditions for methyl exchange at high pH

Table VI. NMR studies of methionine residues in proteins

Protein	Reference	Comment
β-Galactosidase	Cohen *et al.* (1979)	^{13}C enrichment and oxidation to sulfoxide
Basic pancreatic trypsin inhibitor	Harina *et al.* (1980)	^{13}C (25.2 MHz) Met-52 diastereomeric S-[^{13}C]CH_3 BPTI
C-I apolipoprotein	Chen *et al.* (1980)	^{13}C-enriched; incorporation into liposomes
α-Chymotrypsin	Matta *et al.* (1981)	S-[^{13}C]Met-192
Cytochrome *c*	Eakin *et al.* (1975b);	S-[2-^{13}C]carboxymethyl-Met-80
	Eakin *et al.* (1975a);	S-[^{13}C]methyl, Met-80 and Met-98
	Schejter *et al.* (1978);	S-[^{13}C]methyl, Met-55 and Met-131, Fe(II) and Fe(III)
	Wooten *et al.* (1981)	^{13}C- and ^{2}H-enriched Met-65 and Met-80 pH-induced conformations
Myelin basic protein	Deber *et al.* (1978)	S-[^{13}C]methyl at Met-20 and Met-167 (bovine) and Met-22 and Met-169 (human)
Myoglobin (sperm whale)	Jones *et al.* (1976)	C^{13}-enriched Met-55 and Met-131, T_1's at three fields
Ribonuclease	Jaeck and Benz (1979);	^{13}C-enriched Met-29, unfolding study
	Niu *et al.* (1979)	[ε-^{13}C]Met-13 RNase (1–15) and complex

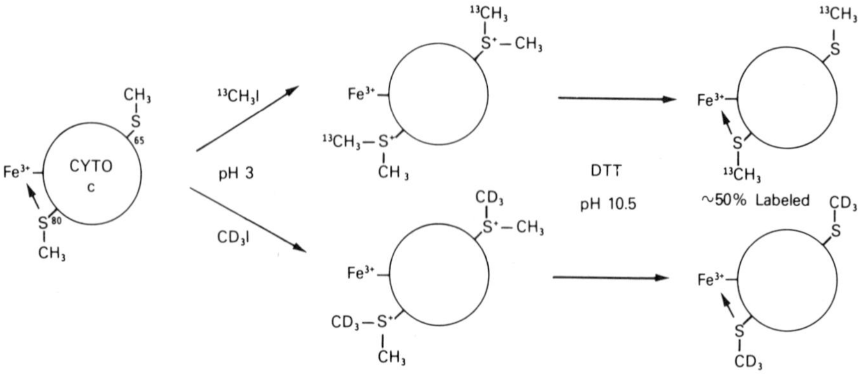

Figure 48. Schematic of ε-methylmethionine isotope substitution in ferricytochrome c, using either $^{13}CH_3I$ or CD_3I.

(~10) could lead to irreversible (or very slowly reversible) processes for cytochrome c. Attempts to carry out the alkylation at lower pH (~9) or with cysteine in place of DTT or for shorter time periods (less than 22 h) were relatively unsuccessful (Wooten *et al.*, 1981).

London (1980) points out that for methionine, parallel variations of NT_1, NOE, and $\Delta_{1/2}$ are frequently observed. After consideration of appropriate current models, he ascribes the variations among methionines to differences in amplitude rather than internal rotation rates. He argues persuasively that available methionine relaxation data suggest essentially free rotation about the $S—CH_3$ bond and restricted amplitude diffusion about the $CH_2—S$ bond.

The ^{13}C chemical shifts for amino acid residues in a series of small peptides and proteins have been tabulated by Howarth and Lilley (1978). They point out that chemical-shift values for residues beyond the penultimate unit are neighbor independent except when preceding proline (which shifts C_α ~0.20 ppm and C_β ~0.80 ppm). Thus in the denatured chain, the resonances from individual amino acids are position independent (except as noted), and observed differences in native proteins are considered to arise from secondary or tertiary structure.

Eakin and co-workers (1975a) incorporated [S-Me-^{13}C]methionine biosynthetically into position 98 in the cytochrome c of the fungus *Neurospora crassa* and considered ring-current effects of the heme. A transmethylation reaction also resulted in formation of the N-trimethyl cationic species of Lys-72. The isolated purified cytochrome c had a singlet at 53.6 ppm for the Fe(II) form and a second upfield resonance at 13.8 ppm for the Fe(III) form. The ^{13}C NMR spectrum of the harvested colonies suspended in water showed a predominant triplet centered at 55 ppm (assigned to $N^+(CH_3)$-Lys-72), [^{13}C]Met at 15.0 ppm, and hydroxymethyl carbons at 63.7 ppm. From the relative areas of these peaks, the relative degrees of chemical modification of these residues could be estimated. Eakin *et al.* (1975b) also examined the S-[2-^{13}C]carboxymethyl derivative of Met-80 in horse cytochrome *c*. Of course, carboxymethylation is a substantial chemical modification of protein structure, and later studies have tended to emphasize the isotopic replacement reaction for methyl groups on methionine. In one of the pioneer studies, Met-55 and Met-131 of sperm whale myoglobin were ^{13}C enriched, and relaxation measurements were reported at three frequencies (15.1, 25.2, and 67.9 MHz; Jones *et al.*, 1976). The T_1 values show the expected increase with frequency, based on the assumed relaxation predominantly by the ^{13}C–^{1}H dipolar mechanism for protonated carbons. The T_1's for C_ε of both methyl residues have the same value. The separate chemical shifts are sensitive to temperature, pH, ligand, and spin state of the heme iron. The resonance of Met-55 is downfield from that of Met-131 ($\Delta\delta \approx$ 1 ppm).

The dynamic studies were interpreted as showing substantial rotational motion in the side chain in addition to that normally expected about the S—C_ε bond; such mobility apparently serves the function of conformational adaptability for the protein mediated by the heme and its ligands.

Schejter *et al.* (1978) used $^{13}CH_3I$ to enrich the two methionine residues in horse cytochrome *c* and examined both oxidized, Fe(III), and reduced, Fe(II), states in various solvents (at 67.9 MHz). The chemical shifts are sensitive to the oxidation state of the iron atom; the signal for Met-80 disappears in the oxidized form because of its direct coordination to the paramagnetic Fe(III) atom. For the diamagnetic Fe(II) state, the Met-80 and Met-65 resonances are at 13.1 and 14.45 ppm, respectively (at 15°C). For the cyano-Fe(III) form, cyanide displaces Met-80 as a coordinating ligand and the latter shows its characteristic position. At high pH (~11) the data also suggested that Met-80 is displaced from the iron atom.

More recently, Wooten *et al.* (1981) extended the ^{13}C NMR studies (enriched and natural abundance) of horse ferricytochrome *c*, as well as ^{2}H NMR of ^{2}H-enriched methionine, including pH-dependent unfolding of the protein through a series of conformational states. Below pH 2, ferricytochrome *c* has a loosely folded tertiary structure. Refolding to the native state under acidic conditions corresponds to the bonding of the Met-80 sulfur to the heme iron. An alkaline transition also occurs, and at high pH, Fe(III)—S bonding is disrupted. In accord with numerous studies of methionine dynamics, appreciable internal motion was found about both S—CH_3 and S—CH_2 for Met-65. However, coordination to Fe severely restricts the CH_2—S rotation for Met-80.

Blakley and co-workers (1978) incorporated [methyl-^{13}C]methionine by *in vivo* synthesis into all seven methionines of the dihydrofolate reductase (DHFR) isoenzyme 2 of *S. faecium*. The denatured enzyme (6 *M* aqueous urea) shows a single resonance peak at 15.32 ppm (15°C, 25.2 MHz), whereas the enriched sample has three resonances (15.35, 14.99, and 14.66 ppm). Detailed studies included the free enzyme as well as complexes with various substrates (e.g., NADPH, $NADP^+$, dihydrofolate, 5-formyltetrahydrofolate) and the cancer drug methotrexate. The $NADP^+$ and the inhibitors show the most pronounced effects. Chemical shifts for C_ε ranged from 17.1 (for $NADP^+$) to 13.7 ppm, with 15.3 ppm corresponding to the denatured enzyme. The T_1's measured range from 0.35 to 0.72 s, and NOEs from 1.5 to 2.1. Using an overall molecular correlation time of 2×10^{-8} s, it was necessary to postulate not only rapid rotation for the C_ε-methyl group but considerable restricted rotation about $CH_2—S$. This appears to be the general picture emerging for Met.

The use of ^{13}C enrichment and oxidation to sulfoxide were both investigated (Cohen *et al.*, 1979) for the methionines in β-galactosidase. The sample was prepared by adding [methyl-^{13}C]methionine to the growth medium of *E. coli* K-12. The normal $^{13}C_\varepsilon$ peak for this enzyme (20°C, 67.9 MHz, aqueous in HCl) is at 14.19 ppm. The enzyme has 23 methionines for each of four subunits. The signals of both Met and Met sulfoxide were readily assigned, since C_ε-CH_3 for the latter has a characteristic downfield shift of 22.6 ppm resulting from oxygen deshielding. Measurement of peak areas allowed quantitative determination of the amount of oxidation undergone by methionine residues, and this roughly correlates with activity measurements. The average correlation time τ_c for the methionyl methyl groups is 2.1×10^{-9} s.

Two studies involving liposomes and methylmethionine markers were reported (Deber *et al.*, 1978; Chen *et al.*, 1980). The single methionine residue of apolipoprotein C-I has been both ^{13}C enriched and spin-labeled with nitroxide. The protein (57 residues) is of interest because its single methionine (at position 38) is in a helical segment (residues 33–53) considered to participate in phospholipid binding. The following results were noted: (1) a rotational correlation time τ_c of 0.22 ns was calculated from the nitroxide BPR data; (2) $T_1 = 320$ ms (in 10 m*M* phosphate buffer, 24°C, 25.2 MHz); (3) linewidth $W_{1/2} = 6.0$ Hz. Relaxation studies of the apo C-I protein plus multilamellar liposomes of dimyristoylphosphatidylcholine (DMPC) were also made in 1.6 *M* guanidine hydrochloride (to mitigate self-association of protein), and in 2-chloroethanol–phosphate buffer (to examine the effect of a helix-promoting solvent). Among the interesting results observed were the increase of linewidth of the thiomethyl peak (from 2.6 to 4.7 Hz) in the presence of the liposomes and the decrease in T_1 from ~1 to ~0.4 s, indicating restricted mobility of the methionine in the complex. The chloroethanol increased the linewidth but did not affect the spin-lattice relaxation time. The motions of the methionine were judged to be complex and not analyzable by simple models (Chen *et al.*, 1980).

Deber and co-workers (1978) reported data on ^{13}C-enriched human and bovine myelin basic protein (MBP) including incorporation of the protein into liposomes. The enriched residues correspond to positions 20 and 167 for bovine and 22 and 169 for the human protein. Both methylated samples give a singlet at 167.9 ppm (from external CS_2 or 25.9 ppm from TMS) whereas, each "native" or reduced form has a single upfield resonance at 178.5 ppm. This difference could result from removal of positive charge from near the methionines. The corresponding T_1's are 0.5 and 0.6 s for the respective cationic species, and 0.74 and 0.83 s for the native species. The authors concluded that there is substantial local mobility; note, for example, the nonresolution of the two $^{13}CH_3$ groups. The results are suggestive of a loose or open structure in which residues are exposed to solvent. The liposomes used were 50:50 phosphatidylcholine and phosphatidic acid. The addition of the enriched proteins appreciably broadens the liposome resonances. The methylated form shows one [^{13}C]methyl signal (167.8 ppm) but the reduced or native form shows what appears to be two peaks (176.77 and 178.15 ppm), possibly reflecting different environments for the two methionines.

Matta *et al.* (1980) reported the ^{13}C S-methylation of Met-192, which lies near the active site of α-chymotrypsin (Met-180 is buried and presumed to be nonreactive). The initial ^{13}C NMR spectrum is a doublet (of equal intensity peaks with δ's of 23.82 and 24.11 ppm), but upon equilibration the upfield peak increases. Their tentative explanation attributes the spectral differences to changes in counterions and conformational mobility accompanying dissolution of the enzyme. The S-methylated species was converted, presumably at Ser-195, to the phenylmethanesulfonyl derivative [—S(O)CH_2-ϕ] so that active site–inhibitor interactions could be examined. The PMS sample induces an 0.28 ppm downfield shift in the ^{13}C enriched methyl resonance, but does not exhibit the second peak. Diastereomeric effects were not reported but may provide an explanation for the doublet. Matia *et al.* (1981) also used 2-mercaptoacetyl-4′-methoxyanilide (pH 8.6; 5°C) to convert the S-methylsulfonium cation to $^{13}C_{\varepsilon}$ Met-192 α-chymotrypsin without the denaturing disruption of disulfide bonds caused by demethylations involving mercaptoethanol and dithiothreitol. The ^{13}C NMR (22.5 MHz, 25°C) results were pH dependent; at pH 5 three peaks were noted, 14.7, 16.3, and 13.3 ppm, with the latter being minor. The peak previously reported at 23.8 ppm was absent, and changes in peak intensities occurred with time. A full interpretation of the data was not forthcoming.

The single methionine residue of BPTI has been ^{13}C enriched (Harina *et al.*, 1980) and the two diastereomers resolved ($\Delta\delta$ = 0.2 ppm, invariant over the pH range 1–11). The Met-52 is "close to a hydrophobic region containing several aromatic residues (Tyr-21, Tyr-23, Phe-4, Phe-22 and Phe-45)." This derivatized BPTI was used to examine conformational changes in the protein, which carries an extra positive charge but is still biologically active. The nonequivalence of the peaks was noted to be sensitive to experimental conditions and was considered conformation dependent. Complete coalescence was found only upon complete denaturation above pH 13.

Preliminary studies on the unfolding of RNase were carried out by Jaeck and Benz (1979) using the ^{13}C-enriched Met-29 sulfonium derivative (δ = 26.7 ppm). Denaturation by 2.5 *M* guanidine deuterochloride shows an equilibrium between native and denatured states with chemical shifts of 26.9 and 26.3 ppm, respectively. Niu *et al.* (1979) reported semisynthetic complexes of RNase S′ with (1–15) ^{13}C-enriched peptides. Specifically, of the three synthetic peptides synthesized, one was [ε-^{13}C]Met-13 RNase (1–15). It was found that the midpoint of the temperature and acid denaturation of the RNase S′ complex is lower in temperature (5°C) and higher in pH (0.5 pH units) for the Met residue than for the His-12, Asp-14, or Ala-5 ^{13}C-enriched residues. This may indicate a weaker hydrophobic interaction for the methionine side chain.

5.6.2. Cysteine and Cystine. The sulfhydryl group (SH) in cysteine and the disulfide link (—S—S—) in cystine, formed by elimination of hydrogen (oxidation) between two cysteine residues, have significant catalytic and structural roles. The chemistry of the thiol moiety is reviewed by Torchinsky (1981).

In the thermal-unfolding study of RNase A observed by ^{13}C NMR (45.2 MHz) over the pH range 1–7, cysteine resonances (actually eight residues in four cystine cross-links) were examined by Howarth (1980). Linewidth comparisons, including those at pH 1.1 and 70°C, suggest that the disulfide (and also proline residues) is not in the fully denatured state. Packer *et al.* (1977) assigned the Cys ^{1}H resonances using ^{13}C enrichment and decoupling in *Clostridial* ferredoxins. Man and Bryant (1974) have used ^{35}Cl NMR relaxation studies for the quantitative determination of sulfhydryl groups in the protein rhodanese.

The isotope ^{33}S would be an ideal means to observe cystine and cysteine residues in proteins were it not for the very low sensitivity and prohibitively broad linewidths due to quadrupolar broadening (Cohen, 1978). But ^{77}Se may be a suitable substitute in view of its much greater sensitivity and narrow linewidths. (Luthra *et al.*, 1982).

5.7. Aliphatic Side Chains

Reports of resolved individual carbon resonances of aliphatic side chains in proteins are still not numerous (see Table VII) and involve predominantly low-molecular-weight proteins such as BPTI, lysozyme, cytochrome *c*, myoglobin, and a notable exception, hemoglobin. It should be noted that these residues are not observable by other spectroscopic methods. The increasing availability of high-field spectrometers should allow substantial progress on the assignments and molecular dynamics of alkyl moieties including the study of larger multisubunit proteins.

The typical aliphatic side-chain region in ^{13}C NMR spectra (~10 to 50 ppm; see Figure 1) shows a broad envelope corresponding to the many protonated side chains overlying occasional highly resolved narrow resonances, now understood to arise from internal motions of side-chain fragments (e.g., methyl and methylene) that are fast relative to the rate of overall

Table VII. NMR studies of aliphatic residues in proteins

Protein	References	Comment
Basic pancreatic trypsin inhibitor	DeMarco *et al.* (1977);	Ala-16; ^{13}C, active site
	Richarz and Wüthrich (1978);	Methyl-group dynamics
	Wüthrich *et al.* (1980);	
	Richarz *et al.* (1980a)	Ala, Leu, Ile; ^{13}C, two fields
Collagen (chick embryo)	Torchia and VanderHart (1976); Jelinsky and Torchia (1979)	Gly; ^{13}C, collagen fibrils, solid state
Cytochrome *c* (horse)	Moore and Williams (1980b);	Ala, Leu, Ile; ^{1}H, assignment
	Oldfield *et al.* (1975);	^{1}H, aliphatic region
	Wilbur and Allerhand (1977a)	^{13}C, comparison of several species
Cytochrome *c* (*Rhodospirillum*)	Crespi *et al.* (1968)	Leu; ^{1}H, deuterated protein hybrids
α-Elastin (aortic)	Urry and Mitchell (1976)	Ala, Val; ^{13}C, coacervate and fibrous states; methyl-group dynamics
Insulin (porcine)	Saunders and Offord (1972)	Gly; [1-^{13}C] or [2-^{13}C] substitution for N-terminal Phe
Lysozyme (hen)	Norton *et al.* (1977)	^{13}C, two fields
Myoglobin (sperm whale)	Jones *et al.* (1976);	Ile; ^{13}C, relaxation data
	Wittebort *et al.* (1979)	Ile; ^{13}C, dynamics
Parvalbumin (carp)	Nelson *et al.* (1976)	Ala; ^{13}C, dynamics, T_1
Proteoglycan core protein (chick limb bud)	Torchia and VanderHart (1976); Jelinsky and Torchia (1979)	Gly; ^{13}C, collagen fibrils, solid state
Rhodopsin (*Halobacterium halobium*)	Kinsey *et al.* (1981a,b)	[γ-$^{2}H_6$]Val, [^{13}C]Leu; membrane-bound protein, solid-state ^{2}H NMR
Ribonuclease	Chaiken (1974);	[1-^{13}C]Gly and [2-^{13}C]Gly; substituted for Ala
	Niu *et al.* (1979)	Ala; ^{13}C, semisynthetic complex

molecular tumbling, as discussed previously (see Section 3.2). The lack of alkyl chemical reactivity naturally has limited chemical-modification studies, with a corresponding emphasis on molecular dynamics and theoretical treatments.

As high-quality NMR relaxation data become available, they have been compared with theoretically calculated values using increasingly sophisticated models. For example, Richarz *et al.* (1979,1980a) reported methyl-group assignments, T_1's, and NOEs at both 25.1 and 90.5 MHz for BPTI (experimental values of T_1's were 45 and 260 ms, and those of NOEs, 1.4 and 1.2, respectively, at 39°C). Lipari and Szabo (1982) then calculated that the observed motional flexibility is consistent with an interpretation allowing free methyl rotation plus wobbling of the C_α—C_β methyl group within a cone (see Section 3.2). Such studies fully support the concept of mobile, but sometimes restricted, side chains, although the full implications with respect to enzyme mechanisms and the possible participation of protein "breathing" implied by local chain mobility remain to be fully explored. The seeming paradox of the essential correspondence between dynamic protein NMR structures and close-packed crystal diffraction structures has been considered by several authors (see Jones *et al.*, 1976; Hugli and Gurd, 1970; Wittebort *et al.*, 1979). However, the theoretical treatments thus far apply most accurately to low-molecular-weight, approximately spherical proteins (i.e., those isotropic in their molecular tumbling). An advance has recently been made by Lipari and Szabo (1982), who have developed a two-parameter theory to treat molecular dynamics (see Section 3.2).

Having the smallest possible side chain, a proton, makes glycine unique in the following respects: no optical isomerism is possible; lack of steric hindrance permits close approach of protein chain segments; and conformational flexibility of the chain is promoted. In fact, glycine has been observed to have hinging functions in, for example, trypsinogen (Huber and Bode, 1978; Richardson, 1981). Specific studies that feature glycine are (1) the semisynthetic ^{13}C synthesis of Saunders and Offord (1972) in which either [1-^{13}C]glycine or [2-^{13}C]glycine was substituted for the N-terminal phenylalanine in chain B of porcine insulin (51 amino acids) and (2) a study in which Ala-6 of synthetic S-peptide of RNase A was replaced by these same two isotopic substitutions (Chaiken, 1974). However, such substitutions or additions are of limited value compared with utilization of the correct enriched amino acid.

Torchia *et al.* (1981) have studied by ^{13}C NMR (15.1 MHz) the chick limb bud proteoglycan core protein enriched with [2-^{13}C]glycine and [3-^{13}C] serine. The motional characteristics of this branched copolymer species, a protein core with polysaccharide side chains, were determined from T_1 values, linewidths, NOE values, and correlation times, and were analyzed in detail by calculations and comparison to theoretical models. Heterogeneous peptide-backbone motion and flexibility of the side chains was noted; these serines serve as covalent branch points for the glycosaminoglycan side chains and show hindered rotation.

In their ^{13}C NMR study of muscle calcium-binding protein, Nelson *et al.* (1976) reported that NMR parameters of the β carbons of alanine do not fit the isotropic-diffusion model or the single internal rotation model, and Howarth and Lilley (1978) warned that existing models for calculating alanine methyl T_1's do not fit the data well. They also suggested the possibility of nonexponential T_1 relaxation from cross-correlation studies, for example, in tetragastrin.

Studies of alanine often occur in conjunction with studies of other methyl-containing residues. Thus the 270 MHz NMR study of a total of 49 methyl groups of horse cytochrome *c* involves six alanines (Moore and Williams, 1980b). This paper has an extended discussion of the difficulties of differentiating aliphatic proton peaks, for example, those of alanine, from those of valine and leucine. After establishing multiplet structures and coupling constants, they supplemented their double-resonance procedure by known crystallographic structures and the known sequence of homologous cytochrome *c*'s; Ala-15 was definitively assigned with chemical-shift values for the Fe(II) form, CH_3 = 2.07 ppm and $C_{\alpha}H$ = 5.63 ppm at 57°C. The advantage of working with well-characterized systems is striking. Complete assignment was not possible, but by careful analysis of chemical-shift effects (e.g., heme ring-current effects) and known X-ray structure, 15 resonances of 28 aliphatic protons of ferrocytochrome *c* and 10 of 24 of ferricytochrome *c* were assigned. The six heme methyls were previously assigned. The heme proteins are an excellent example not only of effects on chemical shifts of folding into globular form but of the pronounced ring-current effects of moieties such as the heme group (Dobson *et al.*, 1974).

The rather striking rotational freedom of methyl groups of several fibrous proteins was recently reported. Two such studies are ideal models for NMR of these nonglobular proteins with repetitive amino acid sequences. These include the studies of collagen fibrils by Torchia and co-workers (1976; Jelinski *et al.*, 1980) and measurements reported by Urry and Mitchell (1976) on aortic α-elastin and its model peptides. By finding identifiable sharp peaks for valine and alanine, they confirmed the appreciable motion of methyl groups even in the condensed phases of fibrous elastin.

As part of their study of protein-membrane dynamics, Kinsey *et al.* (1981a,b) examined [γ-$^2H_6$]Val bacteriorhodopsin in the purple membrane. In contrast to bulky aromatic residues, fast motions ($>10^6$ s^{-1}) around Val C_{β}—C_{γ} bonds extends to as low as -75°C (see Section 2.4.1). Crespi and co-workers (1968,1969) in early studies biosynthetically prepared isotopic hybrids with [^1H]-L-leucine incorporated into the otherwise fully deuterated proteins c-phycocyanin, c-phycoerythrin, and cytochrome *c*; the C_{α} protons of Leu were found to exchange for ^2H during synthesis, and of course normally exchanging protons equilibrate with ^1H_2O during the NMR measurements. Peak assignments were thus readily made in the absence of other aliphatic-side-group signals. The narrow, sharper resonance signals shown by cytochrome *c* were interpreted as showing appreciable mobility for many of the leucines.

Levy *et al.* (1981) did molecular-dynamics simulations (25 MHz) of BPTI in order to examine how the T_1 values for buried, specifically assigned ^{13}C resonances are perturbed by fluctuations in the picosecond range. Their detailed analysis examines 62 interior protonated carbons for BPTI at 300°K, subdivided into various carbon types. The non-C_α's of Leu and Ileu are especially susceptible to increases of their ^{13}C relaxation times by the picosecond dynamics. The authors caution that for some carbons, slower motions (e.g., nanosecond) may also affect T_1.

The Leu-32 of horse ferrocytochrome *c* has been assigned (Moore and Williams, 1980b) at 57°C, 270 MHz; the protein contains five other leucines. Horse cytochrome *c* also contains six Ile residues that have been studied by Moore and Williams (1980b). These authors used homologous proteins to assist in peak assignments (only three of the six Ile were noted to be conserved), and specifically they used ring-current calculations, pH, and temperature variations to distinguish Ile-57 from Ile-81.

The nine Ile residues of myoglobin have been well characterized by NMR (Jones *et al.*, 1976; Wittebort *et al.*, 1979). Eight of the resonances observed correspond to $C_{\delta 1}$ individual carbons. The most favorable dynamic model is a restricted-diffusion model allowing free rotation of terminal methyls plus hindered, possibly concerted, rotations of appreciable amplitude (± 50 degrees) about C_α—C_β and C_β—$C_{\gamma 1}$ bonds. The NMR analysis corroborated the view that aliphatic side chains possess considerable mobility.

As an example of ^{13}C-enriched noncovalently bound species, the study of the binding of [1-^{13}C]ethyl and [2-^{13}C]ethyl isocyanide to myoglobin and hemoglobin by Gilman (1979) can be cited. Chemical shifts, relaxation times, and NOEs were measured and analyzed by the restricted-diffusion model. The data analysis led to the conclusion that the ethyl moiety in the bound ethyl isocyanide shows virtually the same motional characteristics in sperm whale myoglobin, harbor seal myoglobin, and the R form of human hemoglobin.

5.8. Proline

Although there have been extensive NMR studies of Pro-containing peptides, there have been very few studies of Pro in proteins. This arises largely from the poor resolution in the aliphatic region of ^1H and ^{13}C spectra, where Pro resonances occur. It was shown in ^1H NMR studies of oligoprolines and poly-L-prolines that it is possible to resolve signals from the C_α H of *cis*- and *trans*-peptide conformations (Deber *et al.*, 1970). More recently it has been demonstrated that similar resonances can be resolved in ^{13}C NMR spectra of cyclic hexapeptides containing Pro, that is (X-Pro-Y)$_2$. These resonances correspond to the *cis* and *trans* backbone carbonyls of the Pro residues. The relative signal intensities allow delineation of the proportions of these two conformations, if present, and enable deductions to be made about the characteristic sequences one would expect to encounter around Pro residues in β turns (Gierasch *et al.*, 1981).

Although there is a dearth of NMR studies of proline resonances in spectra of proteins, this is somewhat surprising given the role the *cis–trans* isomerization of Pro apparently plays in slow equilibration during refolding of the polypeptide chain (Brandts *et al.*, 1977) and also the fact that some biologically important proteins, such as the hormone human chorionic gonadotropin, contain large proportions of Pro residues (Bell *et al.*, 1969).

6. CONCLUSIONS AND PROGNOSIS

It is 25 years since the first NMR spectrum of a protein was published (Saunders *et al.*, 1957). A cursory comparison of this spectrum of RNase at 40 MHz with that obtained recently at 500 MHz (Figure 49) is enough to demonstrate how far we have come. Clearly a large measure of this advance results from the purely technological developments, notably in magnets,

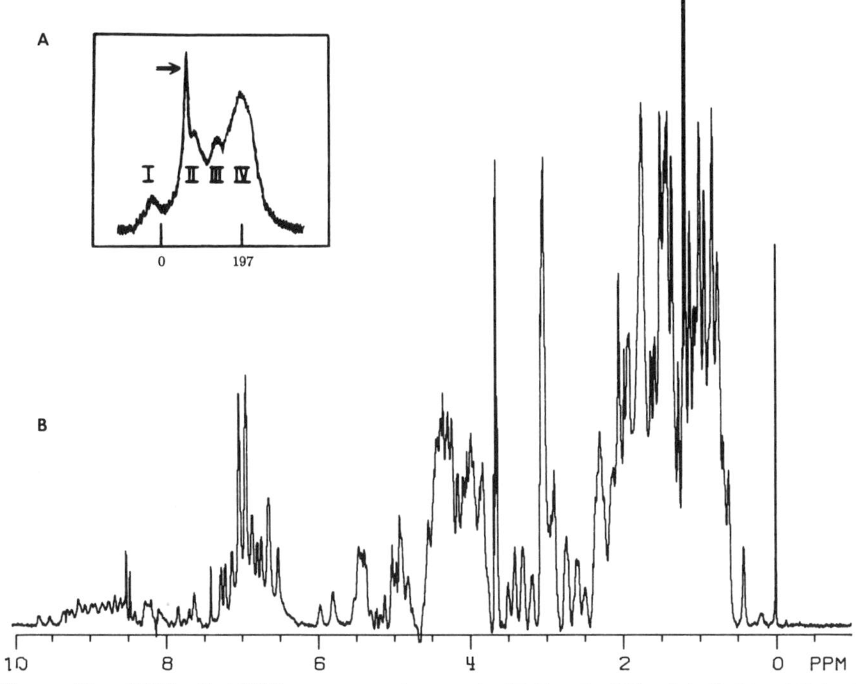

Figure 49. (*A*) The first NMR spectrum of a protein (RNAse A; 20% w/v in D_2O) published, a single slow scan at 40 MHz (Saunders *et al.*, 1957.) The arrow indicates the HDO resonance. [Reprinted with permission from Saunders *et al.* (1957). Copyright 1957, American Chemical Society.] (*B*) 500-MHz proton NMR spectrum of RNAse A (23 mg in 0.5 ml of 0.2 *M* deuteroacetate-D_2O, pH 6.8) recorded on an NT500 spectrometer at 40°C. The sample was not lyophilized from D_2O; HDO was continuously decoupled (0.4 W), resulting in the minimum at 4.35 ppm. The standard is TSP; the sharp peak at 2.8 ppm arises from acetate; the sharp resonances at 9 ppm are His C_ε protons, and the other resonances from 7.5 to 9 ppm are unexchanged N—H. Spectral parameters were 90° pulse (9.5 μs); repetition rate, 2.4 s; line broadening, 1 Hz.

electronics, and computers. But as far as the subject matter is concerned, in the intervening years a great deal of valuable information has been gleaned about proteins in solution.

The first unambiguous assignment of a resolved resonance of a single atom in a protein molecule, that of the C_ε proton of the single histidine residue at position 15 in hen egg white lysozyme (Meadows *et al.*, 1967), was reported a scant 15 years ago. The first resolution of individual resonances in the ^{13}C NMR spectra of a protein, also hen egg white lysozyme, was reported only 9 years ago (Allerhand *et al.*, 1973). The ability to study individual side-chain groups in many proteins has in recent years become essentially routine. The range of the NMR method was recently demonstrated by Mildvan and his associates in their resolution of C_ε proton resonances of histidine residues in very high molecular weight proteins such as pyruvate kinase (MW 237,000) (Meshitsuka *et al.*, 1981).

The multivariate uses of stable isotopes (^{13}C, ^{2}H, ^{15}N, ^{19}F), incorporated both chemically and biosynthetically, have extended the range of the NMR method in many elegant experiments. The recent introduction of sophisticated two-dimensional and multipulse NMR techniques (see Section 2.3) may result in a new era in the application of NMR methods to proteins.

We commented elsewhere (Cohen and Wlodawer, 1982) on the need for NMR spectroscopists not to seek to "confirm" structures of proteins determined by X-ray crystallography. It is also true that protein chemists must learn to deal with the results of NMR experiments as willingly, but perhaps not as uncritically, as they have dealt with X-ray structures of proteins in the past.

This review represents the state of the art when NMR spectroscopy, in all its various forms, can be considered to have become a standard technique in advancing our knowledge of protein structure and functional correlates. In 25 years hence, perhaps, the spectrum recorded at 500 MHz (Figure 49B) may appear as limited as that at 40 MHz (Figure 49A) appears today. At that juncture, any side chain in any protein, even at very low concentrations, may be a candidate for the NMR spectroscopist's molecular magnifying glass.

ACKNOWLEDGMENTS

We thank Elmer Leininger for his outstanding work in typing this manuscript, and Chi-wan Chen for running the 500 MHz spectrum in Figure 49.

SELECTED BIBLIOGRAPHY

Books

Becker, P. D. (1980), *High Resolution NMR*, 2nd ed., Academic Press, New York.
Dwek, R. (1973), *NMR in Biochemistry*, Clarendon Press, Oxford.

Farrar, T., and E. D. Becker (1971), *Pulse and Fourier Transform NMR*, Academic Press, New York.

Fukushima, E., and S. B. W. Roeder (1981), *Experimental Pulse NMR*, Addison-Wesley, Reading, Mass.

Jardetzky, O., and G. C. K. Roberts (1981), *NMR in Molecular Biology*, Academic Press, New York.

Levy, G., R. Lichter, and G. Nelson (1980), *Carbon-13 NMR*, 2nd ed., Wiley, New York.

Shaw, D. (1976), *Fourier Transform NMR Spectroscopy*, Elsevier, New York.

Wüthrich, K. (1976), *NMR in Biological Research: Peptides and Proteins*, Elsevier, New York.

Reviews

Allerhand, A. (1979), *Meth. Enzymol.* **61B,** 458.

Egan, W., H. Shindo, and J. S. Cohen (1977), *Ann. Rev. Biophys. Bioeng.* **6,** 383.

Howarth, I. W., and D. M. J. Lilley (1978), *Progr. NMR Spect.* **12,** 1.

London, R. E. (1980), *Magn. Reson. Biol.* **1,** 1.

REFERENCES

Aguiar, A., G. H. DeMaas, E. R. Jansen, A. J. Slotboom, and R. J. P. Williams (1979), *Eur. J. Biochem.* **100,** 511.

Allerhand, A., R. F. Childers, and E. Oldfield (1973), *Biochemistry* **12,** 1335.

Allerhand, A. (1979), *Meth. Enzymol.* **61B,** 458.

Allerhand, A., R. S. Norton, and R. F. Childers (1977), *J. Biol. Chem.* **252,** 1786.

Allerhand, A., D. W. Cochran, D. Doddrell (1970), *Proc. Natl. Acad. Sci. USA* **67,** 1093.

Allerhand, A., D. Doddrell, V. Glushko, D. W. Cochran, E. Wenkert, P. J. Lawson, and F. R. N. Gurd (1971), *J. Am. Chem. Soc.* **93,** 544.

Alma, N. C. M., B. J. W. Harmsen, C. W. Hilbers, G. Van der Marel, and J. H. van Boom (1981a), *FEBS Lett.* **135,** 15.

Alma, N. C. M., B. J. M. Harmsen, W. E. Hull, G. van der Marel, J. H. van Boom, and C. W. Hilbers (1981b), *Biochemistry* **20,** 4419.

Arata, Y., S. Kimura, H. Matsuo, and K. Narita (1976), *Biochem. Biophys. Res. Comm.* **73,** 133.

Arndt, K. T., F. Boschelli, P. Lu, and J. H. Miller (1981), *Biochemistry* **20,** 6109.

Arus, C., L. Paolillo, R. Llovens, R. Napolitano, X. Pares, and C. M. Cuchillo (1981), *Biochim. Biophys. Acta* **660,** 117.

Bachovchin, W. W., and J. D. Roberts (1978), *J. Am. Chem. Soc.* **100,** 8041.

Bachovchin, W. W., R. Kaiser, J. H. Richards, and J. D. Roberts (1981), *Proc. Natl. Acad. Sci. USA* **78,** 7323.

Baillargeon, M. W., M. Laskowski, Jr., D. E. Naves, M. A. Porubcan, R. E. Santini, and J. L. Markley (1980), *Biochemistry* **19,** 5703.

Ball, J. C., and D. L. Vander Jagt (1981), *Biochemistry* **20,** 899.

Batchelder, I. W., C. E. Sullivan, L. W. Jelinski, and D. A. Torchia (1982), *Proc. Natl. Acad. Sci. USA* **79,** 386.

Becker, P. D. (1980), *High Resolution NMR,* 2nd ed., Academic Press, New York.

Bell, J. J., R. E. Canfield, and J. J. Sciarra (1969), *Endocrinology* **84,** 298.

Bendall, M. R., D. M. Doddrell, and D. T. Pegg (1981), *J. Am. Chem. Soc.* **103,** 4603.

Berliner, I. J., and R. Kaptein (1980), *Biochemistry* **20,** 799.

Blackburn, P., and S. Moore (1982), *The Enzymes* **15B**, 317.

Blake, C. C. P., R. Cassels, C. M. Dobson, F. H. Poulsen, R. J. P. Williams, and K. S. Wilson (1981), *J. Mol. Biol.* **147**, 73.

Blakley, R. I., L. Cocco, R. E. London, T. E. Walker, and N. A. Matwiyoff (1978), *Biochemistry* **17**, 2284.

Blomberg, F., W. Maurer, and M. Ruterjans (1977), *J. Am. Chem. Soc.* **99**, 8149.

Blow, D. M., J. J. Birktoft, and B. S. Hartley (1969), *Nature* **221**, 337.

Bolton, P. H. (1981), *J. Magn. Reson.* **45**, 418.

Boswell, A. P., G. R. Moore, R. J. P. Williams, J. C. Chien, and L. C. Dickinson (1980), *J. Inorg. Chem.* **13**, 347.

Botelho, L. H., and F. R. N. Gurd (1978), *Biochemistry* **17**, 5188.

Bradbury, E. M., and H. W. E. Rattle (1972), *Eur. J. Biochem.* **27**, 270.

Bradbury, E. M., P. D. Cary, C. Crane-Robinson, H. W. E. Rattle, M. Boublik, and P. Sautiere (1975), *Biochemistry* **14**, 1876.

Bradbury, J. H., and B. E. Chapman (1972), *Biochem. Biophys. Res. Commun.* **49**, 891.

Bradbury, J. H., and L. R. Brown (1973), *Eur. J. Biochem.* **40**, 565.

Bradbury, J. H., and R. S. Norton (1973), *Biochim. Biophys. Acta* **328**, 10.

Bradbury, J. H., and R. S. Norton (1975), *Eur. J. Biochem.* **53**, 387.

Bradbury, J. H., and H. A. Scheraga (1966), *J. Am. Chem. Soc.* **88**, 4240.

Bradbury, J. H., and J. S. Teh (1975), *Biochem. Biophys. Res. Commun.* **52**, 936.

Bradbury, J. H., M. W. Crompton, and J. S. Teh (1977), *Eur. J. Biochem.* **81**, 411.

Bradbury, J. H., S. D. M. Deacon, and M. D. Ridgway (1979), *Chem. Commun.* 997.

Bradbury, J. H., V. Ramesh, and G. Dodson (1981), *J. Mol. Biol.* **150**, 609.

Brandts, J. F., M. Brennan, and L. N. Liu (1977), *Proc. Natl. Acad. Sci. USA* **74**, 4178.

Brauer, M., and F. W. Benz (1978), *Biochim. Biophys. Acta* **533**, 186.

Brown, F. F., T. F. Parsons, D. S. Sigman, and J. G. Pierce (1979), *J. Biol. Chem.* **254**, 4335.

Brown, L. R., and J. H. Bradbury (1975), *Eur. J. Biochem.* **54**, 219.

Brown, L. R., A. De Marco, G. Wagner, and K. Wüthrich (1976), *Eur. J. Biochem.* **62**, 103.

Browne, W. J., A. C. T. North, D. C. Phillips, K. Brew, T. C. Vanaman, and R. L. Hill (1969), *J. Mol. Biol.* **42**, 56.

Burns, P. D., and G. N. La Mar (1979), *J. Am. Chem. Soc.* **101**, 5844.

Burger, A. R., S. J. Lippard, M. W. Pantolino, and J. S. Valentine (1980), *Biochemistry* **19**, 4139.

Bystrov, V. F., A. S. Argeniev, Y. D. Gavrilov (1978), *J. Magn. Reson.* **30**, 151.

Campbell, I. D., C. M. Dobson, R. J. P. Williams, and A. V. Xavier (1973), *J. Magn. Reson.* **11**, 172.

Campbell, I. D., S. Lindskog, and A. I. White (1974), *J. Mol. Biol.* **90**, 469.

Campbell, I. D., C. M. Dobson, and R. J. P. Williams (1975a), *Proc. Royal Soc. London* **B189**, 503.

Campbell, I. D., S. Lindskog, and A. J. White (1975b), *J. Mol. Biol.* **98**, 597.

Campbell, I. D., C. M. Dobson, R. J. P. Williams, and P. B. Wright (1975c), *FEBS Lett.* **57**, 96.

Campbell, I. D., C. M. Dobson, G. R. Moore, S. J. Perkins, and R. J. P. Williams (1976), *FEBS Lett.* **70**, 96.

Canioni, P., P. J. Cozzone, and L. Sardi (1980), *Biochim. Biophys. Acta* **621**, 29.

Carver, J. P., B. H. Barber, and B. J. Fuber (1977), *J. Biol. Chem.* **252**, 3141.

Cassels, R., C. M. Dobson, F. M. Poulson, and R. J. P. Williams (1978), *Eur. J. Biochem.* **92**, 81.

Cave, A., C. M. Dobson, J. Parello, and R. J. P. Williams (1976), *FEBS Lett.* **65,** 190.

Chaiken, I. M. (1974), *J. Biol. Chem.* **249,** 1247.

Chaiken, I., M. H. Freedman, J. R. Lyerla, and J. S. Cohen (1973), *J. Biol. Chem.* **248,** 884.

Chaiken, I. M., J. S. Cohen, and E. Sokoloski (1974), *J. Am. Chem. Soc.* **96,** 4703.

Chapman, G. E., P. G. Hartman, and E. M. Bradbury (1976), *Eur. J. Biochem.* **61,** 69.

Chapman, G. E., F. J. Aviles, C. Crane-Robinson, and E. M. Bradbury (1978), *Eur. J. Biochem.* **90,** 287.

Chen, T. C., R. D. Knapp, M. F. Rohde, J. R. Brainard, A. M. Gotto, Jr., J. T. Sparrow, and J. D. Morrisett (1980), *Biochemistry* **19,** 5140.

Chien, J. C. W., and W. B. Wise (1973), *Biochemistry* **12,** 3418.

Cocco, I., R. L. Blakley, T. E. Walker, and R. E. London (1977), *Biochem. Biophys. Res. Commun.* **76,** 183.

Cocco, I., R. L. Blakley, T. E. Walker, R. E. London, and N. A. Matwiyoff (1978), *Biochemistry* **17,** 4285.

Cohen, J. S. (1969), *Nature* **223,** 43.

Cohen, J. S. (1971), *Biochim. Biophys. Acta* **229,** 603.

Cohen, J. S. (1978), *Crit. Rev. Biochem.* **5,** 25.

Cohen, J. S., and M. B. Hayes (1974), *J. Biol. Chem.* **249,** 5472.

Cohen, J. S., and O. Jardetzky (1968), *Proc. Natl. Acad. Sci. USA* **60,** 93.

Cohen, J. S., and H. Shindo (1975), *J. Biol. Chem.* **250,** 8874.

Cohen, J. S., and A. Wlodawer (1982), *Trends Biochem. Sci.,* **7,** 389.

Cohen, J. S., R. I. Shrager, M. McNeel, and A. N. Schechter (1970a), *Biochem. Biophys. Res. Commun.* **40,** 144.

Cohen, J. S., A. N. Schechter, R. Shrager, and M. McNeel (1970b), *Nature* **228,** 642.

Cohen, J. S., M. Feil, and I. Chaiken (1971), *Biochim. Biophys. Acta* **236,** 468.

Cohen, J. S., C. T. Yim, M. Kandel, A. G. Gornall, S. I. Kandel, and M. H. Freedman (1972a), *Biochemistry* **11,** 327.

Cohen, J. S., H. P. Hagenmaier, H. Pollard, and A. N. Schechter (1972b), *J. Mol. Biol.* **71,** 513.

Cohen, P., J. H. Griffin, M. Camier, M. Caizergues, P. Fromagect, and J. S. Cohen (1972c), *FEBS Lett.* **25,** 282.

Cohen, J. S., J. Griffin, and A. N. Schechter (1973), *J. Biol. Chem.* **248,** 4305.

Cohen, J. S., W. R. Fisher, and A. N. Schechter (1974), *J. Biol. Chem.* **249,** 1113.

Cohen, J. S., J. Yariv, A. J. Kalb, L. Jacobson, and Y. Schechter (1979), *J. Biochem. Biophys. Meth.* **1,** 145.

Cohen, J. S., C. H. Niu, S. Matsuura, and H. Shindo (1980), in *Frontiers in Protein Chemistry*, T. Y. Liu, G. Mamiya, and K. T. Yasunobu, Eds., Elsevier, New York, p. 3.

Coleman, J. E., and I. M. Armitage (1978), *Biochemistry* **17,** 5038.

Craik, D. J., A. Kumar and G. C. Levy (1982), *J. Chem. Inf. Comp. Sci.*, in press.

Crespi, H. L., and J. J. Katz (1969), *Nature* **224,** 560.

Crespi, H. L., R. M. Rosenberg, and J. J. Katz (1968), *Science* **161,** 795.

Crestfield, A. M., W. H. Stein, and S. Moore (1963), *J. Biol. Chem.* **238,** 2413.

Cross, T. A., and S. J. Opella (1981), *Biochemistry* **20,** 290.

Cutnell, J. D. (1982), *J. Am. Chem. Soc.* **104,** 362.

Moscarello, and D. D. Wood (1978), *Biochemistry* **17,** 898.

Deber, C. M., F. A. Bovey, J. P. Carver, and E. R. Blout (1970), *J. Am. Chem. Soc.* **92,** 6191.

Deber, C. M., M. A. Moscarello, and D. D. Wood (1978), *Biochemistry* **17,** 898.

Deese, A. J., E. A. Dratz, F. W. Dahlquist, and M. R. Paddy (1981), *Biochemistry* **20,** 6420.

Deisenhofer, J., and W. Steigemann (1975), *Acta Cryst.* **B31,** 238.

Deslauriers, R., W. H. McGregor, D. Sarantakis, and I. C. P. Smith (1974), *Biochemistry* **13,** 3443.

De Marco, A., M. Llinas, and K. Wüthrich (1978), *Biopolymers* **17,** 617.

De Marco, A., H. Tschesche, G. Wagner, and K. Wüthrich (1977), *Biophys. Struct. Mech.* **3,** 303.

Dobson, C. M. (1975), *Biochemistry* **14,** 4905.

Dobson, C. M. (1977), in *NMR in Biology*, L. A. Dwek, R. A. Campbell, I. D. Richards, Eds., Academic Press, New York, p. 63.

Dobson, C. M., N. J. Hoyle, C. F. Geraldes, M. Bruschi, J. Legall, P. E. Wright, and R. J. P. Williams (1974), *Nature* **249,** 425.

Dobson, C. M., S. J. Ferguson, F. M. Poulsen, and R. J. P. Williams (1978), *Eur. J. Biochem.* **92,** 99.

Doddrell, D. M., and D. T. Pegg (1980), *J. Am. Chem. Soc.* **102,** 6390.

Dooijewaard, G., F. F. Roossien, and G. T. Robillard (1979), *Biochemistry* **18,** 2996.

Dwek, R. (1973), *NMR in Biochemistry,* Clarendon Press, Oxfrod.

Dwek, R. (1973), *NMR in Biochemistry*, Clarendon Press, Oxford.

Dwek, R. A., J. C. A. Knott, D. Marsh, A. C. McLaughlin, E. M. Press, N. C. Price, and A. I.

Eakin, R. T., L. O. Morgan, and N. A. Matwiyoff (1975a), *Biochem. J.* **152,** 529.

Eakin, R. T., L. O. Morgan, and N. A. Matwiyoff (1975b), *Biochemistry* **14,** 4538.

Edsall, J. T., R. E. Martin, and B. R. Hollingworth (1958), *Proc. Natl. Acad. Sci. USA* **44,** 505.

Edwards, B. F. P., and B. D. Sykes (1978), *Biochemistry* **17,** 684.

Egan, W., H. Shindo, and J. S. Cohen (1977), *Ann. Rev. Biophys. Bioeng.* **6,** 338.

Egan, W., H. Shindo, and J. S. Cohen (1978), *J. Biol. Chem.* **253,** 16.

Epstein, H., A. N. Schechter, and J. S. Cohen (1971), *Proc. Natl. Acad. Sci. USA* **68,** 2042.

Feeney, J., G. C. K. Roberts, B. Birdsall, D. V. Griffiths, R. W. King, P. Scudder, and A. S. V. Burgen (1977), *Proc. Roy. Soc. (London)* **B196,** 267.

Findlay, D., D. G. Herries, A. P. Mathias, B. R. Rabin, and C. A. Ross (1962), *Biochem. J.* **85,** 152.

Forsen, S., and R. A. Hoffman (1963), *J. Chem. Phys.* **39,** 2892.

Freedman, M., J. S. Cohen, and I. Chaiken (1971), *Biochem. Biophys. Res. Comm.* **42,** 1148.

Freedman, M. H., J. R. Lyerla, Jr., I. M. Chaiken, and J. S. Cohen (1973), *Eur. J. Biochem.* **32,** 215.

Freeman, R. (1980), *Proc. Royal Soc. London* **A373,** 149.

Freeman, R., and G. A. Morris (1979), *Bull. Magn. Res.* **1,** 5.

Fujii, S., K. Akasaka, and H. Hatane (1980), *J. Biochem. (Tokyo)* **88,** 789.

Fung, L. W., K. C. Liu, and C. Ho (1975), *Biochemistry* **14,** 3424.

Furie, E. C., M. Blumenstein, and B. Furie (1979), *J. Biol. Chem.* **254,** 12521.

Gall, C. M., J. A. DiVerdi, and S. J. Opella (1981), *J. Am. Chem. Soc.* **103,** 5039.

Gall, C. M., T. A. Cross, J. A. DiVerdi, and S. J. Opella (1982), *Proc. Natl. Acad. Sci. USA* **79,** 101.

Gassner, M., O. Jardetzky, and W. Conover (1978), *J. Magn. Reson.* **30,** 141.

Gerken, T. A., E. Joyce, J. B. Jentoft, N. Jentoft, and D. C. Dearborn (1982), personal communication.

Gettins, P., and R. A. Dwek (1977), in *NMR in Biology*, R. A. Dwek, I. D. Campbell, R. E. Richards, and R. J. P. Williams, Eds., Academic Press, London, p. 125.

Gettins, P., J. Boyd, C. P. J. Glaudemans, M. Potter, and R. A. Dwek (1981), *Biochemistry* **20,** 7463.

Gierasch, L. M., C. M. Deber, V. Madison, C-H. Niu, and E. R. Blout (1981), *Biochemistry* **20,** 4730.

Gilman, J. G. (1979), *Biochemistry* **18,** 2273.

Glasoe, P. K., and F. A. Long (1960), *J. Phys. Chem.* **64,** 188.

Glick, R. E., D. F. Kates, and S. J. Ehrenson (1959), *J. Chem. Phys.* **31,** 567.

Glickson, J. D., C. C. McDonald, and W. D. Phillips (1969), *Biochem. Biophys. Res. Commun.* **35,** 492.

Glickson, J. D., W. D. Phillips, and J. A. Rupley (1971), *J. Am. Chem. Soc.* **93,** 4031.

Glushko, V., P. J. Lawson, and F. R. N. Gurd (1972), *J. Biol. Chem.* **247,** 3176.

Gordon, S. L., and K. Wüthrich (1978), *J. Am. Chem. Soc.* **100,** 7094.

Greenfield, N. J., and N. M. Williams (1972), *Biochim. Biophys. Acta* **257,** 187.

Griffin, J. H., A. N. Schechter, and J. S. Cohen (1973), *Ann. N.Y. Acad. Sci.* **222,** 693.

Gupta, R. K., and J. M. Pesando (1975), *J. Biol. Chem.* **250,** 2630.

Gupta, R. K., J. A. Ferretti, and E. D. Becker (1974), *J. Magn. Reson.* **13,** 275.

Gurd, F. R. N., and T. M. Rothgeb (1979), *Adv. Prot. Chem.* **33,** 73.

Haffner, P. H., and J. H. Wang (1973), *Biochemistry* **12,** 1608.

Hagen, D. S., J. H. Weiner, and B. D. Sykes (1978), *Biochemistry* **17,** 3860.

Harina, B. M., D. F. Dyckes, R. M. Willcott, III, and W. C. Jones, Jr. (1980), *J. Am. Chem. Soc.* **102,** 1120.

Hartman, P. G., G. E. Chapman, and E. M. Bradbury (1977), *Eur. J. Biochem.* **77,** 45.

Hayes, M. B., H. Hagenmaier, and J. S. Cohen (1975), *J. Biol. Chem.* **250,** 7461.

Heinrickson, R. L., W. H. Stein, A. M. Crestfield, and S. Moore (1965), *J. Biol. Chem.* **240,** 2921.

Hentschel, R., and H. W. Spiess (1979), *J. Magn. Reson.* **35,** 157.

Hill, A. V. (1910), *J. Physiol.* **40,** 4.

Hilton, B. D., K. Trudeau, and C. K. Woodward (1981), *Biochemistry* **20,** 4697.

Hoffman, R. A., and S. Forsen (1966), *Prog. NMR Spectrosc.* **1,** 15.

Howard, B. B., B. Linder, and M. T. Emerson (1962), *J. Chem. Phys.* **36,** 485.

Howarth, O. W. (1980), *Biochem. Soc. Trans.* **8,** 634.

Howarth, O. W., and D. M. J. Lilley (1978), *Progr. NMR Spect.* **12,** 1.

Huber, R., and W. Bode (1978), *Acc. Chem. Res.* **11,** 114.

Hugli, T. B., and F. R. N. Gurd (1970), *J. Biol. Chem.* **245,** 1939.

Hull, W. E., and B. D. Sykes (1975), *J. Mol. Biol.* **98,** 121.

Hull, W. E., and B. D. Sykes (1976), *Biochemistry* **15,** 1535.

Hunkapillar, M. W., S. H. Smallcombe, D. R. Whitaker, and J. H. Richards (1973), *Biochemistry* **12,** 4732.

Hunkapillar, M. W., M. D. Forgac, E. H. Yu, and J. H. Richards (1979), *Biochem. Biophys. Res. Commun.* **87,** 25.

Irving, C., and A. Lapidot (1976), *Chem. Commun.*, 43.

Isenberg, I. (1979), *Ann. Rev. Biochem.* **48,** 159.

Jaeck, G., and F. W. Benz (1979), *Biochem. Biophys. Res. Commun.* **86,** 885.

Jardetzky, O. (1981), *Accts. Chem. Res.* **14,** 291.

Jardetzky, O., and G. C. K. Roberts (1981), *NMR in Molecular Biology*, Academic Press, New York.

Jarema, M. A. C., K. T. Arndt, M. Savage, P. Lu, and J. H. Miller (1981), *J. Biol. Chem.* **256,** 6544.

Jelinski, L. W., and D. A. Torchia (1979), *J. Mol. Biol.* **133,** 45.

Jelinski, L. W., and D. A. Torchia (1980), *J. Mol. Biol.* **138,** 255.

Jelinski, L. W., C. B. Sullivan, and D. A. Torchia (1980), *Nature* **284,** 531.

Jentoft, J. B., N. Jentoft, T. A. Gerken, and D. C. Dearborn (1979), *J. Biol. Chem.* **254,** 4366.

Jentoft, J. B., T. A. Gerken, N. Jentoft, and D. C. Dearborn (1981), *J. Biol. Chem.* **256,** 231.

Johnson, R. N., J. H. Bradbury, and C. A. Appleby (1978), *J. Biol. Chem.* **253,** 2148.

Jones, W. C., T. M. Rothgeb, and F. R. N. Gurd (1976), *J. Biol. Chem.* **251,** 7452.

Jordan, F., and L. Polgar (1981), *Biochemistry* **20,** 6366.

Kalbitzer, H. R., J. Deutscher, W. Hengstenberg, and P. Rosch (1981), *Biochemistry* **20,** 6178.

Kalk, A., and H. J. C. Berendsen (1976), *J. Magn. Reson.* **24,** 343.

Kanamori, K., A. H. Cain, and J. D. Roberts (1978), *J. Am. Chem. Soc.* **100,** 4979.

Karplus, M., and J. A. McCammon (1981), *CRC Crit. Rev. Biochem.* **9,** 293.

Keim, P., R. A. Vigna, J. S. Morrow, R. C. Marshall, and F. R. N. Gurd (1973), *J. Biol. Chem.* **248,** 7811.

Kennelly, P. J., R. Timkovich, and M. A. Cusanovich (1981), *J. Mol. Biol.* **145,** 583.

Kilmartin, J. V., J. J. Breen, G. C. K. Roberts and C. Ho (1973), *Proc. Natl. Acad. Sci. USA* **70,** 1246.

Kimber, B., D. V. Griffiths, B. Birdsall, R. W. Kind, P. Scudder, J. Feeney, G. C. K. Roberts, and A. S. V. Burgen (1977), *Biochemistry* **16,** 3492.

Kimber, B. J., J. Feeney, G. C. K. Roberts, B. Birdsall, D. V. Griffiths, A. S. V. Burgen, and B. D. Sykes (1978), *Nature* **271,** 184.

King, N. L. R., and J. H. Bradbury (1971), *Nature* **229,** 404.

King, R. W., and G. C. K. Roberts (1971), *Biochemistry* **10,** 558.

Kinsey, R. A., A. Kintanar, and E. Oldfield (1981a), *J. Biol. Chem.* **256,** 9028.

Kinsey, R. A., A. Kintanar, M.-D. Tsai, R. D. Smith, N. James, and E. Oldfield (1981b), *J. Biol. Chem.* **256,** 4146.

Kleier, D. A., and G. Binsch (1970), *J. Magn. Reson.* **3,** 146.

Klein, P. (1974), *J. Biol. Chem.* **249,** 4149.

Kumar, A., G. Wagner, R. R. Ernst, and K. Wüthrich (1981), *J. Am. Chem. Soc.* **103,** 3654.

La Mar, G. N., D. L. Budd, and H. Goff (1977), *Biochem. Biophys. Res. Commun.* **77,** 104.

La Mar, G. N., R. R. Anderson, D. L. Budd, K. M. Smith, K. C. Langry, K. Gersonde, and H. Sick (1981), *Biochemistry* **20,** 4429.

Lauterbur, P. C. (1970), *Appl. Spectrosc.* **24,** 450.

Leipert, T. K., and J. H. Noggle (1975), *J. Am. Chem. Soc.* **97,** 269.

Lenstra, J. B., B. G. J. M. Bolscher, S. Stob, J. J. Beintema, and R. Kaptein (1979), *Eur. J. Biochem.* **98,** 385.

Levine, B. A., D. Mercola, D. Coffman, and J. M. Thornton (1977), *J. Mol. Biol.* **115,** 743.

Levy, R. M., M. Karplus, and J. A. McCammon (1981), *J. Am. Chem. Soc.* **103,** 994.

Lewis, S. D., F. A. Johnson, and J. A. Shafer (1981), *Biochemistry* **20,** 48.

Lilley, D. M. J., O. W. Howarth, V. M. Clark, J. F. Pardon, and B. M. Richards (1975), *Biochemistry* **14,** 4590.

Lipari, G., and A. Szabo (1982), *J. Am. Chem. Soc.* **104,** 4546, 4559.

London, R. E. (1980), *Magn. Reson. Biol.* **1,** 1.

London, R. E., T. E. Walker, V. H. Kollman, and N. A. Matwiyoff (1978), *J. Am. Chem. Soc.* **100,** 3723.

London, R. E., J. P. Groff, and R. L. Blakley (1979), *Biochem. Biophys. Res. Commun.* **86,** 779.

Luthra, N. P., R. C. Costello, J. D. Odom, and R. B. Dunlap (1981), *J. Biol. Chem.* **257,** 1142.

Maghuin-Rogister, G., J. Degelaen, and G. C. K. Roberts (1979), *Eur. J. Biochem.* **96,** 59.

Man, M., and R. G. Bryant (1974), *J. Biol. Chem.* **249,** 1109.

Markley, J. L. (1973), *Biochemistry* **12,** 2245.

Markley, J. L. (1975a), *Acc. Chem. Res.* **8,** 70.

Markley, J. L. (1975b), *Biochemistry* **14,** 3546.

Markley, J. L. (1975c), *Biochemistry* **14,** 3554.

Markley, J. L., and W. R. Finkenstadt (1975), **14,** 3562.

Markley, J. L., and I. B. Ibanez (1978), *Biochemistry* **17,** 4627.

Markley, J. L., and O. Jardetzky (1970), *J. Mol. Biol.* **50,** 223.

Markley, J. L., and M. A. Porubcan (1976), *J. Mol. Biol.* **102,** 487.

Markley, J. L., M. N. Williams, and O. Jardetzky (1970), *Proc. Natl. Acad. Sci. USA* **65,** 645.

Markley, J. L., W. R. Finkenstadt, D. Dugas, P. Ledue, and G. R. Drapeau (1975), *Biochemistry* **14,** 998.

Martin, M. L., G. J. Martin, and J.-J. Delpuech (1980), *Practical NMR Spectroscopy*, Heyden, London, p. 139.

Matta, M. S., M. E. Landis, T. B. Patrick, P. A. Henderson, M. W. Russo, and R. L. Thomas (1980), *J. Am. Chem. Soc.* **102,** 7151.

Matta, M. S., P. A. Henderson, and T. B. Patrick (1981), *J. Biol. Chem.* **256,** 4172.

Matthews, D. A., R. A. Alden, J. J. Birktoft, S. T. Freer, and J. Kraut (1977), *J. Biol. Chem.* **252,** 8875.

Maurer, W., W. Haar, and H. Ruterjans (1974), *Z. Phys. Chem.* **93,** 119.

Mayer, R., F. Toulme, T. Monteney-Garestier, and C. Helene (1979), *J. Biol. Chem.* **254,** 75.

McCammon, J. A., P. J. Wolynes, and M. Karplus (1979), *Biochemistry* **18,** 927.

McCammon, J. A., and M. Karplus (1980), *Biopolymers* **19,** 1375.

McDonald, C. C., and W. D. Phillips (1967), *J. Am. Chem. Soc.* **89,** 6332.

Meadows, D. B., J. L. Markley, J. S. Cohen, and O. Jardetzky (1967), *Proc. Natl. Acad. Sci. USA* **58,** 1307.

Meadows, D. B., O. Jardetzky, R. M. Epand, H. H. Ruterjans, and H. A. Scheraga (1968), *Proc. Natl. Acad. Sci. USA* **60,** 766.

Meadows, D. B., G. C. K. Roberts, and O. Jardetzky (1969), *J. Mol. Biol.* **45,** 491.

Meshitsuka, S., G. M. Smith, and A. S. Mildvan (1981), *J. Biol. Chem.* **256,** 4460.

Migchelsen, C., and J. J. Benteima (1973), *J. Mol. Biol.* **79,** 25.

Miyamoto, K., Y. Arata, H. Matsuo, and K. Narita (1981), *J. Biochem. (Tokyo)* **89,** 49.

Moore, R., and R. J. P. Williams (1980a), *Eur. J. Biochem.* **103,** 493.

Moore, R., and R. J. P. Williams (1980b), *Eur. J. Biochem.* **103,** 503.

Moore, R., and R. J. P. Williams (1980c), *Eur. J. Biochem.* **103,** 513.

Moore, R., and R. J. P. Williams (1980d), *Eur. J. Biochem.* **103,** 533.

Morris, G. A., and R. Freeman (1979), *J. Am. Chem. Soc.* **101,** 760.

Munowitz, M., W. W. Bachovchin, J. Herzfeld, C. M. Dobson, and R. G. Griffin (1982), *J. Am. Chem. Soc.* **104,** 1192.

Nagayama, K. (1981), *Adv. Biophys.* **14,** 139.

Nagayma, K., and K. Wüthrich (1981a), *Eur. J. Biochem.*, **114,** 365.

Nagayama, K., and K. Wüthrich (1981b), *Eur. J. Biochem.* **115,** 653.

Nagayama, K., K. Wüthrich, and R. R. Ernst (1979), *Trends Biochem. Sci.* **4,** N178.

Nagayama, K., A. Kumar, K. Wüthrich, and R. R. Ernst (1980), *J. Magn. Reson.* **40,** 321.

Nakano, A., T. Miyazawa, S. Nakamura, and Y. Kaziro (1979), *Arch. Biochem. Biophys.* **196,** 233.

Nelson, D. J., S. J. Opella, and O. Jardetzky (1976), *Biochemistry* **15,** 5552.

Niu, C-H., H. Shindo, J. S. Cohen, and M. Gross (1977), *J. Am. Chem. Soc.* **99,** 3161.

Niu, C-H., S. Matsuura, H. Shindo, and J. S. Cohen (1979), *J. Biol. Chem.* **254,** 3788.

Norton, R. S., and A. Allerhand (1976), *J. Biol. Chem.* **251,** 6522.

Norton, R. S. and A. Allerhand (1977), *J. Biol. Chem.* **252,** 1795.

Norton, R. S., and J. H. Bradbury (1974), *Chem. Commun.*, 870.

Norton, R. S., A. O. Clouse, R. Addleman, and A. Allerhand (1977), *J. Am. Chem. Soc.* **99,** 79.

Ohms, J. P., H. Hagenmaier, M. B. Hayes, and J. S. Cohen (1979), *Biochemistry* **18,** 1599.

Oldfield, E., and A. Allerhand (1975), *J. Biol. Chem.* **250,** 6403.

Oldfield, E., and T. M. Rothgeb (1980), *J. Am. Chem. Soc.* **102,** 3635.

Oldfield, E., R. S. Norton, and A. Allerhand (1975), *J. Biol. Chem.* **250,** 6368, 6381.

Opella, S. J., D. J. Nelson, and O. Jardetzky (1974), *J. Am. Chem. Soc.* **96,** 7157.

Opella, S. J., T. A. Cross, J. A. DiVerdi, and C. F. Sturm (1980), *Biophys. J.* **10,** 531.

Otvos, J. D., and I. M. Armitage (1980), *Biochemistry* **19,** 4021.

Otvos, J. D., and D. T. Browne (1980), *Biochemistry* **19,** 4011.

Otvos, J. D., I. M. Armitage, J. F. Chebowski, and J. E. Coleman (1979), *J. Biol. Chem.* **254,** 4707.

Packer, E. L., H. Sternlicht, and J. C. Rabinowitz (1972), *Proc. Natl. Acad. Sci. USA* **69,** 3278.

Packer E. L., H. Sternlicht, and J. C. Rabinowitz (1973), *Am. N.Y. Acad. Sci.* **222,** 824.

Packer, E. L., W. V. Sweeney, J. C. Rabinowitz, H. Sternlicht, and E. N. Shaw (1977), *J. Biol. Chem.* **252,** 2245.

Patel, D. J., L. L. Canuel, and F. A. Bovey (1975), *Biopolymers* **14,** 987.

Patt, S. L., and B. D. Sykes (1972), *J. Chem. Phys.* **56,** 3182.

Pesando, J. M. (1975), *Biochemistry* **14,** 675, 681.

Pletcher, C. H., E. F. Bouhoutsos-Brown, R. G. Bryant, and G. L. Nelsestrian (1981), *Biochemistry* **20,** 6149.

Poe, M., K. Hoogsteen, and D. A. Matthews (1979), *J. Biol. Chem.* **254,** 8143.

Porubcan, M. A., D. E. Neves, S. K. Rausch, and J. L. Markley (1978), *Biochemistry* **17,** 4640.

Putter, I., A. Baretto, J. L. Markley, and O. Jardetzky (1969), *Proc. Natl. Acad. Sci. USA* **64,** 1396.

Quirt, A. R., J. R. Lyerla, Jr., I. R. Peat, J. S. Cohen, W. F. Reynolds, and M. H. Freedman (1974), *J. Am. Chem. Soc.* **96,** 570.

Rabenstein, D. L., M. S. Greenberg, and O. A. Evans (1977), *Biochemistry* **16,** 977.

Ramirez, J. E., J. R. Cavanaugh, K. S. Schweizer, and P. D. Hoagland (1977), *Anal. Biochem.* **75,** 130.

Redfield, A. G. (1978), *Meth. Enzymol.* **49,** 253.

Reynolds, W. F., I. R. Peat, M. H. Freedman, and J. R. Lyerla, Jr. (1973), *J. Am. Chem. Soc.* **95,** 328.

Ribeiro, A. A., P. King, C. Restivo, and O. Jardetzky (1980), *J. Am. Chem. Soc.* **102,** 4040.

Rice, D. M., R. J. Wittebort, R. G. Griffin, E. Meirovitch, E. R. Stimson, Y. C. Meinwald, J. H. Freed, and H. A. Scheraga (1981), *J. Am. Chem. Soc.* **103,** 7077.

Richards, T. M., and P. J. Vithayathil (1959), *J. Biol. Chem.* **234,** 1459.

Richards, F. M., and H. W. Wyckoff (1973), in *Atlas of Molecular Structures in Biology 1; Ribonuclease S* (Phillips, D. C., and F. M. Richards, eds.) Clarendon, Oxford.

Richardson, J. S. (1981), *Adv. Prot. Chem.* **34,** 167.

Richarz, R., and K. Wüthrich (1977), *FEBS Lett.* **79,** 64.

Richarz, R., and K. Wüthrich (1978), *Biochemistry* **17,** 2263.

Richarz, R., P. Sehr, G. Wagner, and K. Wüthrich (1979), *J. Mol. Biol.* **130,** 19.

Richarz, R., K. Nagayama, and K. Wüthrich (1980a), *Biochemistry* **19,** 5189.

Richarz, R., H. Tschesche, and K. Wüthrich (1980b), *Biochemistry* **19,** 5711.

Roberts, G. C. K., D. H. Meadows, and O. Jardetzky (1968), *Biochemistry* **8,** 2053.

Roberts, G. C. K., J. Feeney, B. Birdsall, B. J. Kimber, D. V. Griffiths, R. W. King, and A. S. V. Burgen (1977), in *NMR in Biology*, R. A. Dwek, I. D. Campbell, R. E. Richards, and R. T. P. Williams (Eds.), Academic Press, Oxford, pp. 95–109.

Robillard, G., and R. G. Shulman (1974), *J. Mol. Biol.* **86,** 519.

Rosevear, P. R., P. Desmeules, G. L. Kenyon, and A. S. Mildvan (1981), *Biochemistry* **20,** 6155.

Rothgeb, T. M., and E. Oldfield (1981), *J. Biol. Chem.* **256,** 1432.

Rothgeb, T. M., R. D. England, B. N. Jones, and R. S. Gurd (1978), *Biochemistry* **17,** 4564.

Russu, I. M., N. T. Ho, and C. Ho (1982) *Biochemistry* **21,** 5031.

Russu, I. M., N. T. Ho, and C. Ho (1980), *Biochemistry* **19,** 1043.

Russu, I. M., and C. Ho (1982) *Biochemistry* **21,** 5044.

Ruterjans, H., and H. Witzel (1969), *Eur. J. Biochem.* **9,** 118.

Ruterjans, H., H. Witzel, and O. Pongs (1969), *Biochem. Biophys. Res. Commun.* **37,** 247.

Sachs, D. H., A. N. Schechter, and J. S. Cohen (1971), *J. Biol. Chem.* **246,** 6576.

Santoro, J., H.-P. Juretschke, and H. Ruterjans (1979), *Biochim. Biophys. Acta* **578,** 346.

Saunders, D. J., and R. E. Offord (1972), *FEBS Lett.* **26,** 286.

Saunders, M., A. Wishnia, and J. Kirkwood (1957), *J. Am. Chem. Soc.* **79,** 3289.

Schechter, A. N., D. H. Sachs, S. R. Heller, R. T. Shrager, and J. S. Cohen (1972), *J. Mol. Biol.* **71,** 39.

Scheek, R. M., R. Kaptein, and J. W. Verhoever (1979), *FEBS Lett.* **107,** 288.

Schejter, A., A. Lanir, I. Vig, and J. S. Cohen (1978), *J. Biol. Chem.* **253,** 3768.

Seelig, J. (1977), *Quart. Rev. Biophys.* **16,** 353.

Sheinblatt, M. (1966), *J. Am. Chem. Soc.* **88,** 2845.

Shindo, H. (1981), *Eur. J. Biochem.* **120,** 309.

Shindo, H., and J. S. Cohen (1976), *Proc. Natl. Acad. Sci. USA* **73,** 1979.

Shindo, H., M. B. Hayes, and J. S. Cohen (1976), *J. Biol. Chem.* **251,** 2644.

Shindo, H., W. Egan, and J. S. Cohen (1978), *J. Biol. Chem.* **253,** 6751.

Shrager, R., J. S. Cohen, S. Heller, D. H. Sachs, and A. N. Schechter (1972) *Biochemistry* **11,** 327.

Smith, G. M., and A. S. Mildvan (1981), *Biochemistry* **20,** 4340.

Snyder, G. H., R. Rowan, III, S. Karplus, and B. D. Sykes (1975), *Biochemistry* **14,** 3765.

Sperling, R., and E. J. Wachtel (1981), *Adv. Prot. Chem.* **34,** 1.

Stenflo, J., P. Fernlund, W. Egan, and P. Roepstorff (1974), *Proc. Natl. Acad. Sci. USA* **71,** 2730.

Sternlicht, H., and D. Wilson (1967), *Biochemistry* **6,** 2881.

Stoesz, J. D., D. P. Malinowski, and A. G. Redfield (1979), *Biochemistry* **18,** 4669.

Sykes, B. D., and W. E. Hull (1978), *Meth. Enzymol.* **49,** 270.

Sykes, B. D., and J. H. Weiner (1980), *Magn. Reson. Biol.* **1,** 171.

Sykes, B. D., H. I. Weingarten, and M. J. Schlesinger (1974), *Proc. Natl. Acad. Sci. USA* **71,** 469.

Sykes, B. D., W. E. Hull, and G. H. Snyder (1978), *Biophys. J.* **21,** 137.

Tanokura, M., T. Mitsuo, and T. Miyazawa (1976), *Biopolymers* **15,** 393.

Torchia, D. A., M. A. Hasson, and V. C. Hascall (1981), *J. Biol. Chem.* **256,** 7129.

Torchia, D. A., and A. Szabo (1982), *J. Magn. Reson.*, **49,** 107.

Torchia, D. A., and D. L. VanderHart (1976), *J. Mol. Biol.* **104,** 315.

Torchinsky, I. M. (1981), *Sulfur in Proteins*, Permagon Press, New York.

Tran-Dinh, S., S. Fermandjian, E. Sala, R. Mermet-Bouvier, M. Cohen, and P. Fromageot (1974), *J. Am. Chem. Soc.* **96,** 1484.

Ugurbil, K., R. S. Norton, A. Allerhand, and R. Bersohm (1977), *Biochemistry* **16,** 886.

Urry, D. W., and L. Mitchell (1976), *Biochem. Biophys. Res. Commun.* **G8,** 1153.

Usher, D. A. (1969), *Proc. Natl. Acad. Sci. USA* **62,** 661.

Usher, D. A., D. I. Richardson, Jr., and Eckstein (1970), *Nature* **228,** 663.

Veloso, D., W. W. Cleland, and J. W. Porter (1981), *Biochemistry* **20,** 887.

Wagner, G., and K. Wüthrich (1979), *J. Mol. Biol.* **130,** 31.

Wagner, G., A. Kumar, and K. Wüthrich (1981), *Eur. J. Biochem.* **114,** 375.

Walker, E. J., G. B. Ralston, and I. G. Darvey (1976), *Biochem. J.* **153,** 329.

Wang, F. C. H., and C. H. W. Hirs (1979), *J. Biol. Chem.* **254,** 1090.

Wasylishen, R., and J. S. Cohen (1974), *Nature* **249,** 847.

Wemmer, D., H. Shvo, A. Ribeiro, R. P. Bray, and O. Jardetzky (1981), *Biochemistry* **20,** 3351.

Westmoreland, D. G., C. R. Matthews, M. B. Hayes, and J. S. Cohen (1975), *J. Biol. Chem.* **250,** 7456.

Wider, G., R. Baumann, K. Nagayama, R. R. Ernst, and K. Wüthrich (1981), *J. Magn. Reson.* **42,** 73.

Wilbur, D. J., and A. Allerhand (1976), *J. Biol. Chem.* **251,** 5187.

Wilbur, D. J., and A. Allerhand (1977a), *J. Biol. Chem.* **252,** 4968.

Wilbur, D. J., and A. Allerhand (1977b), *FEBS Lett.* **74,** 272.

Wilbur, D. J., and A. Allerhand (1977c), *FEBS Lett.* **79,** 144.

Wittebort, R. J., T. M. Rothgeb, A. Szabo, and F. R. N. Gurd (1979), *Proc. Natl. Acad. Sci. USA* **76,** 1059.

Wittebort, R. J., A. Szabo, and F. R. N. Gurd (1980), *J. Am. Chem. Soc.* **102,** 5723.

Witzel, H. (1963), *Prog. Nucl. Acids Res.* **2,** 221.

Wlodawer, A., and L. Sjölin (1981), *Proc. Natl. Acad. Sci. USA* **78,** 2853.

Wlodawer, A., R. Bott, and L. Sjölin (1982), *J. Biol. Chem.* **257,** 1325.

Wodak, S. Y., M. Y. Liu, and H. W. Wyckoff (1977), *J. Mol. Biol.* **116,** 855.

Wooten, J. B., and J. S. Cohen (1979), *Biochemistry* **18,** 4188.

Wooten, J. B., J. S. Cohen, I. Vig, and A. Schejter (1981), *Biochemistry* **20,** 5394.

Wüthrich, K. (1970), *Struct. Bond.* **8,** 53.

Wüthrich, K. (1976), *NMR in Biological Research: Peptides and Proteins*, Elsevier, New York.

Wüthrich, K., and G. Wagner (1975), *FEBS Lett.* **50,** 265.

Wüthrich, K., and G. Wagner (1978), *Trends Biochem. Sci.* **3,** 227.

Wüthrich, K., and G. Wagner (1979), *J. Mol. Biol.* **130,** 1.

Wüthrich, K., K. Nagayama, and R. R. Ernst (1979), *Trends Biochem. Sci.* **4,** N178.

Wüthrich, K., G. Wagner, R. Richarz, and W. Braun (1980), *Biophys. J.* **32,** 549.

Wyckoff, H. W., D. Tsernoglu, A. W. Hanson, J. R. Knox, E. Lee, and F. M. Richards (1970), *J. Biol. Chem.* **245,** 305.

Wyeth, P., A. Gronenborn, B. Birdsall, G. C. K. Roberts, J. Feeney, and A. S. V. Burgen (1980), *Biochemistry* **19,** 2608.

Yeh, H. C., K. L. Kirk, L. A. Cohen, and J. S. Cohen (1975), *J. Chem. Soc.* **II,** 928.

York, J. L., F. S. Millett, and L. B. Minor (1980), *Biochemistry* **19,** 2583.

Five

Water Relaxation in Heterogeneous and Biological Systems

L. J. Lynch

CSIRO Physical Technology Unit
Ryde, NSW, 2112 Australia

1. **Introduction** 248
2. **Relaxation Theory** 251
3. **Applications of NMR to Studies of Water in Heterogeneous Systems** 255
   3.1. General, 255
   3.2. The Scope of NMR Measurements, 257
      3.2.1. Temperature Scanning, 257
      3.2.2. Dispersion, 257
      3.2.3. Multinuclei Studies, 258
      3.2.4. Tests for Anisotropy, 258
      3.2.5. Proton Local Field–Multiwindow Analysis, 259
      3.2.6. Self-Diffusion Measurements, 260
   3.3. Models Used, 261
   3.4. Cross-Relaxation Effects at the Water–Substrate Interface, 267
   3.5. Surface Water Proton Exchange Kinetics, 278
   3.6. Assessment of "Bound" or "NonFreezable" Water, 280
   3.7. ^{1}H NMR Discrimination of Physiological States of Tissues, 290
4. **Concluding Remarks** 294
   List of Symbols 296
   References 300

1. INTRODUCTION

The intention here is to focus on some of the more recent developments in the use of nuclear magnetic resonance (NMR) techniques to study the nature of water contained in biological and other heterogeneous systems. Two particular topics are discussed in some detail: (1) developments in methods of analysis for taking account of the interfacial proton–proton coupling that leads to cross-relaxation and magnetization transfer between protons of water and those of other molecules of the system and (2) the application of temperature scanning to detect and quantify "bound" or "nonfreezing" water and related phenomena of parts of the water.

Most NMR studies of water in heterogeneous systems have been based on NMR relaxation measurements. Relaxation phenomena are described by theory that takes account of aspects of the structure and molecular dynamics of the system. The theory is effective when applied to simple systems, but its extension and application to heterogeneous systems are complicated. There are many factors to be considered and hence parameters necessary to define suitable dynamic structural models. This complexity can be countered to some extent by the experimental flexibility of NMR in being able to provide a number of independent observations. The availability of versatile spectrometers allows comprehensive studies to be made so that dispersion, multinuclei, and combined relaxation-time studies are now commonplace. However, as pointed out by Resing *et al.* (1977), the present situation with respect to the interpretation of NMR data on biological water is still largely one of model building and hypothesis testing.

Although NMR relaxation theory (BPP theory: Bloembergen, Purcell, and Pound, 1948; Kubo and Tomita, 1954; Solomon, 1955) is successful in predicting the spin-lattice relaxation of water, it has been shown that NMR relaxation is insensitive to details of the molecular motion in the liquid (Sharma and Joshi, 1963). Chiarotti *et al.* (1955) pointed out that the apparent success of the BPP theory, incorporating the Debye macroscopic-diffusion model of a liquid, in predicting the NMR behavior of water nevertheless revealed little about the microstructure of water. This is not surprising when it is considered that NMR relaxation is sensitive to molecular motions down to a time scale of perhaps 10^{-10} s, whereas the microscopic-diffusion processes on which water models are based have time scales of the order of 10^{-12}–10^{-11} s, so that the NMR measurements reflect only the gross average behavior. The implied expectation that the microdynamic processes of water of hydration are much slower than those for the bulk liquid has encouraged the widespread application of NMR to the problem of determining the nature of water in heterogeneous systems.

It is usually assumed that at least part of the water sorbed to surfaces or associated with macromolecules in complex systems is different from bulk water. This modified water is considered to be the result of either specific interactions of an electrostatic nature such as hydrogen bonding, or of the water dipole moment in the potential field gradients associated with charge distributions of the substrate molecules, or nonspecific perturbations of interfacial water. It is further considered that because of the nonuniform charge distribution of the water molecules and the inherent asymmetry of the surface environment, these interactions can lead to preferential orientations of the water molecules with respect to the interface and to anisotropy in the molecular reorientation processes. Because of the perturbations to the water structure and restrictions in its movements, it is unlikely that the strong correlation between rotational and translational motions that exists for bulk water (Krynicki, 1966) is sustained within the interacting water phase.

For the purposes of this discussion, the water in heterogeneous systems is considered to consist of interacting and noninteracting components. Interacting water is conceived of as having physical properties modified to some extent from bulk water as a consequence of interaction with the substrate materials, as distinct from the noninteracting or essentially bulk water. These definitions parallel the concepts of "bound" and "free" water that have been widely used in studies of water–interface systems. The term *bound* has been of variable definition related both to the method of its assessment and to the nature of the system under study. These many definitions have been discussed elsewhere (Kuprianoff, 1958; Kuntz and Kauzmann, 1974; Berendsen, 1975). Its most general definition equates bound water to that water in the system that is *not "free"* (Kuprianoff, 1958)—a concept equivalent to the present definition of interacting water. The terms *bound* and *interacting* are therefore used interchangeably in this chapter.

For NMR studies two classes of heterogeneous water systems can be conveniently distinguished. The first is systems containing low concentrations of water, that is, not significantly greater than the saturation water content (SWC) for insoluble-substrate systems. For these systems the water comprises essentially sorbed water and is thus, prima facie, mostly interacting water. A second class comprises systems containing large quantities of water in excess of that which could conceivably be bound to the substrate molecules or surfaces. This class includes macromolecular solutions and high-water-content biological tissues.

To explain the NMR observations of water associated with macromolecular systems and surfaces, models ascribing various forms of anisotropy and heterogeneity in the structure and/or microdynamics are used. These complexities can be encountered in the anisotropic motions of the molecules and/or in the structural heterogeneity of the system whereby the microdynamics are different for each class of molecules (Section 3.3). Thus a system might be considered to contain a mixture of water molecular states distinguished by different correlation functions (Section 2) and/or different spin interactions and therefore relaxation behaviors. How this heterogeneity manifests itself in the net observed NMR relaxations depends largely on the degree of microdynamic mixing that occurs.

Nuclear magnetic resonance investigations of biological systems have been influenced by the controversy between those who consider water in such systems to be essentially in the liquid state and those who hold that the water is modified by long-range coupling to the macromolecular constituents

(Foster *et al.*, 1976). Similar NMR data have been used to support, on the one hand, models in which a very small fraction of the water is highly modified in its properties and, on the other, models in which the bulk of the water is slightly modified (Resing *et al.*, 1977; Foster *et al.*, 1976; Chang and Woessner, 1977; Bryant and Shirley, 1980).

2. RELAXATION THEORY

Nuclear magnetic resonance relaxation processes are determined by the specific magnetic interactions of the resonant species and the nature of the fluctuations in these interactions. The proton (^{1}H), with a nuclear spin quantum number $I = \frac{1}{2}$ and a magnetogyric ratio $\gamma_{^1\mathrm{H}} = 2.673 \times 10^8$ C/kg, deuterium (^{2}H), with $I = \frac{3}{2}$ and $\gamma_{^2\mathrm{H}} = 0.4103 \times 10^8$ C/kg, and ^{17}O, with $I = \frac{5}{2}$ and $\gamma_{^{17}\mathrm{O}} = 0.3623 \times 10^8$ C/kg, have all been used as NMR probes in water studies. All have magnetic dipole moments, and ^{2}H and ^{17}O also have electric quadrupole moments. Because the dipole interaction energy between two nuclei i and j is proportional to $\gamma_i^2\gamma_j^2$, this interaction is much stronger for ^{1}H–^{1}H interaction than for any other combination. For both ^{2}H and ^{17}O, the Coulomb interactions experienced by their quadrupole moments in the local electric-field gradients are strong compared with their nuclear dipolar interactions. For water, both these quadrupole interactions are intramolecular; for ^{2}H the gradient is directed along the O—^{2}H bond, and for ^{17}O the major component of the field-gradient tensor is along the diad axis of the water molecule.

The theory considers the interactions experienced by a single resonant nucleus (Bloembergen *et al.*, 1948) or a spin-$\frac{1}{2}$ pair (Solomon, 1955) and how fluctuations determined by the microdynamics of the molecular system modulate the interactions.

As outlined here, the theory relates strictly to magnetic relaxation determined by rotational diffusion for ^{1}H dipole–dipole interactions and ^{2}H quadrupolar interactions. It is also considered sufficiently correct to account for dipolar interactions in general and for the nuclear quadrupolar spin-lattice relaxation rate R_1 of ^{17}O (Rubenstein *et al.*, 1971; Civan *et al.*, 1978). Also within certain limits, the expression obtained for the spin-spin relaxation rate R_2 is appropriate for ^{17}O (Civan *et al.*, 1978).

Because this basic theory is essentially a bulk theory, molecular translations are assumed to be restricted in neither dimensions nor distance. For heterogeneous systems the possibility of restricted diffusion arises, and also for surface phases the possibility that translations are confined to two dimensions requires consideration (Tabony, 1980).

Each microdynamic mode contributing to the fluctuations is characterized by a correlation time parameter τ_c specified by the correlation function $k_a(\tau)$ that describes the evolution with elapsed time τ of that random molecular process in terms of particular position functions $F_a(t)$ fluctuating in time t.

For rotational motions, correlation functions are assumed to be of the form

$$k_a(\tau) = \overline{\langle F_a^*(t)\, F_a(t + \tau)\rangle} \exp(-\tau/\tau_c) \tag{1}$$

where the average is over both population and time as indicated. The correlation function $k_a(\tau)$ contains all the information known of the random process in that it measures the rate of change induced in the system by it.

The functions $F_a(t)$ represent the different components of the fluctuating interaction-energy function. These functions, which are all associated with the time-dependent second-order spherical harmonics are (Bloembergen *et al.*, 1948; Woessner, 1962)

$$F_0(t) = (1 - 3N^2) \tag{2a}$$

$$F_1(t) = N(L + iM) \tag{2b}$$

$$F_2(t) = (L + iM)^2 \tag{2c}$$

where L, M, and N are the direction cosines in the laboratory frame that describe the interaction vector with respect to the applied magnetic field H_0 aligned with the z axis. This is the line drawn between interacting nuclei for dipolar interaction and the direction of the local electric-field gradient for quadrupolar interactions.

Analysis of the power spectra of the fluctuations described by $k_a(\tau)$ enables specification of the relaxation rates.

Thus the spectral densities or Fourier intensities $J_a(\omega)$ at the frequency ω given by the Weiner-Khintchine theorem, are

$$J_a(\omega) = \int_{-\infty}^{\infty} k_a(\tau) \exp(i\omega\tau)\, d\tau \tag{3}$$

Components of the interaction that contain the orientation functions $F_1(t)$ and $F_2(t)$ couple states differing in energy by $\hbar\omega_0$ and $2\hbar\omega_0$ and therefore contribute to the spin-lattice relaxation by inducing transitions at the resonant frequencies ω_0 and $2\omega_0$, respectively, so that the general expression for the spin-lattice relaxation rate R_1 is

$$R_1 = C[J_1(\omega_0) + J_2(2\omega_0)] \tag{4}$$

The quantity C is determined by the strength of the static nuclear spin interactions involved. In particular for a proton spin population, C is contributed to by all pair-wise dipolar interactions involving the average spin i such that

$$C = \sigma_0^2 = \frac{9}{20}\left(\gamma_{^1H}^4 \hbar^2 \sum_j r_{ij}^{-6}\right) \tag{5}$$

where r_{ij} is the distance between the interacting spins and σ_0^2 is termed the rigid lattice second moment. For the much stronger quadrupolar interactions experienced by ^2H and ^{17}O in water, C is proportional to the square of the nuclear quadrupole coupling constants. In that the relaxation rate is directly proportional to the strength of the nuclear spin interaction, the ^2H relaxation rate should be about 10 times that of the ^1H rate in (^1H_2O) water. For similar reasons the ^{17}O relaxation rate should be about 1000 times that of ^1H.

Spin-spin relaxation can be understood as a dephasing of the induced coherence, which determines the net transverse magnetization of the spin system M_x resulting from a spread in the static local interaction. Therefore, all those spectral components with motions more rapid than the dephasing are averaged out so that only those near zero frequency contribute to the dephasing or spin-spin relaxation process. Here the spectral density component $J_0(0)$ is important in determining the spin-spin relaxation. As the molecular motions shift to higher frequencies and $J_0(0)$ becomes smaller, the spin-spin relaxation becomes dependent on the lifetime of the nuclei in a given state, as does the spin-lattice relaxation. The spin-spin relaxation rate R_2 is given by

$$R_2 = \frac{C}{4}\,[J_0(0) + 10\,J_1(\omega_0) + J_2(2\omega_0)] \tag{6}$$

Intramolecular interactions are modulated by rotational motions, whereas intermolecular interactions are modulated also by translational diffusion. If the translational motions are ascribed correlation functions similar to those of the rotational motions, relaxation associated with both types of interaction is of the same form. In fact, because bulk water is a highly associated liquid in that its rotational and translational diffusions are correlated, a single correlation time has been used to describe the microdynamics of the bulk liquid. For such an isotropic and homogeneous system the relaxation rates are

$$R_1 = C\left[\frac{\tau_c}{1 + (\omega_0\tau_c)^2} + \frac{4\tau_c}{1 + 4(\omega_0\tau_c)^2}\right] \tag{7}$$

and

$$R_2 = \frac{C}{2}\left[3\tau_c + \frac{5\tau_c}{1 + (\omega_0\tau_c)^2} + \frac{2\tau_c}{1 + 4(\omega_0\tau_c)^2}\right] \tag{8}$$

The spin-lattice relaxation measured under the influence of a resonant spin-locking field $H_{1\rho}$ differs from that in the absence of such a field if the molecular motions are not rapid [i.e., $\omega_0^2\tau_c^2 \geqslant 1$ (Jones, 1966)]. This spin-lattice relaxation is influenced by a mechanism in which the spin system is relaxed toward thermal equilibrium in the effective magnetic field of the rotating frame. Spectral-density components at the frequency $\omega_1 = \gamma H_{1\rho}$ affect this relaxation, and the rotating-frame spin-lattice relaxation rate is given by

$$R_{1\rho} = \frac{C}{2}\left[\frac{3\tau_c}{1 + 4(\omega_1\tau_c)^2} + \frac{5\tau_c}{1 + (\omega_0\tau_c)^2} + \frac{2\tau_c}{1 + 4(\omega_0\tau_c)^2}\right] \tag{9}$$

In considering the applications of these equations to water, one should note that for the realizable states of pure liquid water and the experimentally attainable resonance frequencies ω_0, $\tau_c \ll 1/\omega_0$, when $R_1 = R_2$. This is well beyond the conditions under which the NMR relaxation processes are especially sensitive to the microdynamics, that is, $\tau_c^{-1} \approx \omega_0$ for R_1; $\tau_c^{-1} \approx 10^5$ Hz for R_2, and $\tau_c^{-1} \approx \omega_1 \approx 10^3$ Hz for $R_{1\rho}$. Also Equation (8) does not hold for the "rigid lattice" state when the molecular fluctuations are too slow to time average the nuclear spin interactions, and the relaxation is nonexponential, as is the case of ice. This occurs when $\tau_c > C^{-\frac{1}{2}}$, and as an approximation,

$$R_2 \approx C^{\frac{1}{2}} (0.5\pi) \tag{10}$$

The range of mobility above the rigid-lattice value and until the nuclear spin interaction is effectively time averaged is termed the motional narrowing region, and here $R_1 < R_2$. An analytic relationship between R_2 and τ_c covering the transition from motional narrowing to rigid lattice conditions is due to Kubo and Tomita (1954).

$$R_2 = 4\pi^{-1} \ln 2\, C \tan^{-1}\left[\pi\tau_c\left(\frac{4 \ln 2}{R_2}\right)^{-1}\right] \tag{11}$$

Equations (7), (8), (10), and (11) are represented in Figure 1.

Recent developments in the theory and application of NMR relaxation in systems with two-dimensional rather than bulk geometries have been reviewed by Tabony (1980). It is reasonable that two-dimensional theories could be appropriate for adsorbed systems, and Tabony quotes the success of their application to experimental data in several studies of nonwater systems (Silbernagel and Gamble, 1974; Avogadro and Villa, 1977; Riekel *et al.*, 1979). Avogadro and Villa (1977) compared the time-correlation and spectral-density functions for two- and three-dimensional translation diffusions and showed that they coincide only at short times or high frequencies but diverge rapidly at long times or low frequencies. The relaxation rates thereby specified for the two-dimensional fluid differ significantly from those of the bulk in that there is enhanced intermolecular contribution, and the transverse relaxation rate is always much greater than the spin-lattice relaxation rate. Also the frequency dispersion for the intermolecular spin-lattice relaxation is logarithmic for the two-dimensional case compared with the functions of the form given in Equation (7).

The effects of restricted diffusion on the relaxation have been considered by Tabony (1980); he states that a characteristic frequency determined by the diffusion rate and boundary distance introduces an extra step in the spectral-density function and hence the relaxation behavior. Presumably this treatment (unpublished) has much in common with that which Woessner (1977) used to describe the effects of substrate order on the relaxation of water in heterogeneous systems (Section 3.3).

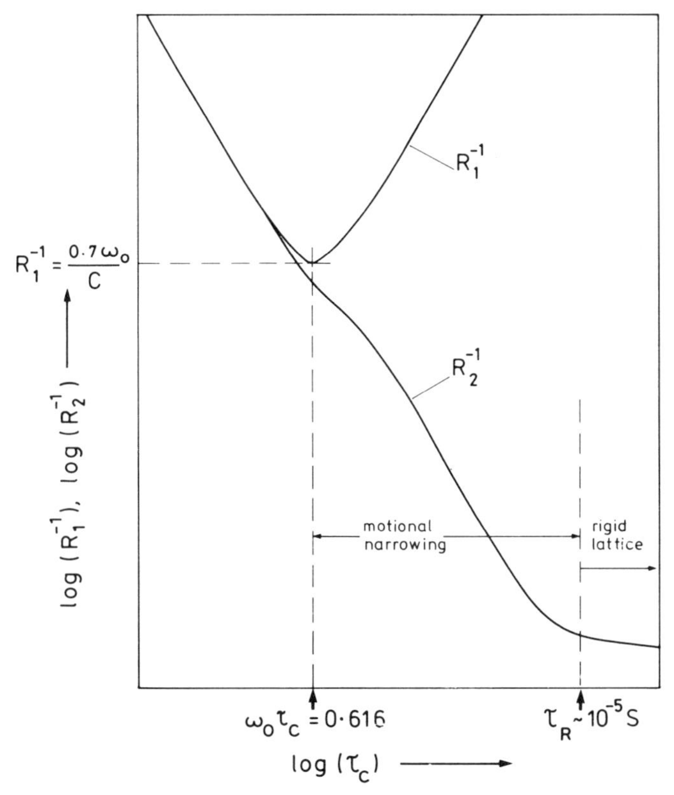

Figure 1. The dependence of the spin–lattice R_1^{-1}, Equation (7), and spin-spin R_2^{-1}, Equations (8), (10) and (11), relaxation times on the correlation time τ_c for a homogeneous and isotropic nuclear spin system.

3. APPLICATIONS OF NMR TO STUDIES OF WATER IN HETEROGENEOUS SYSTEMS

3.1. General

As outlined above, a single molecular correlation function or microdynamic parameter τ_c is found adequate to specify the observed NMR properties of bulk water. This parameter has an estimated value of $\sim 10^{-11}$ s (Krynicki, 1966) for the liquid and a value of $\sim 10^{-5}$ s when attributed to ice near 273K (Bloembergen *et al.*, 1948). Whereas the NMR properties are insensitive to molecular reorientation processes at the rates for liquid water, it is the expectation that these rates are much slower for at least some of the water in heterogeneous systems that has encouraged the use of NMR for their study.

Nuclear magnetic resonance relaxation measurements generally support this expectation: the room-temperature rates R_1 and R_2 are invariably greater than those for bulk water, and $R_2 > R_1$, so the application of Equations (7) and (8) indicates values of $\tau_c > 1.6/\omega_0$. This simplistic approach is discarded, however, both because of the expected heterogeneous properties of the associated water and because extended NMR data are incompatible with it. In particular, studies of R_1 and R_2 by temperature scanning (Lynch and Marsden, 1969; Resing, 1965; Clifford and Sheard, 1966), which detected maximum spin-lattice relaxation rates, showed, compared with the predictions of Equations (7) and (8) that (1) the maximum values of R_1 are low and (2) $R_1/R_2 < 1$ at room temperature above the R_1 maximum.

The spectral-density functions, which describe the water states in the NMR theory, are not directly measured by NMR experiments. Rather NMR measures relaxation processes, which are determined by the lifetimes of the spin states, so there is, more often than not, no simple correspondence between the structural details of heterogeneous systems and the observed NMR properties. This is first of all apparent in the relationships [Equations (7) and (8)] between the relaxation rates R_1 and R_2 and τ_c over most of the mobility range, such that $R_1^{-1} > R_2^{-1} \gg \tau_c$, which imposes limitations on the time resolution of simple NMR measurements for directly revealing heterogeneity in a system.

It is important to keep in mind the separate factors involved in the analysis of data of heterogeneous systems. The molecular system is described in terms of a distribution of correlation functions reflecting a multiplicity of anisotropic reorientations and/or a distribution of the molecules over a range of states, and including functions describing the exchange processes that determine the dynamic equilibrium. These exchange processes, determined by the lifetimes of protons or molecules in particular phase states τ_{ei}, play an important role in how the gross NMR relaxation behavior reveals the structural detail. If exchange is sufficiently rapid that spin migration occurs between the phases within the lifetime of the spin state, all spins experience the same average reorientational history and relax homogeneously regardless of the structural heterogeneity.

A further consideration is that whereas the molecular structure is determined by the force fields that govern molecular interactions, the behavior of the nuclear spin system is determined by specific nuclear–spin interactions that contribute nothing to the molecular interactions but are themselves influenced by the molecular structure and dynamics. Hence there are factors, such as isotopic composition (e.g., $^1H:^2H$), that markedly affect the spin interactions but are incidental to the dynamic structural model of the molecular system. Some such factors have often been overlooked in the analysis of NMR data. An important example of this in biological water systems has been the neglect of interfacial spin–spin interactions between the proton spin systems of the water molecules and those of the macromolecules. These interactions can affect the relaxation processes and lead to magnetization transfer. Many recent studies have shown these effects to be prevalent and have demonstrated their considerable influence on the analysis of NMR data (Edzes and Samulski, 1977; Koenig *et al.*, 1978; Fung and McGaughy, 1980). These developments are discussed in Section 3.4.

The ultimate objective of NMR studies of heterogeneous systems is to extract from observations of the relaxation processes an unambiguous population or time-weighted distribution of spectral functions that describe the microdynamic heterogeneity of the system. However, because of the paucity of data available with respect to the apparent complexity of the systems and the failure to take account of some factors that affect the behavior of the spin system, few studies have resulted in unambiguous and/or unchallenged conclusions in the analyses.

3.2. The Scope of NMR Measurements

Data can be obtained by measurements of the basic NMR relaxation parameters and other parameters that are derived from them, as functions of temperature and resonance frequency. This applies to each of the three resonant species, ^1H, ^2H, and ^{17}O. The isotopic ratio $^1H:^2H$ is another useful variable for ^1H resonance studies. Further data are provided by NMR signal intensity measurements as a function of temperature.

3.2.1. Temperature Scanning. The NMR monitoring of water systems during temperature change is useful both as a means of detecting phase transitions (Section 3.6) and for investigating the molecular dynamics and phase structure. Analysis of temperature-scanning data requires the assumption of thermal-activation laws governing shifts in the motional spectra. Sometimes the magnitude of the molecular correlation time can be estimated at the temperature of a maximum in the spin-lattice relaxation rates, where $\omega_{0,1}\tau \approx 0.5$ [Equations (7) and (9)], but this is a questionable practice for complex systems (Lynch *et al.*, 1969). Thermal-activation coefficients result from such analyses, but the use of these coefficients to extrapolate data to different temperatures has been criticized (Resing *et al.*, 1977).

3.2.2. Dispersion. Nuclear magnetic resonance theory relates the observable relaxation rates to the dynamic molecular structure via the spectral-density functions $J_a(\omega)$ of the microdynamic modes of the system. Maximum information, therefore, can only be obtained if the system is observed over as wide a range of resonance frequencies as possible. Variable-frequency spectrometers enable dispersion studies over a wide range of frequencies, and this range is extended to very low values by magnetic-field-cycling methods available in some laboratories (Hallenga and Koenig, 1976; Kimmich, 1980). Dispersion studies are usually confined to R_1 and $R_{1\rho}$ measurements. The two dispersive terms in Equation (7) are not resolvable and can be approximated by the single term $\tau_c/[1 + 3(\omega_0\tau_c)^2]$ for the range $0.1 < \omega_0\tau_c < 2.0$ (Civan *et al.*, 1978; Lindstrom and Koenig, 1974; Koenig and Schillinger, 1969). Ideally a dispersion scan would reveal dispersive regions and inflection frequencies for each mode of the system, those at high frequency being apparent in the conventional R_1 scan and those at low frequency in the $R_{1\rho}$ scan or R_1 field-cycling scan at low field strengths.

3.2.3. Multinuclei Studies. The three nuclides of water useful for NMR studies, ^1H, ^2H, and ^{17}O (^3H is avoided because of its radioactivity), have different nuclear spin interactions. These differences have been exploited in a number of ways. Thus for pure water, the relaxation rates for the three nuclides can be ranked, R_1, ^{17}O; $\approx 100R_1$, ^2H; $\approx 1000R_1$, ^1H, where it is assessed that the dominant quadrupolar spin interactions of ^2H and ^{17}O are intramolecular and virtually independent of the concentration of the other nuclides. The dipolar spin–spin interaction involving ^1H can have a significant intermolecular component and is affected by the $^1H:^2H$ ratio. This is because of the greater strength of the $^1H–^1H$ dipolar interaction compared with that of a $^1H–^2H$ pair. The much faster relaxation of ^{17}O and ^2H nuclides compared with ^1H enables better time resolution for assessing heterogeneous relaxation in complex systems (Koenig *et al.*, 1975). Comparative dispersion studies have been used to distinguish the intermolecular contribution to the interactions that affect the ^1H but not the ^2H or ^{17}O relaxations (Civan *et al.*, 1978; Civan and Shporer, 1975). Further, since the relaxations of the three nuclides depend on details of the time-averaged orientations of different vectors with respect to the external field ($^1H–^1H$ direction for ^1H, $O–^2H$ for ^2H, and the water diad axis for ^{17}O) it is possible to test for the possible influence of anisotropic effects on the relaxation (Walmsley and Shporer, 1978; Section 3.3). The ^1H relaxation as a function of ^2H concentration has been used to separate intermolecular and intramolecular contributions to the relaxations and to test particular models (Civan *et al.*, 1978; Resing *et al.*, 1976; Edzes and Samulski, 1978).

3.2.4. Tests for Anisotropy. There are many examples, for water sorbed to macroscopically ordered substrates such as fibrous proteins and layered silicates, of NMR line splitting and dependence of the NMR properties on the orientation of the specimens in the applied magnetic field H_0. These are direct evidence of anisotropy in the orientational probability distribution of the water molecules, which can be modeled in terms of an ordered water structure or in terms of preferred axes of molecular reorientation. Because this is such a regularly occurring phenomenon for oriented specimens, it has been suggested (Woessner, 1977) that the interpretation of relaxation data of water in all heterogeneous systems should take account of these anisotropic effects.

Woessner (1977) has given a theory and demonstrated how surface-induced anisotropic effects can be detected and evaluated in terms of the microdynamic structure of heterogeneous systems that are not necessarily macroscopically oriented. He has shown how structural information can be obtained for such systems from the comparison of experimental results with the theoretical behaviors for specific models.

3.2.5. Proton Local Field–Multiwindow Analysis. It has been proposed by Peemoeller and Pintar (1979) that better NMR resolution of the structure of aqueous heterogeneous systems can be obtained by studying the proton local field H_l', in the rotating frame together with analyses of the various relaxation processes R_2, R_1, and $R_{1\rho}$. This is done by assessing the contributions to the magnetization of different proton fractions at different time windows during the free-precession decay of the ^1H NMR signal for each experiment. Consistency in the estimations from the separate NMR measurements is taken as an overall confirmation of the specific model chosen to describe the heterogeneity.

The ^1H local field H_l', which is a measure of the static proton spin interaction, is assessed by plotting the observed transverse magnetization M_x as a function of the strength of the spin-locking field H_1. The relationship used to interpret the measurements includes an assumed distribution function that specifies the structure of the heterogeneous system in terms of the static local field $g(H_l')$. Thus the magnetization $M_x(H_1)$ dependence on the local field H_l' is given by

$$M_x(H_1) = M_0 \int_0^\infty g(H_l') \left(\frac{H_1^2}{H_1^2 + H_l'^2} \right) dH_l' \tag{12}$$

This equation is reduced to a discrete summation for a particular model, and sufficient data to determine the model are obtained by measurements at several times during the free-induction decay (multiwindow analysis).

The local field H_l', which is proportional to R_2, is negligible for proton fractions that are subject to rapid isotropic motional averaging, so that the contributions of those fractions to $M_x(H_1)$ are independent of H_1 and the time of the observation window. Peemoeller and Pinter (1979) chose time windows at 16 and 200 μs in their analysis. At 200 μs it is only those fractions with zero or very small local fields that contribute to the total magnetization. Closer in (at 16 μs), the solidlike fractions with large local fields contribute. Hence by using a discrete form of Equation (12) and successive approximations from the measurements at the different windows, it is possible to obtain a fit to the data so that estimates of the fractions contributing to the magnetization are obtained. These authors noted that the accuracy of local-field analysis is limited by the conflicting field pulse requirements of solid and semisolid components, on the one hand, to achieve equilibrium, and on the other, to avoid significant magnetization decay via the $R_{1\rho}$ process. Although they point out that this multiwindow approach enhances the structural analysis of complex aqueous systems by the accumulation of more-extensive data "at a considerable cost in time," it would appear that, with a suitable data-acquisition method, the multiwindow approach could be used to analyze the complete decay to extract maximum information.

3.2.6. Self-Diffusion Measurements. At any instant the resonant nuclei of a system are labeled by the value of the effective magnetic field H_{eff} at their locations in space, and this provides a means for the accurate measurement of the self-diffusion coefficient D of water in homogeneous systems. Considered in two dimensions, a linear gradient applied perpendicular to H_0 in the X direction, $G(X)$, results in an effective field at a point X of

$$H_{eff}(X) = H_0 + H_l(X) + G(X) \cdot X \tag{13}$$

where $H_l(X)$ is the contribution of the intrinsic local field. It is negligible for regions of high molecular mobility, and when significant, ideally should be independent of position for homogeneous systems. Isotropic diffusion during a measurement time interval of t_d results in a change of effective field experienced by the nuclei of $(2Dt_d)^{\frac{1}{2}}G(X)$ and thereby an apparent enhancement of the transverse-relaxation process. This enhancement can be measured by spin-echo methods (Simpson and Carr, 1958; Stejskal and Tanner, 1965) as an attenuation A in the echo amplitude by the applied-linear-field gradient, and the self-diffusion coefficient calculated using the relation

$$A = \exp\left(-\frac{\gamma^2 G^2(X) D t_d^3}{12}\right) \tag{14}$$

Thus D can be obtained directly from NMR measurements and is not dependent on a particular relaxation model. From this equation it can be seen that the magnitude of A and therefore the sensitivity of the method for assessing D is greater for larger values of diffusion time t_d, field gradient G, and self-diffusion coefficient D. The maximum value of t_d depends on the pulse sequence used for the echo measurements, and for reported studies of aqueous systems, it is of the order of R_2^{-1} or R_1^{-1}. More-complex multiple-pulse decoupling sequences have been used recently to extend the range of t_d in studies of other systems (Silva Crawford *et al.*, 1980). Maximum values of G and accurate definition of t_d can be obtained by pulsed magnetic-field-gradient techniques (Stejskal and Tanner, 1965; Webster and Marsden, 1974). Measurements of D for water in systems with large water contents (Abetsedarskaya *et al.*, 1968; Walter and Hope, 1971) are made with little difficulty, and values of D are found to differ little from that of bulk water. It is much more difficult to measure D for low-water-content sorbed-water systems, which are found to have much smaller D values (Webster, 1971) and are also limited to shorter diffusion times t_d, because of their small R_1^{-1} and R_2^{-1} values; these measurements are further restricted by the small quantity of water in the specimens. Values of D as low as 10^{-12} m^2 s^{-1} have been measured for such systems (Webster, 1971).

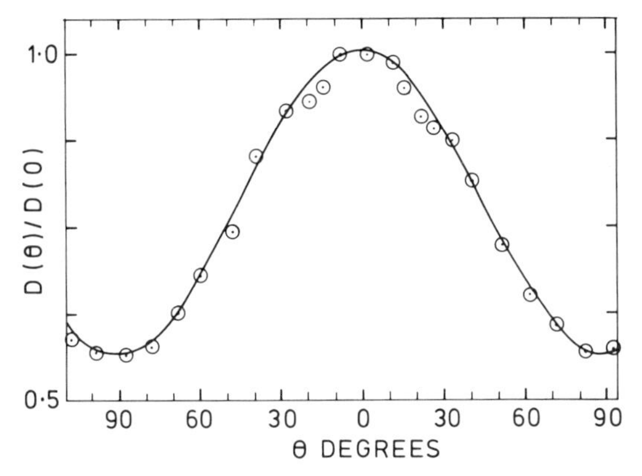

Figure 2. Orientation dependence of the diffusion coefficient of water adsorbed by aligned keratin fibrils; θ is the angle between the fibrillar axis of the keratin and the direction in which diffusion is measured. The curve was calculated assuming D_θ to be related to the principal components $D_\parallel$ and $D_\perp$, parallel and perpendicular to the fibrillar axis, respectively, by $D_\theta = \cos^2\theta\, D_\parallel + \cos^2(90 - \theta)D_\perp$.

Because of the directionality of the applied field gradient, it is possible to measure D as a function of the orientation of a specimen in the applied field and thereby detect the presence of anisotropy in the translational diffusion. Plotted in Figure 2 are the results of such a study of water sorbed to aligned keratin; it can be seen that the self-diffusion is twice as large in the axial or fiber direction as in the radial direction (Webster, 1971).

Using pulsed magnetic-field gradients, it is possible to accurately vary the diffusion time t_d. This is useful for detecting the presence of barriers restricting diffusion within the specimen and for estimating the size of the domains enclosed by these barriers. When the diffusion distance $(2Dt_d)^{\frac{1}{2}}$ becomes comparable to the separation of these barriers, application of Equation (14) no longer yields the true value of D, but rather a value that is dependent on the time t_d. Theories to account for this effect for some simple geometries have been developed (Wayne and Cotts, 1965; Tanner and Stejskal, 1968; Tanner, 1978).

3.3. Models Used

Dynamic structural models of heterogeneous water systems are formulated using variations of three themes: (1) a distribution of motionally isotropic phases or phase states that are often identified with structural or physicochemical features of the system; these phases are ultimately identified by their different NMR relaxation behavior and are usually assumed to have approximately the same chemical shift; (2) the molecular or proton exchange processes that occur between the phases; and (3) reorientational anisotropy of interfacial water molecules.

The formulation for models based on themes 1 and 2 is due to Zimmerman and Britten (1957) and Miyake (1957), and developments of the theory are to be found in Zimmerman and Lasater (1958), Woessner (1961), Woessner and Zimmerman (1963), Odajima (1959), and Resing (1965). The formulation of theories to account for anisotropy are due mainly to Woessner and his co-workers (Woessner, 1961,1962,1974a,b,1977; Woessner and Snowden, 1969a,b), and more recently to the two-dimensional work of Avogadro and Villa (1977). The reader is referred to these papers and the review articles by Pfeifer (1972) and Resing (1967,1972) for treatments of relaxation analysis.

For systems with large concentrations of bulk water, the concern has mainly been to assess the existence of, and if so the amount of, "hydration" or "bound" water. A model widely used and applied variously to the analysis of both R_1 and R_2 relaxation data is simply that of a fast-relaxing fraction p_b bound to the substrate or macromolecule for a time τ_b undergoing rapid exchange with a free fraction p_a. The homogeneous relaxation predicted by the model has a relaxation rate R_i' (i = 1, 2) given by

$$R_i' = p_a R_{ia} + p_b R_{ib} \tag{15}$$

where R_{ib} and R_{ia} are the bound and free relaxation rates. When $R_{ib} \gg R_{ia}$, the R_i' relaxation can be influenced by the residence time τ_b, and if $p_a > p_b$, it is expressed conveniently as (Woessner and Zimmerman, 1963; Luz and Mieboom, 1963; Lindstrom and Koenig, 1974)

$$R_i' = p_b (R_{ib}^{-1} + \tau_b)^{-1} + R_{ia} p_a \tag{16}$$

Resing and co-workers (Foster *et al.*, 1976; Resing *et al.*, 1976; Murday *et al.*, 1975; Resing, 1976) have systematically examined the more general possibilities when the condition of fast exchange does not necessarily hold, so two component relaxations can be observed (R_{ia}' and R_{ib}'). The approximate equations for the observed transverse relaxation rates (this approach is more directly applicable to the transverse relaxation than the slower spin-lattice relaxation) for a two-phase system (Murday *et al.*, 1975) with residence times τ_a and τ_b are

$$\begin{aligned} R_{2a}' &= R_{2a} + p_b[(1 - p_b)\tau_b + R_{2b}^{-1}]^{-1} \\ &= R_{2a} + [\tau_a + (R_{2b} p_b)^{-1}]^{-1} \end{aligned} \tag{17}$$

$$R_{2b}' = R_{2a} + \tau_b^{-1} (1 - p_b)^{-1} \tag{18}$$

and

$$p_a \tau_a^{-1} = p_b \tau_b^{-1} \tag{19}$$

Here, depending on the temperature, the less mobile phase can be either in the rigid-lattice condition [Equation (10)] or in the motional-averaging condition ($\tau_{cb} < 10^{-5}$ s) such that

$$R_{2b} \approx C\tau_{cb} \gg R_{2a} \tag{20}$$

where τ_{cb} is the molecular correlation time in the bound phase.

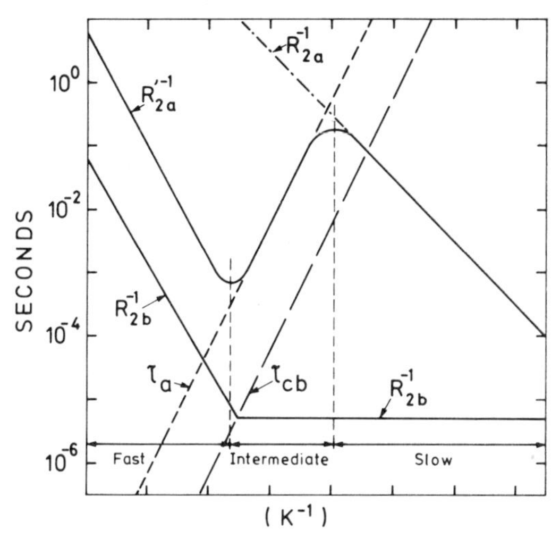

Figure 3. Variation with reciprocal temperature (k^{-1}) of the transverse relaxation time of the bulk phase (R'^{-1}_{2a}) for a model of irrotationally bound water molecules in exchange with bulk water for the conditions $p_b = 0.01$ and $\tau_b = \tau_{cb}$ after the calculation of Resing *et al.* (1976). The plot shows the three distinct regions of fast-, intermediate-, and slow-exchange-rate behavior. The assumed temperature dependences of the transverse relaxation times of the two phases unaffected by exchange (R^{-1}_{2a} and R^{-1}_{2b}), the residence time τ_a, and the correlation time τ_{cb} are also shown.

Three regions of different behavior are predicted for R'_{2a} during change of temperature. This results from the separate temperature dependences of the microdynamic parameters τ_{ca} and τ_{cb} (the molecular correlation times in the free and bound phases that determine R_{2a} and R_{2b} respectively), and τ_a. These all decrease with temperature according to their respective thermal-activation laws. At low temperatures the slow-exchange condition holds, so that $R'^{-1}_{2a} = R^{-1}_{2a}$ and increases with temperature until, combined with the decrease in τ_a, exchange mixing becomes effective and a maximum occurs in R'^{-1}_{2a}, after which $R'^{-1}_{2a} \approx \tau_a$. In this intermediate or exchange-rate-limited region, R'^{-1}_{2a} decreases with temperature. At still higher temperatures where Equation (20) applies for R_{2b}, exchange averaging is fully effective $[(R_{2b}p_b)^{-1} > \tau_a]$, so the relaxation is homogeneous and R'^{-1}_{2a} is the weighted average of R^{-1}_{2a} and R^{-1}_{2b}. This range of behavior is shown in Figure 3 for the conditions $p_b = 0.01$ and $\tau_b = \tau_{cb}$ (see below) (Resing *et al.*, 1976). Experimental evidence for systems exhibiting such thermal behavior has been observed in a number of hydrated systems. However, in most it has been attributed to proton exchange between labile protons of the substrate and water protons (Murday *et al.*, 1975; Resing, 1976; Lynch and Marsden, 1973; Woodhouse *et al.*, 1975; Derbyshire and Duff, 1974; Resing, 1974; Pfeifer, 1975; Freude *et al.*, 1978; Basler and Maiwald, 1979). These observations have been used to study the kinetics of surface reactions (Section 3.5).

Resing *et al.* (1976) suggested that room-temperature R_2 and $R_{1\rho}$ data for muscle water could be explained in terms of this generalized exchange model. They proposed that a small fraction p_b of irrotationally bound water molecules exchanges with the bulk water at the intermediate or rapid rate. The only motion allowed the bound molecules is to jump out of the bound state so that, for this phase, the correlation and exchange times are identical, that is, $\tau_{cb} = \tau_b$. As a consequence, the terms in Equation (17) dominating the intermediate $[p_b/(1 - p_b)\tau_b]$ and the fast $(p_b R_{2b})$ exchange regions vary with temperature at rates equal in magnitude but opposite in sign (Figure 3).

For this model, when $R_{2a}^{-1} \gg R_{2b}^{-1}$, Equation (17), accounting for the intermediate- and rapid-exchange conditions, can be written

$$R_{2a}'^{-1} = \tau_a + (R_{2b} p_b)^{-1} \tag{21}$$

This equation is identical with Equation (16) within the range of approximations made, but features the lifetime in the free (τ_a) rather than the bound state (τ_b). This equation has been modified to account for intermolecular contributions to the relaxations; this is taken up in Section 3.4.

Resing *et al.* (1976, Foster *et al.*, 1976) tested this model against proton relaxation rates for a range of isotopic dilution ratios and the deuterium rate for muscle water. The model equation is written

$$R_{2a}'^{-1} = \tau_a + x(R_{2b} p_b)^{-1} \tag{22}$$

where the experimental variable x is the isotopic ratio $^1H:^2H$ for ^1H relaxation or the enhancement factor with respect to ^1H relaxation for ^2H relaxation. Thus for intermediate exchange, $R_{2a}'^{-1}$ is independent of x, but for fast exchange, $R_{2a}'^{-1}$ is linearly dependent on x. Agreement between theory and experiment was marginal but indicated that the fast-exchange condition is more likely.

A model in contrast with those assuming that a small quantity of highly immobilized water determines the bulk relaxation behavior by exchange is that in which all associated water molecules are considered to be in continuous, long-range interaction with the substrate such that the average relaxation rate is increased with respect to bulk water (Koenig *et al.*, 1975; Civan and Shporer, 1975). This has been used to explain extensive dispersion data (Hallenga and Koenig, 1976; Kimmich, 1980; Lindstrom and Koenig, 1974; Koenig and Schillinger, 1969; Koenig *et al.*, 1975; Civan and Shporer, 1975) on protein solutions, and it has been suggested that the coupling mechanism is related to some nonspecific hydrodynamic effect.

How anisotropic molecular motions could account for the observed deviations of NMR data of sorbed-water systems from the behavior of the liquid in general has been theoretically developed by Woessner *et al.* (Woessner, 1962,1974a,b; Woessner and Snowden, 1969a,b). Woessner expressed the reorientational functions $F_a(t)$ in terms of angles describing the interaction vector with respect to a nonspherical body reorientating in the applied field. Depending on the symmetry of the body, the fluctuations are described by multiple correlation times τ_{ci}, and hence multiple contributions to the Fourier intensities. Thus

$$R_1 = \sum_i K_i R_{1i}(\tau_{ci}) \tag{23}$$

and

$$R_2 = \sum_i K_i R_{2i}(\tau_{ci}) \tag{24}$$

where the $R_{1i}(\tau_{ci})$ and $R_{2i}(\tau_{ci})$ functions are given by Equations (7) and (8), and

$$\sum_i K_i = 1 \tag{25}$$

The constants K_i depend on the details of the anisotropic motion. These equations show that for overall rapid anisotropic molecular reorientation $[(\omega_0\tau_{ci})^2 \ll 1$ for all $i]$, $R_1 = R_2$, but allow the possibility for the R_2 relaxation to be dominated by a slow reorientation about a particular axis s such that $\omega_0\tau_{cs} > 1$ and $R_2/R_1 > 1$. It is therefore often suggested, by implication, that observed NMR properties of aqueous heterogeneous systems can be accounted for by exchange between such a motionally anisotropic surface phase, whose R_2 relaxation is dominated by a slow reorientation process, and the bulk of the associated water. Woessner (1974b,1977) explicitly demonstrated such influence and showed that NMR measurements can be used to detect and evaluate it in some model systems. Whereas the long correlation time τ_{cs} can be the result of slow reorientation about a particular axis because of specific interaction at the interface, it is also possible, when a preferred orientation of the water molecule with respect to the surface exists, for a long correlation time to result, even when all local molecular reorientations are rapid. This correlation time would be the time associated with a change in the direction of the preferred orientation, brought about by translational diffusion of the water molecules throughout the domain structure of the system. Its value therefore would be determined by the extent of the ordered domains and the translational-diffusion rate. The theory has been extended to enable evaluation of specific model systems including "powder" samples in which there is an overall random distribution of the orientation of the reference surfaces or structural domains in the applied field.

Structural information can be obtained for model systems by comparison of the transverse relaxations stimulated by 90–180° (R_2) and 90–90° (R_3) pulse sequences with the corresponding computed theoretical behaviors. Two types of random diffusion have been evaluated: rotational diffusion, in which consecutive angular changes are small, and step diffusion restricted to the four orientations of fixed tetrahedrons. For this procedure to work it is necessary that the rate of change of orientation τ_{cs}^{-1} be greater than the characteristic doublet splitting constant B of the individual domains, that is, $B\tau_{cs}^{-1} > 1$. When this is the case, differences observed in the R_2 and R_3

relaxations are direct evidence of the presence of preferential orientation. It is possible to distinguish the more likely type of ordering by discrimination between the fits to the data obtained for rotational and tetrahedral diffusion. Ultimately a value of τ_{cs} is provided by the analysis, which together with an estimate of the self-diffusion coefficient, gives a value for the domain size.

Woessner and Snowden (1973) speculated that since the ratio of the ^2H and ^1H splitting constants for interfacial water is the same for a wide range of substrate substances of vastly different physical and chemical properties, the structural and dynamic characteristics of water might be determined by the *existence* of an interface rather than the nature of the substrate. They suggested that preferential orientation of water molecules is determined by interactions between interfacial water molecules rather than interactions between water and surface (Woessner, 1974b). Support for these ideas is given by Glasel and Lee's (1974) finding that ^2H relaxation of associated water is unaffected by the character of the substrate surface. Their particular conclusion was that the enhanced R_2 relaxation with respect to bulk water is due to local magnetic-field inhomogeneities created by the heterogeneity in the bulk diamagnetic susceptibilities of the system materials.

Walmsley and Shporer (1978) were concerned with the difficulty posed by the equality of ^1H and ^2H splittings when normalized with respect to intrinsic interaction strengths, as this implies a symmetry axis perpendicular to the plane of the molecule that contains both the proton–proton interaction vector and the electric-field gradient direction at the ^2H nucleus. Also the extensive dispersion results of Koenig *et al.* for protein solutions show that after correction for intermolecular contributions for proton relaxation, the normalized longitudinal relaxation rates at all frequencies are the same, not only for ^1H and ^2H but also for ^{17}O (Koenig *et al.*, 1975). This introduces further geometric complexities because the field gradient tensor for the ^{17}O interaction is different from that of ^2H. Walmsley and Shporer (1978) have been able to reconcile these data by a physically simple model that invokes "a single average interaction of the water molecules with its inherently anisotropic liquid surface," and a fast exchange of the molecules between this anisotropic surface layer and the isotropic bulk environment. For the surface phase they specified a molecular-orientational probability function with cylindrical symmetry around the normal to the surface. The expansion of this in spherical harmonics enables specification of the probability surface in terms of the relative magnitudes of two finite second-order expansion or anisotropy coefficients r_{22} and r_{20}. They defined a scaling factor F to relate the average perturbation Hamiltonians for each nucleus in the surface to the Hamiltonians for each of ^1H, ^2H, and ^{17}O in a water molecule of reference orientation. This scaling factor can be related to the anisotropy coefficients and tensor elements describing the relevant field gradients for the ^2H and ^{17}O interactions. The treatment allows a direct comparison of the reduced relaxation rates for the three nuclei for a full range of molecular-orientational-probability distributions specified by the ratio $r_{22}:r_{20}$. This comparison reveals that if the water is described by the solid-state experimental values of the field-gradient anisotropies at the ^2H and ^{17}O nuclei, the reduced relaxation rates are almost equal for all three nuclei, for a value of $r_{22}/r_{20} = 0.2$ and only for this value. This remarkable observation can be used to explain both the augmented relaxation rates in protein solutions and the line-splitting phenomena of ordered systems. Any proposed model specifying the orientational distribution of water in the surface phase can therefore be tested by the value of r_{22}/r_{20} it predicts. They were therefore able to demonstrate how the observed second-order anisotropy can be the result of an intrinsic property of water surfaces and independent of the nature of the substrate.

Walmsley and Shporer (1978) have modified the two-phase, fast-exchange model to take account of the augmentation of the spin-lattice relaxation rates that results from modulation of the nuclear spin interaction of the surface-oriented water molecules; this modulation is caused by reorientation of the surface-water domains by such phenomena as macromolecular tumbling. The relaxation rate is given by

$$R_1 = p_a R_{1a} + p_s F^2 R_{1s} \tag{26}$$

where s designates the surface fraction, and the scaling factor F can be small compared with 1 because of the anisotropic constraints of the surface. It has been pointed out that this changes the criterion for fast exchange in that the upper limit for the residence time of molecules in the surface phase is now increased by the factor F^{-2}.

Halle *et al.* (1981) analyzed ^{17}O transverse and spin-lattice relaxation dispersion data for protein solutions in terms of such a fast-exchange two-phase model with local anisotropy. They concluded that approximately two layers of water molecules are hindered in their reorientation with respect to bulk water by a factor of about 8. This, they point out, is intermediate between the hydration levels predicted by models based on irrotationally bound molecules and the concept of extended zones of polarized water.

3.4. Cross-Relaxation Effects at the Water–Substrate Interface

It has long been recognized that magnetic interaction of protons acting across phase boundaries could affect the measured spin-lattice relaxation behavior of water protons in heterogeneous systems (Kruger and Helcke, 1967). Cross-relaxation occurs between the proton populations if significant dipolar interaction occurs across the interface. If the two populations differ in their intrinsic spin-lattice relaxation rates, and/or the spin-lattice relaxation induced at the phase boundary is different for each population, the observed relaxation of both populations will differ from the intrinsic rates. Kruger and Helcke attributed the observed change in the proton spin-lattice relaxation of dry lysozyme on hydration, so as to parallel that of the sorbed water over a wide temperature range, to such an effect. Kimmich and Noack (1971) found the same proton spin-lattice relaxation behavior in

^1H_2O and ^2H_2O protein solutions and concluded that all effective relaxation mechanisms reside within the protein molecules. They concluded that there is a "nonmaterial" exchange process between solvent and solute protons because of the identical simple exponential relaxation functions of both protein signal and water signal components observed in all investigated solutions. Further strong evidence for the influence of interfacial proton–proton coupling on the spin-lattice relaxation comes from multinuclear studies. Civan *et al.* (1978) found that $R_1(^2H)/R_1(^{17}O)$ in muscle water is close to that of pure water, whereas $R_1(^1H)/R_1(^{17}O)$ is twice as great in muscle water as in pure water.

Evidence for the occurrence of cross-relaxation effects between protons of substrate species and interacting water protons has been detected in a number of other systems: water–collagen (Edzes and Samulski, 1977,1978; Fung and McGaughy, 1980), hydrated lysozyme (Bryant and Shirley, 1980; Kruger and Helcke, 1967; Hilton *et al.*, 1977), solutions of biopolymers (Koenig *et al.*, 1978; Kimmich and Noack, 1971; Sykes *et al.*, 1978; Stoesz *et al.*, 1978), red blood cells (Eisenstadt and Fabry, 1978), wood (Hsi *et al.*, 1977), cellulose (Hsi *et al.*, 1979), and brown coal (Lynch and Webster, 1980). Most analyses of these data have considered cross-relaxation to result only in a magnetization exchange, whereas more careful attention to the underlying theory shows that direct spin relaxation is also a consequence, and recently this was demonstrated experimentally (Fung and McGaughy, 1980).

Kimmich and Noack (1971) described a general model for heterogeneous systems incorporating spin-diffusion (magnetization transfer) across phase boundaries. Inherent in this model are many of the assumptions that have subsequently been used to describe the mechanism of cross-relaxation and quantitatively analyze data for hydrated systems. Two bulk proton populations were considered, each maintained essentially isothermal by rapid diffusion of spin magnetization—via molecular or proton exchange in the water phase and via spin diffusion in the substrate phase. (For the proton populations to be described by a spin temperature also implies sufficiently strong spin–spin interactions within each population, a condition more likely to be met in the more rigid substrate.) Coupling by spin interaction across the interface results in cross-relaxation effects including magnetization transfer between the phases.

The mechanisms of magnetization transfer at the interface and diffusion of magnetization within the bulk phases enable the population with the greater intrinsic spin-lattice relaxation rate to act as a sink for the magnetic excitation of the other population. The driving force for this process can be envisaged as the difference in the magnetizations or spin temperatures created and/or maintained by the different intrinsic relaxation rates (Koenig *et al.*, 1978). On the other hand, when the intrinsic spin-lattice relaxation rates of the two populations are similar, there can be no detectable effect unless the populations are selectively excited to different magnetizations. A technique for achieving this has been described by Edzes and Samulski (1978).

A model based on the Solomon theory of relaxation of spin pairs (Solomon 1955), developed by Kalk and Berendsen (1976) to account for cross-relaxation effects between different proton groups in proteins, has been used to analyze spin-lattice relaxation data of hydrated systems (Hilton *et al.*, 1977; Eisenstadt and Fabry, 1978; Hsi *et al.*, 1977,1979; Lynch and Webster, 1980). The inadequacy of this approach has been pointed out by Koenig *et al.* (1978) in that this theory does not account for the macroscopic heterogeneity of hydrated systems. Phenomenological models have been described and applied by Edzes and Samulski (1977,1978) and Koenig *et al.* (1978). The models were formulated simply by including a cross-relaxation term coupling the bulk populations in Bloch equations for the spin-lattice relaxations of the two spin populations designated I and S. The equations written in terms of the reduced magnetization, M_I and M_S, are

$$\frac{dM_I}{dt} = -R^B_{1I}\, M_I - k_I\,(M_I - M_S) \tag{27a}$$

$$\frac{dM_S}{dt} = -R^B_{1S}\, M_S - k_S\,(M_S - M_I) \tag{27b}$$

where R^B_{1S} and R^B_{1I} are considered to be the intrinsic bulk spin-lattice relaxation rates of the two populations free of cross-relaxation (i.e., when $k_I, k_S = 0$). The coefficients k_I and k_S of the cross-relaxation terms are defined as the magnetization transfer rates from their respective populations and are related by the dynamic equilibrium condition

$$n_I k_I = n_S k_S = N R_T \tag{28}$$

where n_S and n_I are the proton populations, N is a constant, and R_T is referred to as the cross-relaxation rate. Edzes and Samulski (1978), noting the similarity in form of Equations (27a and b) with those of the Solomon theory, related R_T to the corresponding rate coefficient of this theory for relaxation of spin pairs [see Equation (33)]. This approach has been criticized by Koenig *et al.* (1978) on the basis that whereas the assumed model requires a flow of magnetization from one bulk phase to another, the explicit form of the Solomon cross-term [Equation (33)] is thermodynamically inconsistent in that it can be either positive or negative depending on the external experimental conditions. However, examination of this term (see below) shows that it relates to both a spin-exchange process and a direct spin-lattice relaxation process, and the sign reflects which of these is dominant under the prevailing conditions. Koenig *et al.* (1978) chose to ignore the microscopic detail and simply considered R_T to be a thermodynamic parameter governing the exchange of magnetization between two populations in thermal contact.

Andree (1978) formulated relationships based on the Solomon theory of dipolar interaction of spin pairs. These relationships describe the behavior of the coupled two-phase system in which three zones of relaxation are considered, those of the bulk phases remote from the interface and that at the interface involving magnetic interaction of protons from both populations across the interface. The equations of motion of the experimentally observable z components of the macroscopic magnetic moments I_z and S_z of a population of dipolar coupled spin pairs I and S in terms of the transition probabilities w_i between the four unperturbed eigenstates [Figure (4)] derived by Solomon (1955) are

$$\frac{dI_z}{dt} = -(w_0 + 2w_1 + w_2)(I_z - I_0) - (w_2 - w_0)(S_z - S_0) \tag{29a}$$

$$\frac{dS_z}{dt} = -(w_2 - w_0)(I_z - I_0) - (w_0 + 2w'_1 + w_2)(S_z - S_0) \tag{29b}$$

where I_0, S_0 are the equilibrium moments. He noted the correspondence of these equations with the Bloch equations with the added terms representing the spin I–spin S interaction. Whereas Solomon's treatment was general and included interactions of unlike spins, we are concerned here only with the proton–proton interactions. The spins I and S are distinguished as belonging to different populations. The transition probabilities for perturbation of the states by dipole–dipole interactions for protons distant r apart are (Solomon, 1955)

$$w_0 = \frac{\hbar^2\gamma^4\tau_c}{10r^6} \tag{30a}$$

$$w_1 = w'_1 = \frac{3\hbar^2\gamma^4}{20r^6}\frac{\tau_c}{(1 + \omega_0^2\tau_c^2)} \tag{30b}$$

$$w_2 = \frac{6\hbar^2\gamma^4}{10r^6}\frac{\tau_c}{(1 + 4\omega_0^2\tau_c^2)} \tag{30c}$$

where w_0 is the transition probability for a mutual spin-flip or magnetization exchange of a spin pair, whereas w_1 and w_2 are probabilities for spin transitions involving exchange with the lattice (Figure 4). To better isolate the spin-exchange contribution w_0 to the overall spin relaxation, Equations (29a and b) can be rearranged thus:

$$(di/dt)_x = -R_{1I}i - \kappa(i - s) \tag{31a}$$

$$(ds/dt)_x = -R_{1S}s - \kappa(s - i) \tag{31b}$$

where i and s are the reduced magnetizations, and the subscript z has been dispensed with. The subscript x designates relaxation in the interface zone. The direct, and cross-relaxation coefficients R_{1I}, R_{1S}, and κ, are

$$R_{1I} = 2(w_1 + w_2) + w_{1e} \tag{32a}$$

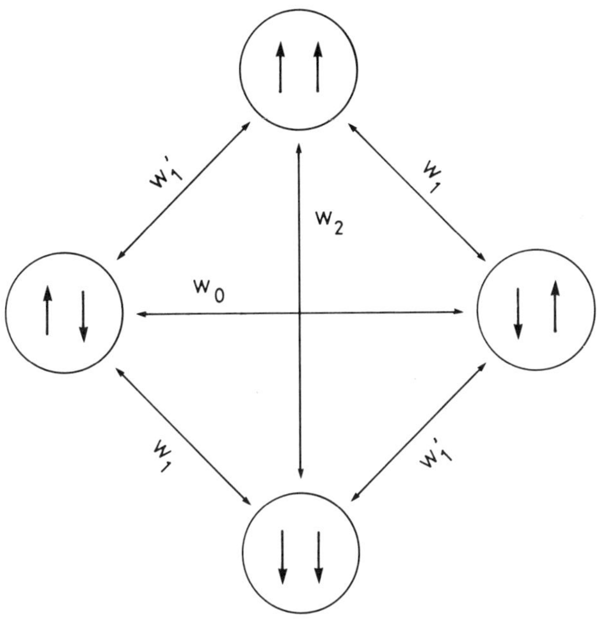

Figure 4. The transition probabilities w_i between the four energy levels of a spin pair for the longitudinal component of the spin magnetization as described in the Solomon theory on the relaxation of spin pairs.

$$R_{1S} = 2(w_1 + w_2) + w_{2e} \tag{32b}$$

and

$$\kappa = w_0 - w_2 \tag{32c}$$

The additional contributions w_{ie} ($i = 1, 2$), attributed to external relaxation mechanisms and included in Equations (32a), and (32b), make $R_{1I} \neq R_{1S}$ and ensure that the magnetizations of the two spin systems are different, and therefore the cross-relaxation or second terms in Equations (31) are finite for $t > 0$. Insofar as the three-zone model is concerned, the effects of any such external mechanisms are not resolvable, and it suffices for the treatment that $w_{ie} = 0$. Equations (31) and (32) therefore describe the relaxation behavior of the protons in the interfacial zone of this three-zone model. The coefficient κ is referred to as the cross-relaxation constant, and it contains both a spin-exchange (w_0) and spin-lattice relaxation (w_2) factor. In terms of the micro-dynamic parameters,

$$\kappa = (w_0 - w_2) = \frac{\hbar^2\gamma^4\tau_c}{10r^6}\left(1 - \frac{6}{1 + 4\tau_c^2\omega_0^2}\right) \tag{33}$$

When the conditions are such that $\omega_0^2\tau_c^2 > \frac{5}{4}$ the spin-spin exchange w_0 dominates and κ is positive. When $\omega_0^2\tau_c^2 < \frac{5}{4}$ the spin-lattice exchange w_2 is dominant and κ is negative. The physical significance of κ when the latter condition prevails is not clear, as w_2 is also represented in the other term of Equations (31). It is assumed that κ is dominated by the spin-spin exchange term w_0 in the following discussion.

The spin exchange between water protons and substrate protons in hydrated systems is enhanced both by intimate contact (small r) and low molecular mobility (large τ_c). For $\omega_0\tau_c > 1$, κ continues to increase with τ_c, whereas the R_{1i} decrease, so that the spin-spin exchange or cross-relaxation term becomes increasingly significant in the net spin-lattice relaxation. The result of effective spin exchange is a tendency for the magnetizations of the two populations to be equalized by a progressive "nonmaterial" transfer of magnetization from the population with the greater magnetization.

In that $\imath$ and s are representations of the magnetizations of separate spin populations that relax at different rates (R_{1I}^B and R_{1S}^B) to equilibrium in isolation, and in that only a small fraction $f_i(i = I, S)$ of the spins of each population is instantaneously represented in the interface zone, the magnitudes and therefore difference between $\imath$ and s are largely determined (ignoring any asymmetrical relaxation terms contributing to the interfacial spin-lattice relaxation) by the relaxation processes that occur in the bulk zones of the two populations. Thus

$$\left(\frac{d\imath}{dt}\right)_B = -R_{1I}^B\imath \tag{34a}$$

$$\left(\frac{ds}{dt}\right)_B = -R_{1S}^B s \tag{34b}$$

An assumption of the model is that each population is effectively maintained isothermal by rapid internal exchange of magnetization, so it is possible to define the total change of magnetization of each population by weighted addition of Equations (31) and (34). Therefore, $\imath$ and s are identical to the reduced magnetizations M_I and M_S defined for Equations (27). If fractions f_I and f_S of total populations n_I and n_S are in the interface zone, the summations result in net relaxation equations for the magnetizations M_I and M_S of each population:

$$\frac{dM_I}{dt} = (1 - f_I)\left(\frac{d\imath}{dt}\right)_B + f_I\left(\frac{d\imath}{dt}\right)_x \tag{35a}$$

and

$$\frac{dM_S}{dt} = (1 - f_S)\left(\frac{ds}{dt}\right)_B + f_S\left(\frac{ds}{dt}\right)_x \tag{35b}$$

Substituting for Equations (31) and (34) and rearranging terms yields

$$\frac{dM_I}{dt} = -R_{Ix}M_I - k_I'(M_I - M_S) \tag{36a}$$

$$\frac{dM_S}{dt} = -R_{Sx}M_S - k'_S(M_S - M_I) \tag{36b}$$

where

$$R_{Ix} = R^B_{1I}(1 - f_I) + R_{1I}f_I \tag{37a}$$

$$R_{Sx} = R^B_{1S}(1 - f_S) + R_{1S}f_S \tag{37b}$$

and

$$k'_I = \kappa f_I; \quad k'_S = \kappa f_S \tag{38}$$

Equations (36a and b) are identical in form to those of the phenomenological model—Equations (27a) and (27b). However, the relaxation constants have a different significance. For Andree's equations the cross-relaxation or magnetization transfer rates k'_I and k'_S are explicitly defined in terms of Solomon's microscopic theory of spin relaxation [Equation (33)], and the direct relaxation rates R_{Ix} and R_{Sx} are not the intrinsic relaxation rates of the bulk phases. From Equations (37a) and (37b) it can be seen that these rates are determined not only by the bulk relaxation processes of the separate populations but also by the distribution of the populations between the bulk and interface zones and also the spin-lattice relaxation induced by cross-coupling at the interface. The degree to which the terms R_{Ix} and R_{Sx} are influenced by the cross-coupling at the interface of course depends on the relative spin-lattice relaxation rates in the bulk and interface zones. Fung and McGaughy (1980) clearly demonstrated that cross-relaxation is the greater influence for the water population in a hydrated collagen system. They found that the calculated R_{Ix} from the water proton spin-lattice relaxation data is only slightly dependent on the $^2H:^1H$ composition, in contrast to this dependence for pure water, a clear indication of significant contribution to this term from the relaxation induced by the cross-coupling. Furthermore, a direct comparison of the spin-lattice relaxation rate of ^2H with the parameter R_{Ix} for protons under identical conditions revealed a considerable enhancement of the R_{Ix} value with respect to the comparison in liquid water. This indicates an intermolecular contribution to the relaxation involving nonwater protons. On the other hand, in the brown-coal–water system, it was concluded that the intrinsic spin-lattice relaxation of the bulk water is dominant because of the large concentration of paramagnetic species accessible to the water protons (Lynch and Webster, 1980).

A general solution for Equations (36a) and (36b) [allowing for differential excitation of the two spin populations when at time $t = 0$, $M_I(0) \neq M_S(0)$] is, for $i, j = I, S$

$$M_i(t) = C_i^+ \exp(-\lambda^+ t) + C_i^- \exp(-\lambda^- t) \tag{39}$$

where

$$2\lambda^\pm = (R_{Sx} + R_{Ix} + k'_I + k'_S) \pm [(R_{Sx} - R_{Ix} + k'_I - k'_S)^2 + 4k'_Ik'_S]^{\frac{1}{2}} \tag{40}$$

$$C_i^{\mp} = \pm M_i(0) \frac{(R_{ix} - \lambda^{\pm})}{(\lambda^+ - \lambda^-)} \pm [M_i(0) - M_j(0)] \frac{k_i'}{(\lambda^+ - \lambda^-)} \tag{41}$$

The theory therefore predicts that when spin exchange is significant, the magnetic relaxation of both phases is the superposition of two exponentials in such a way that the rate of decay of the logarithm of the magnetization decreases with time for the population with the greater direct relaxation rate and increases with time for the other population.

The parameters $C_i^{\mp}$ and $\lambda^{\pm}$ can be extracted from the measured spin-lattice relaxation data for the water phase, and if the spin-lattice relaxation of the substrate proton phase also can be measured or estimated, then values of rates R_{ix} can be computed from Equation (41) (Edzes and Samulski, 1978). Further, if estimates of both R_{Sx} and R_{Ix} are available, Equation (40) yields values for k_S' and k_I', and hence $k_I'/k_S' = n_S/n_I = \boldsymbol{F}$, the ratio of proton populations, assuming all coupling between the populations is via spin pairs. Values of $\boldsymbol{F}$ obtained from analysis of spin-lattice relaxation data for a series of coal–water specimens of variable water content were in good agreement with the known proton population ratios (Figure 5).

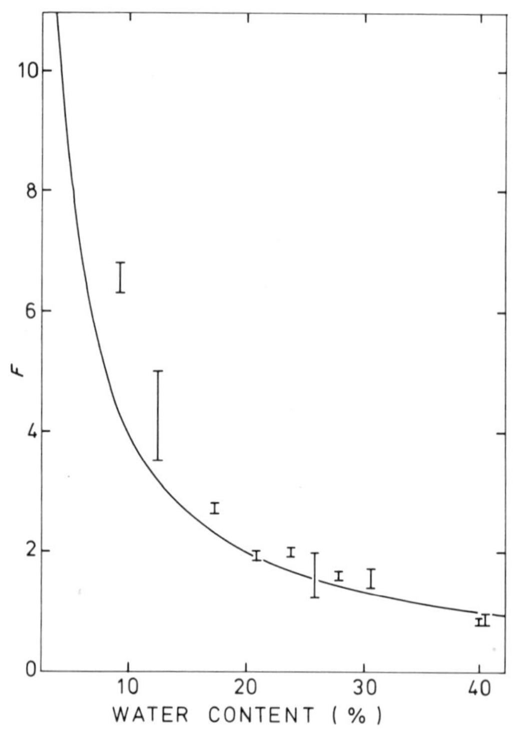

Figure 5. The variation with water content at 290 K of the ratio of the coupled coal and water proton populations, ***F***, derived from NMR relaxation data compared with the independently estimated ratio (solid line).

In an attempt to extract information on the molecular dynamics of the interfacial water from the cross-relaxation analysis, Edzes and Samulski (1978) have considered a model for the mechanics of the transfer of magnetization from one population to the other when interfacial water molecules are irrotationally bound for a time τ_{cb}. They identify three sequential processes: spin diffusion in the solid of rate R_S, cross-relaxation or spin exchange in the phase boundary, κ, and proton exchange between the boundary and bulk water of rate τ_{cb}^{-1} (this is the process that determines the magnetic diffusion rate away from the phase boundary within the water phase).

Thus the effective rate k_e of magnetization transfer across the interface is given by

$$k_e = (\kappa^{-1} + R_S^{-1} + \tau_{cb})^{-1} \tag{42}$$

Each of these processes can be rate determining for k_e depending on the water mobility or dipolar interaction time at the interface, τ_{cb}. These regions are shown in Figure 6 (Edzes and Samulski, 1978). When $\omega_0^{-1} \ll \tau_{cb} < \sigma_0^{-1}$, cross-relaxation at the phase boundary, κ, is dominant, and k_e is directly proportional to τ_{cb}. When $\sigma_0^{-1} < \tau_{cb} < R_S^{-1}$, spin exchange is limited by the

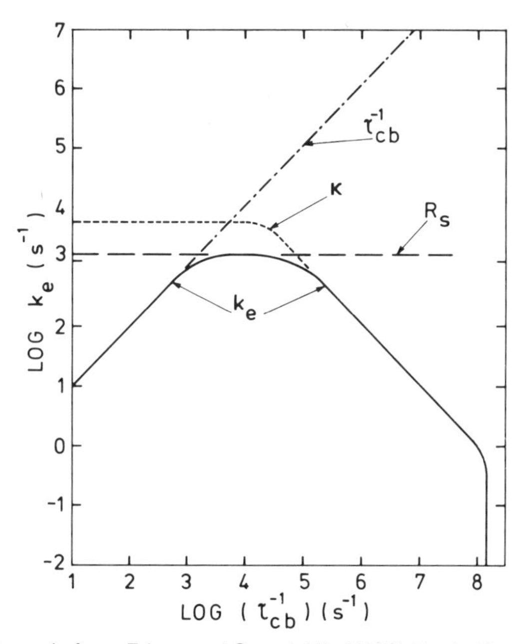

Figure 6. This figure is from Edzes and Samulski's (1978) illustration of their model for the transfer of magnetization between solid and water proton populations of a hydrated system in which interfacial water molecules exchange with the bulk water at a rate τ_{cb}^{-1} (s^{-1}). The calculations are for a resonance frequency of 40 MHz and interproton distance of 2.70 Å. The effective rate of magnetization transfer, k_e, is limited by τ_{cb}^{-1} at lower rates of τ_{cb}^{-1}; by the rate of spin-diffusion within the solid, R_s, at intermediate rates of τ_{cb}^{-1}; and by the cross-relaxation rate of the interface, κ, for higher values of τ_{cb}^{-1}.

spin-diffusion rate of the solid phase, R_S, which is insensitive to τ_{cb} and hence temperature. In this range, magnetization transfer is most effective. For $\tau_{cb} > R_S^{-1}$, k_e is limited by proton exchange between boundary and bulk water and is inversely proportional to τ_{cb}.

Given the existence of dipolar coupling at the interface, a further necessary condition for net magnetization transfer to occur is a difference in magnetization or spin temperature, that is, $|(M_I - M_S)| > 0$. When both populations are equally excited by the preparation pulse [i.e., $M_I(0) = M_S(0)$], this can only evolve with time if the relaxation rates of the two populations differ (i.e., $R_{Ix} \neq R_{Sx}$). This is clearly the case when the bulk relaxation rates R_{1S}^B and R_{1I}^B are different, but Equations (37a) and (37b) show that this condition also can be achieved by a difference in the fractions f_I and f_S of the two spin populations in the interfacial zone, or alternatively by an asymmetry in the direct relaxation induced by the cross-coupling such that $R_{1S} \neq R_{1I}$. For hydrated systems the expectation is that the condition $R_{1S}^B \neq R_{1I}^B$ prevails. When the relaxation rates are such that the conditions for magnetization transfer do not evolve or are insufficient to reveal the existence of cross-relaxation when both populations are equally excited, a method of selective excitation described by Edzes and Samulski (1978) can be used. This method depends on a difference in the spectral width of the proton resonances of the two populations as a consequence of the lesser mobility of the substrate molecules compared with the water molecules. A 180° preparation pulse of suitable duration, t_{180}, determined by the condition R_2^{-1} (substrate) $< t_{180} < R_2^{-1}$ (water) can be selected so that it has a spectral width sufficient to saturate the water proton population [i.e., fully inverted so that for water protons $M_w(0) = 1$] while only partly saturating the substrate proton population, which rapidly becomes isothermal at a different spin temperature than the water population via spin diffusion so that $M_{\text{sub}}(0) < 1$. Edzes and Samulski (1978) have used the fact that the amplitude parameters $C_i^{\pm}$ are dependent both on the degree of saturation and the difference in the saturations of the two populations—$[M_I(0) - M_S(0)]$ in Equation (41)—to enhance and study the effect of cross-relaxation on the spin-lattice relaxation of hydrated collagen and muscle. They were able to demonstrate the occurrence of magnetization transfer by the selective-excitation method. Relaxation curves obtained by varying the 180°-pulse condition revealed that when the macromolecular magnetization is incompletely inverted, the detected magnetization of the macromolecular protons initially increases, directly revealing the transfer of magnetization from the initially more highly magnetized spin population of the water. It was possible from the measured values of λ^+, λ^-, $C_i^{\pm}$, and $M_w(0)$, $M_{\text{sub}}(0)$ to fit the data to Equations (40) and (41) and calculate the direct relaxation rate for water and the cross-relaxation rates k_i'.

Edzes and Samulski (1978) considered the implications of cross-relaxation on the proton transverse relaxation time of hydrated water for Resing's irrotationally bound water model [Equation (22)]. They used the expression for unlike spins to account for the contribution by dipolar interaction between substrate protons and bound-water protons to the spin-spin relaxation, that is,

$$R_{2b}^{\mathrm{u}} = \frac{\hbar^2\gamma^4}{10r^6}\left(\frac{5}{3} + \frac{3}{1 + \omega_0^2\tau_{cb}^2} + \frac{2}{1 + 4\omega_0^2\tau_{cb}^2}\right) \tag{43}$$

Here τ_{cb} is the residence time of the water in the bound state or the effective correlation time for the interaction. When motional narrowing is effective but $\tau_{cb}^{-1} < \omega_0$, comparison of Equation (43) with Equation (33) for the cross-relaxation rate κ gives

$$R_{2b}^{\mathrm{u}} \geqslant \frac{5}{2}\kappa \tag{44}$$

which indicates a significant intermolecular contribution to the spin–spin relaxation of interfacial or bound-water protons when the conditions conducive to cross-relaxation apply. Edzes and Samulski assumed the effective spin-exchange k_e [Equation (42)] to be limited by cross-relaxation at the interface κ and modified the Resing expression [Equation (22)] by adding the intermolecular contribution to the net relaxation for the bound water. This involves replacing R_{2b}/x in Equation (22) with $(R_{2b}/x + R_{2b}^{\mathrm{u}}/n)$, where n is the number of substrate protons with which each bound-water proton is assumed to interact. The modified equation for the proton relaxation is

$$R_{2a}'^{-1} = \tau_a + \frac{xR_{2i}^{-1}}{R_{2b}p_bR_{2i}^{-1} + x} \tag{45}$$

where $R_{2i} = R_{2b}^{\mathrm{u}}/np_b$ is the intermolecular contribution averaged over all the water molecules. This expression enabled a much better fit to the muscle water data than Equation (22) and indicated that the fast-exchange condition was applicable.

Bryant and Shirley (1980) have analyzed proton spin-lattice relaxation measurements of sorbed water–lysozyme powder in terms of the phenomenological model of Edzes and Samulski (1978). They found that the "intrinsic" relaxation rates R_{1w} of the sorbed water are significantly greater than the slower of the measured relaxation rate constants λ^- over the temperature range, and that there appears to be a minimum in R_{1w} near room temperature. From this they concluded that the water molecules at the protein surface have considerable mobility. They estimated that the molecular correlation times are not greater than 100 times that of bulk water and that only a narrow distribution of mobilities, if indeed any, is required to explain the relaxation-time data. Resing (Bryant and Shirley, 1980) criticized this interpretation on the basis of the lack of precision of the estimates of the relaxation times and in particular the assumption that a minimum occurs at room temperature. He demonstrated that the data are consistent with a minimum at +100°C. Resing further observed that according to the Solomon relationship for the coupling of spin pairs [Equation (31)], cross-relaxation should only be significant for $\omega_0^{-1} > 10^{-8}$ s [see Bryant and Shirley's data (1980)]. Bryant, on the other hand, rejected this theoretical basis of the cross-relaxation phenomenon. Resing also suggested that the proton relaxation times of the protein would be influenced by the coupling and molecular dynamics of the sorbed-water protons and therefore could not be determined by the measurement of a specimen hydrated with ^2H_2O. This observation viewed in light of Fung and McGaughy's (1980) demonstration of the strong contribution to the direct relaxation rates of the water and substrate protons for muscle by the cross-coupling at the interface—an effect consistent with the interpretations based on the Solomon theory [Equations (37a and b)]—further emphasizes the difficulty of interpretation of spin-lattice relaxation measurements of these systems. In contrast to Bryant and Shirley, Edzes and Samulski assert that in accordance with the Solomon theory, the presence of cross-relaxation is direct evidence of slowly exchanging water protons at a rate $\tau_b^{-1} \ll \omega_0$.

3.5. Surface Water Proton Exchange Kinetics

There have been many examples in which the exchange-rate-limited transverse-relaxation condition (Section 3.3) has been observed for sorbed-water systems and attributed to a proton-exchange reaction between the water and labile surface groups (—SuH*) of the substrate (Murday *et al.*, 1975; Resing 1974,1976; Lynch and Marsden, 1973; Derbyshire and Duff, 1974; Woodhouse *et al.*, 1975; Pfeifer, 1975; Freude *et al.*, 1978; Basler and Maiwald, 1979) rather than between bound and free water phases. Resing (1974) pointed out the potential of NMR temperature scanning of sorbed systems to study the kinetics of such reactions. He related the exchange probability of the proton from the water phase to the chemisorbed labile hydrogen phase, τ_a^{-1}, as measured by NMR [Equation (17)] to the kinetic constants of an assumed rate-limiting reaction of the type

$$\mathrm{HOH} + \mathrm{SuH}^* \underset{}{\overset{k_f}{\rightleftharpoons}} \mathrm{HOH}^* + \mathrm{SuH} \tag{46}$$

The forward reaction rate constant k_f and the orders of the reaction, a and b, with respect to the thermodynamic activities of water $[H_2O]$, and the substrate species [SuH*], respectively, are related by

$$\tau_a^{-1} = \frac{k_f\,[\mathrm{H_2O}]^a\,[\mathrm{SuH}^*]^{b'}}{N_{\mathrm{H_2O}}} \tag{47}$$

The number density of water molecules, N_{H_2O}, and the water activity $[H_2O]$, which is approximated by the equilibrium water-vapor pressure, are related by the water-vapor-absorption isotherm. If [SuH*] is assumed constant, it is possible to compute estimates of the reaction orders a and b' by fitting Equation (47) to the NMR-determined τ_a^{-1} values. The calculated dependencies of τ_a^{-1} on water content (i.e. $[H_2O]$) reveal great sensitivity to

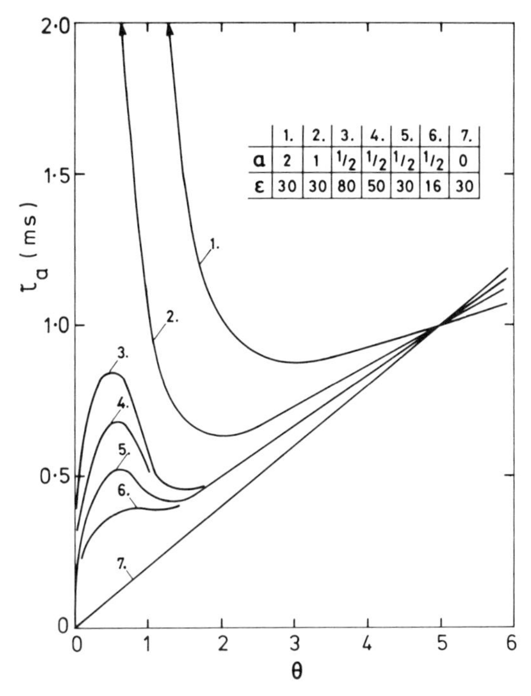

Figure 7. Resing's (1976) diagram showing how the proton exchange time τ_a varies with coverage θ for BET-type adsorption isotherms according to a hypothetical exchange reaction [Equation (46)] with a range of reaction orders of the water, a. Also, for the case of $a = 1/2$, the sensitivity of the relationship to the BET first layer adsorption strength index ε is shown for a range of values of ε.

both the isotherm type and the reaction order. Calculated values for BET-type isotherms (Brunauer, Emmett and Teller, 1938) are shown in Figure 7 (Resing, 1976). The application of this theory to experimental data on the keratin–water system explained an otherwise obscure behavior whereby the NMR-assessed value of τ_a passes through a maximum with increasing water content (Figure 8). This behavior was found to be consistent with a reaction of the form of Equation (46) with a half order for $[H_2O]$ (Resing, 1976).

Extensive NMR data of the water–zeolite 13-X system (Resing, 1974; Murday *et al.*, 1975), which clearly exhibits all the characteristics of the two-phase exchange theory [Equation (17)] including the three limiting conditions for R_2 relaxation (Section 3.3) of no-exchange, intermediate-exchange or exchange-rate-limited, and fast-exchange relaxations, were used to quantitatively determine the kinetics of the proton exchange between the water and the surface complex. Best agreement was achieved with reaction orders of $\frac{1}{2}$ and 1 for the water and surface complex, respectively. This analysis also provided estimates of parameters describing the magnetic interactions and molecular kinetics of the surface complex.

Other studies include those of Freude *et al.* (1978) and Basler and Maiwald

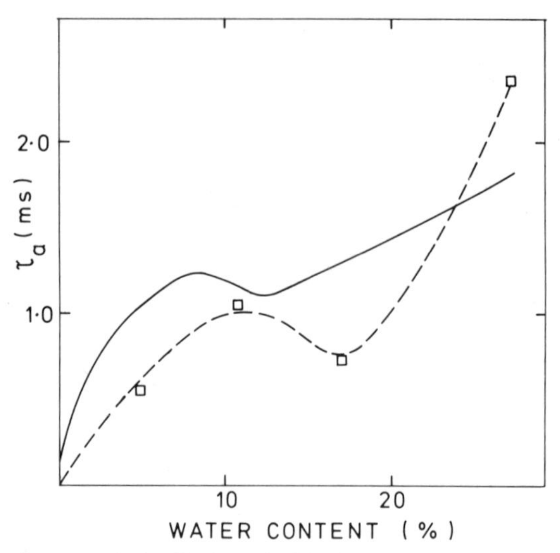

Figure 8. Comparison made by Resing (1976) of the experimental exchange times τ_a between sorbed water and labile keratin protons for a range of water contents, obtained from NMR transverse-relaxation measurements (squares), with a theoretical calculation (solid line) for Resing's hypothetical model [Equation (46)] made using the experimental adsorption isotherm.

(1979) of the interaction of water with zeolites. Freude *et al.* studied a range of synthetic zeolites before and after hydrolysis and hydrothermal treatments. Using the exchange-rate-limited transverse-relaxation-rate values, they were able to assess the proton lifetimes in the surface hydroxyl groups and to demonstrate that the hydrolyzed specimens had a greater population of labile surface hydroxyl groups.

Exchange-rate-limited transverse relaxation for hydrated type A zeolites prepared by different methods was interpreted by Basler and Maiwald (1979) to be the result of proton exchange between water in the α cages and —OH surface groups contained in aluminium compounds in β cages. This, they concluded, indicates defects in the crystalline structure of the zeolite cages.

3.6. Assessment of "Bound" or "Nonfreezable" Water

There have been many NMR studies of the phenomena that occur when hydrated systems are temperature cycled above and below 273K. These studies have often been attempts to assess the quantity of "bound" water, based on its definition as water so modified in its properties that it is nonfreezable. It is therefore important to consider, on the one hand, the accuracy of NMR for estimating nonfreezable water and, on the other, implications from NMR and other studies for the hypothesis that nonfreezable water can be equated with "bound" or interacting water.

Most assessments of nonfreezable water by NMR methods have chosen the intensity of the mobile-water signal at a specified temperature below 273K as being proportional to the amount of "bound" or nonfreezable water (Sussman and Chin, 1966; Toledo *et al.*, 1968; Kuntz *et al.*, 1969; Kuntz and Brassfield, 1971; Kuntz, 1971; Belton *et al.*, 1972; Hsi and Bryant, 1975; Carles and Scallan, 1973; Haschemeyer *et al.*, 1977; Brynjas-Kraljevic and Maricic, 1978; Katayama and Fujiwara, 1979,1980; Leung and Steinberg, 1979). Either the integrated intensity of the absorption signal or the initial intensity of the transverse relaxation of the mobile-water component have been used as the measurement parameter. Andrewartha *et al.* (1978) showed how the sensitivity of the measurement can be greatly enhanced by the use of signal averaging. However, it has been found in most cases that the quantity of mobile water is temperature dependent below any freezing transition (Sussman and Chin, 1966; Belton *et al.*, 1972; Haschemeyer *et al.*, 1977; Katayama and Fujiwara, 1979,1980; Leung and Steinberg, 1979; Cyr *et al.*, 1971; Derbyshire and Parsons, 1972; Oakes, 1976a; Lynch and Webster, 1979a,b; Mrevlishvili and Sharimanov, 1978). Reasons for the temperature dependence of this estimate of nonfreezable water are discussed below.

Mrevlishvili and Sharimanov (1978) questioned the validity of the equation of NMR nonfreezable water and water of hydration because of their finding that this equation indicates a decrease in hydration on the denaturation of collagen, whereas calorimetry indicates an increase. A similar observation was made by Ramirez *et al.* (1974) with respect to their own results and those of Kuntz *et al.* (Kuntz *et al.*, 1969; Kuntz and Brassfield, 1971; Kuntz, 1971).

Resing *et al.* (1977) asserted that nonfreezable water in muscle is not necessarily related to an identifiable phase at room temperature, implying that freezing alters the structure and causes segregation. Evidence for such a structural change is apparent in the NMR data of Narebska and Streich (1980) for gel-type ion-exchange resins. However, there is evidence of close correlation between an identifiable interacting component of water and nonfreezable water in some systems (Kuntz and Kauzmann, 1974; Lynch and Webster, 1979a; Golton *et al.*, 1981; Leung and Steinberg, 1979). Systems that can tolerate wide changes in their level of hydration above and below the saturation water content without denaturation are useful for investigating this connection. The highly cross-linked, insoluble protein wool-keratin falls into this category, and its saturation water content has been estimated by a number of NMR (Lynch and Webster, 1979a) and other methods (Haly and Snaith, 1969; Ashpole, 1952).

The temperature variation of the ^{1}H NMR mobile-water signal intensity I_m (uncorrected for the temperature dependence of the NMR signal) during heating of a wool–water specimen with 38.4% water content (wc) (measured as weight percent of the dry specimen weight) is shown in Figure 9 (Lynch, 1981). If the step transition near 273K is due to fusion of the noninteracting water present, the mobile-water signal strength at 273K before the transition indicates a saturation water content of about 35%. For lower temperatures,

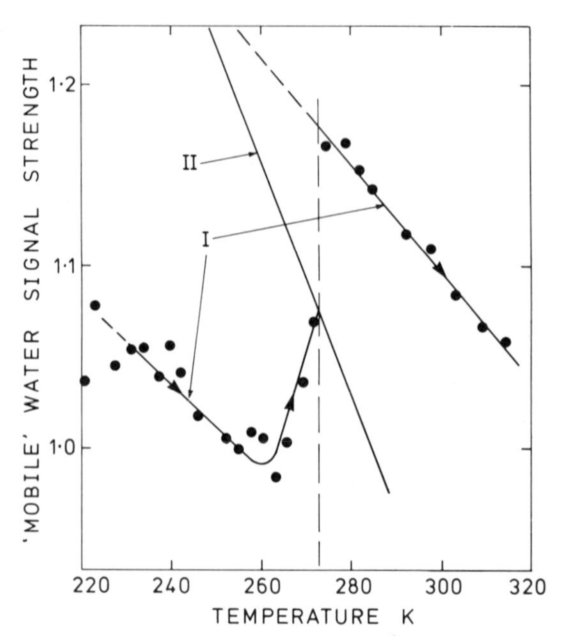

Figure 9. The temperature variation during heating of the ¹H NMR mobile-water signal strength I_m for a wool–water specimen containing 38.4% water (*I*). This is compared with the temperature variation of the ¹H NMR signal strength of a nonfreezing liquid measured with the same apparatus (*II*).

less nonfreezable water is indicated. At temperatures just above the phase transition, for this wool–water system, the mobile-water signal can be resolved into components consistent with a slow or intermediate exchange rate between two water phases. The slower relaxing component is attributed to noninteracting water, and a plot of its intensity against total water content at 280K is shown in Figure 10. The abscissa intercept gives an estimate of ~35% for the saturation water content.

Another NMR method used to distinguish "nonfreezable" or "bound" water is to measure the dependence of the transverse relaxation rate on total water content of the specimen at a temperature below the freezing transition. The relaxation time R_2^{-1} increases with water content until a plateau value is reached (Lynch and Webster, 1979a,b) at a water content that is found to be close to the saturation water content determined by other methods. In Figure 11, such a plot for the wool–water system is shown, from which a value of ~33% for the saturation water content is estimated.

These three independent NMR estimates of the saturation water content of the wool–water system—~35% at 280K, ~35% at 273K, and ~33% at 235K—are to be compared with the value from calorimetry (Haly and Snaith, 1969) of 34% at 273K and a gravimetric sorption value of ~37% at 273K (Lynch and Webster, 1979a). This general agreement suggests a close

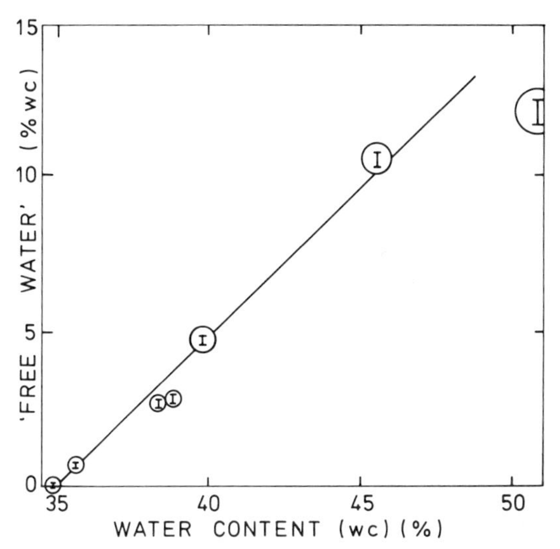

Figure 10. Plot of the relative intensity (in units of percentage water content) of the "noninteracting" water component of the ^{1}H NMR relaxation of wool–water specimens against the total equilibrium water content (% wc). The intercept of the line through the data at 45° to the abcissa indicates a saturation water content of about 35%.

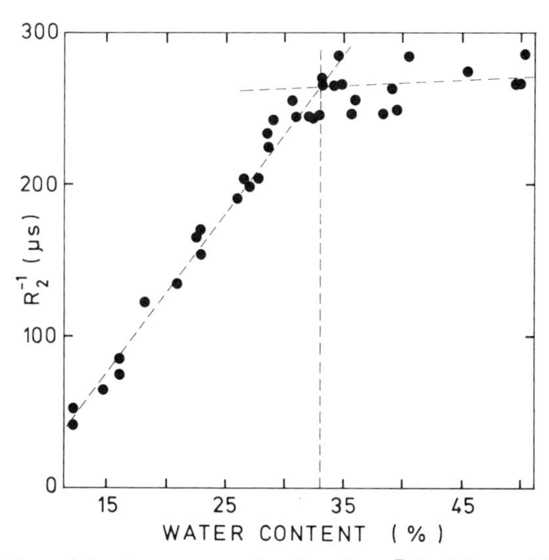

Figure 11. Variation of the transverse relaxation time R_2^{-1} of the mobile-water component versus total water content for wool–water specimens at 235° K. A plateau value is reached near 33% wc.

correlation between nonfreezable water detected by NMR and interacting water in the system.

Leung and Steinberg (1979) found good agreement of NMR nonfreezable-water content of a range of food constituents, with the "bound"-water content determined by a variety of techniques including an NMR assessment by differential radiofrequency (RF) saturation (Shanbhag *et al.*, 1970; Mousseri *et al.*, 1974). In this method the ^1H NMR continuous-wave absorption signal intensity at high RF power is plotted against water content, and the water content at the peak response is taken as the "bound"-water content, since added bulk water greatly enhances the RF saturation because of its slower spin-lattice relaxation.

Dehl (1970) used the low temperature value of the ^2H NMR coupling constant to estimate the quantity of nonfreezable water in collagen. He assumed that the dependence of the coupling constant on water content was the same as at room temperature and his estimates agreed well with those from calorimetry.

The following phenomena are generally observed in NMR temperature scanning of hydrated systems:

1. Sharp phase transitions during cooling for higher-water-content specimens. These occur as step changes in the amount and nature (indicated by relaxation behavior) of the mobile water (e.g., Woodhouse *et al.*, 1975; Lynch and Webster, 1979b; Narebska and Streich, 1980; Bystrov *et al.*, 1973; Fung *et al.*, 1975; Figures 12 and 13).
2. Supercooling of high-water-content specimens indicated by freezing transitions at temperatures below 273K (e.g., Katayama and Fujiwara, 1979; Lynch and Webster, 1979b; Narebska and Streich, 1980; Bystrov *et al.*, 1973; Fung *et al.*, 1975; Figures 12 and 13).
3. A decrease with temperature of the mobile-water content below freezing temperatures (e.g., Sussman and Chin, 1966; Belton *et al.*, 1972; Cyr *et al.*, 1971; Ramirez *et al.*, 1974; Lynch and Webster, 1982; Kvlividze *et al.*, 1978; Deroyane, 1969).
4. Thermal hysteresis in the amount and nature of mobile water (e.g., Haschemeyer *et al.*, 1977; Katayama and Fujiwara, 1980; Lynch and Webster, 1979a,b; Narebska and Streich, 1980; Morariu and Chiricuta, 1976,1977; Chiricuta and Morariu, 1979; Figure 14).

The first two phenomena are consistent with the presence of noninteracting water in the specimens, but the latter two phenomena need consideration in terms of the properties of water in heterogeneous systems and the nature of the NMR measurements.

The following effects have been suggested to explain fully or in part the temperature dependence of the mobile-water content:

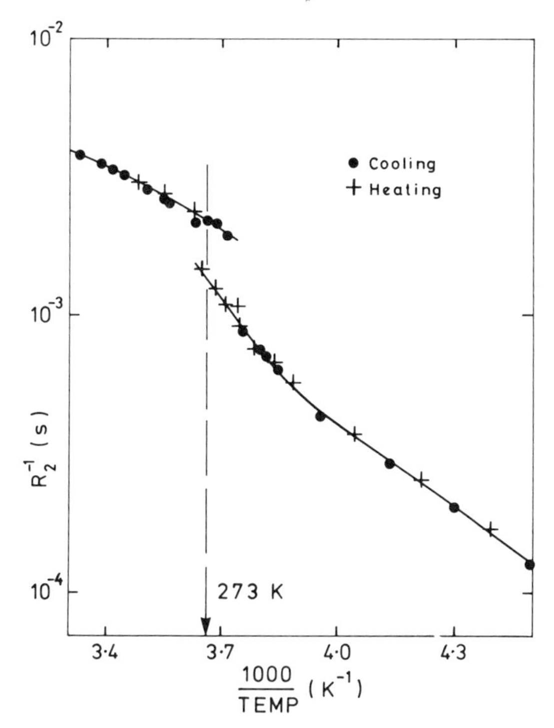

Figure 12. Variation with reciprocal temperature of the ^{1}H NMR transverse relaxation time R_2^{-1} for the mobile-water component of a wool–water specimen of ~ 37% wc during cooling and heating. The extent of supercooling is indicated by the shift in the step transition between heating and cooling.

1. An "apparent phase transition effect" (Resing, 1965), whereby water molecules are progressively transferred from the motional-narrowing to the rigid-lattice state (i.e., where $\tau_c \gtrsim \tau_R \sim 10^{-5}$ s) by thermal deactivation (Woodhouse *et al.*, 1975; Lynch and Webster, 1979a,b; Narebska and Streich, 1980).
2. A capillary effect, whereby water contained in a range of pores of different size progressively freezes according to the Kelvin equation, that is, the depression of the freezing point being inversely proportional to the pore diameter (Oakes, 1976a; Morariu and Chiricuta, 1976,1977; Rennie and Clifford, 1976).
3. A nonspecific effect, whereby the temperature of fusion/freezing of some of the water is lowered depending on the degree of perturbation to the water structure caused by interaction with the substrate materials.

All these explanations involve the concept of a distribution of states for the interacting water. The assumption of a continuous-distribution-of-states model, specified by the molecular correlation time τ_c, is therefore useful for

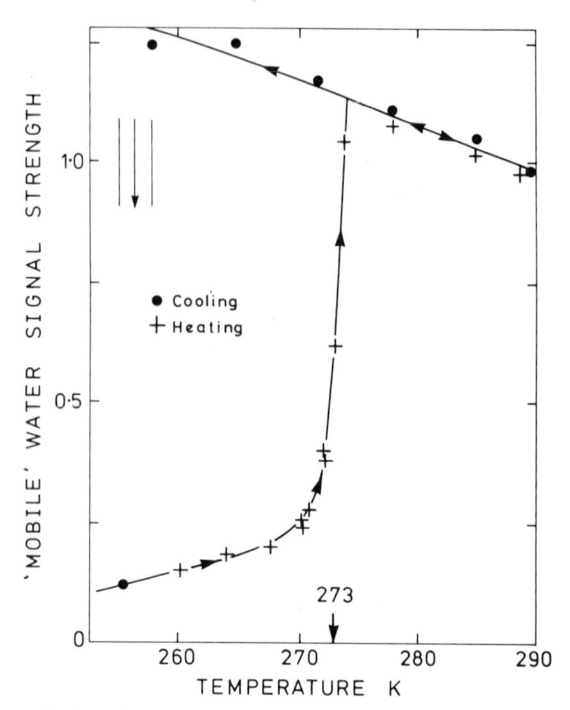

Figure 13. The variation of the mobile-water ^1H NMR signal strength of a brown-coal–water specimen of 141% wc during cooling and heating. The phase transition occurs after supercooling to below 260° K.

further discussion. A schematic representation of such a distribution containing both noninteracting water and a continuous distribution of interacting water states is given in Figure 15. In terms of such a distribution, the first effect involves the least mobile and presumably most strongly perturbed water molecules moving between the mobile and rigid-lattice populations as the temperature is changed, and the latter two effects envisage the least modified water molecules passing between the two populations by a phase transition.

For a continuous distribution of water states, the domain of a state is of molecular dimensions, so the exchange lifetime τ_{ei} effectively equals the correlation time τ_{ci} of the state (Resing, 1965). Thus molecules with τ_c less than about 10^{-5} s satisfy the fast-exchange condition and constitute a time-averaged mobile population. Those with $\tau_c > 10^{-5}$ s contribute to the rigid-lattice population. Because the molecular motions are thermally activated, changes in temperature cause corresponding shifts in the distribution and hence interchange between the mobile and rigid populations. It is therefore possible that this effect could cause an increase in the mobile-water population on heating and cause difficulty in estimating the quantity of water that actually fuses in that temperature range.

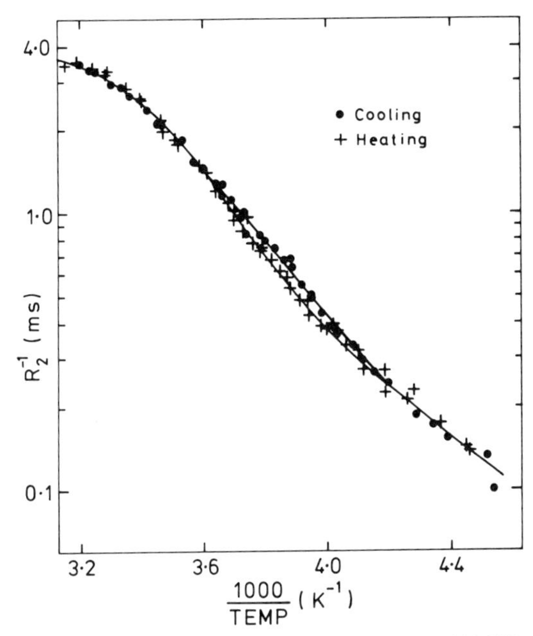

Figure 14. Dependence on the reciprocal temperature of the ¹H NMR transverse relaxation time R_2^{-1} of the water in a keratin–water specimen of near-saturation water content. A hysteresis is apparent in that the relaxation time is greater during cooling than heating over a temperature range below 273° K.

Calorimetric studies have detected diffuse endotherms extending over temperature intervals below 273K for a number of systems containing water at levels near but below the estimated saturation water content (Haly and Snaith, 1969,1971; Dehl, 1970; Mrevlishvili and Privalov, 1969). As the water content of such systems is brought closer to the saturation water content, the endotherms are concentrated closer to 273K. These endotherms indicate the gradual fusion of some of the water that is presumably modified by interaction with the substrate to the extent of lowering the temperature of, but not preventing, the freezing transition. In terms of the continuous-distribution-of-states model for the interacting water, those states containing freezable interacting water would be represented over a range at the short-correlation-time end of the distribution at temperatures above 273K (Figure 15). A correspondence can be envisaged within this range between the molecular mobility and the depression of the fusion transition temperature below 273K. Below the endotherm these states would change to frozen-water states and be represented at the long correlation time, or rigid-lattice end of the distribution (Figures 15*b* and *c*). Thus when such a frozen system was heated, the mobile-water fraction would increase over the range of the endotherm to yield a variable estimate of the nonfreezable water content.

If these two effects occur, the distribution of mobile-water states below

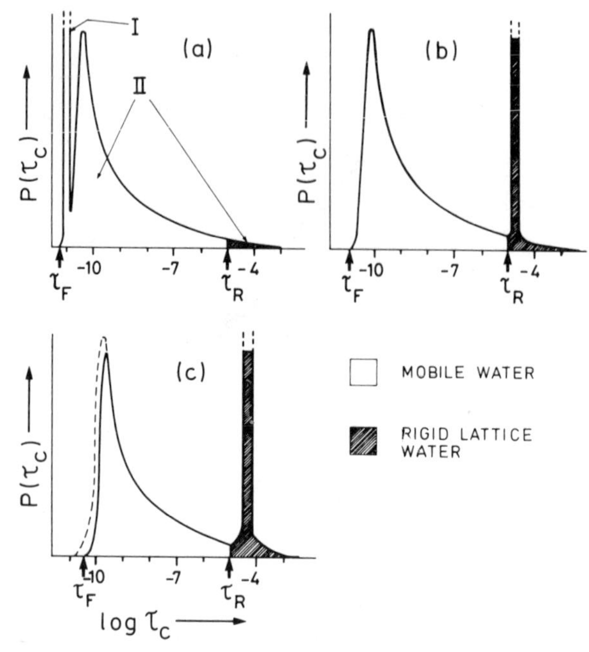

Figure 15. A model represented by the distribution function $P(\tau_c)$ describing the nature of water in a system containing both noninteracting water (*I*) and heterogeneous interacting water (*II*). The distribution of water between the mobile and rigid-lattice ^{1}H NMR populations is shown for (*a*) temperature above 273° K, where only a small fraction of the interacting water with $\tau_c > \tau_R$ is in the rigid-lattice state; (*b*) the condition during heating just before fusion of the noninteracting water at 273° K; here all the noninteracting water (as ice) and a greater fraction of the interacting water than for (*a*) are in the rigid-lattice population; (*c*) temperature below 273° K during heating, where the rigid-lattice population has been further added to both by a shift of the total distribution toward higher values of τ_c because of thermal deactivation and because some of the freezable *interacting* water is in the frozen state.

273K will not correspond with the total interacting-water distribution above 273K in that it will be truncated at both ends.

Nuclear magnetic resonance and calorimetric studies therefore show that there can be two types of freezable water in saturated heterogeneous systems: that which behaves as bulk water and fuses at 273K and that which fuses over a range of temperatures below 273K. This diffuse freezing water falls into the category of interacting water as defined previously. The minimum in the mobile-water signal strength I_m below 273K for the wool–water specimen shown in Figure 9 is consistent with the diffuse fusion of a component of the water. The nonfreezable water in this specimen at 260K is about 31% compared with 35% at 273K. Also shown in Figure 9 is the experimental temperature variation of the ^{1}H NMR signal strength of a constant population of protons. Comparison of the temperature dependence of the measured I_m with this indicates a gradual loss of mobile protons above 273K, probably due to the "apparent phase transition" effect. Thus the estimate of the amount of noninteracting water in the specimen from the NMR mobile-water signal below 273K would be deficient because of both effects.

It is significant to note that whereas calorimetry detects diffuse fusion of water in wool–water specimens with water contents greater than about 23% (Haly and Snaith, 1969), the NMR transverse relaxation at 235K (below the diffuse endotherm) remains sensitive to water content up to about 33% (Lynch and Webster, 1979a; Figure 11), which is in close agreement with the saturation water content estimated by other techniques. These relaxation results strongly suggest that water sorbed up to the saturation water content is interactive with the previously sorbed water and therefore with the substrate, so that at each level there is a unique distribution of water states. They also suggest that whereas beyond 23% water content (determined by calorimetry), some portion of the sorbed water is freezable, it is unlikely that all the water in excess of 23% water content freezes, because the transverse relaxation of the residual mobile water varies with higher total water content.

Ramirez *et al.* (1974) interpreted the gradual disappearance of the mobile water on the cooling of frozen aqueous solutions of amino acids and polypeptides in terms of a complex version of the binary-solution model. For binary solutions, the lowering of the temperature beyond the initial freezing temperature causes ice to be formed, allowing the resultant more concentrated solution to remain liquid until a new equilibrium is reached at a lower temperature, where the cycle is repeated until the eutectic point of the solution is reached, and it freezes completely.

Temperature hysteresis that has been detected in NMR measurements of mobile water has been explained in a number of ways. Morariu and Mills (1972) attributed hysteresis in both R_2 and the amount of mobile water for water sorbed on silica gel to a movement of the phase boundary relating to the gradual transition between frozen and nonfrozen states. The outermost layers of the water are considered to freeze first and the position of the boundary to be further from the substrate surface on the heating than the cooling leg. This explanation is consistent with the mobile-water transverse relaxation rate being faster on heating than cooling, as has been observed for a number of systems (Lynch and Webster, 1979a,b; Narebska and Streich, 1980; Morariu and Chiricuta, 1976). The removal by freezing of molecular states at the short-correlation-time end of the distribution, which are sites least effective in the spin-spin relaxation process, would increase the population-averaged R_2 for the residual mobile water. Hysteresis is therefore accounted for by the incremental supercooling of the diffusely freezing water with respect to its fusion.

Pearson and Derbyshire (1974) also detected temperature-hysteresis effects for the water–silica system, but attributed it to the overlap of supercooling and superheating occurring for water structurally ordered within pores.

A further mechanism to explain hysteresis is that outlined by Barnes (1962). He suggested that the freezing of sorbed water would disturb the sorption equilibrium and cause unfrozen water to move from pores to the surface, where it would freeze because of the lower pressure. On reheating the specimen, fusion of the water would occur at higher temperatures.

Narebska and Streich (1980) recently reported hysteresis in both the transverse relaxation and amount of mobile water for macroporous and nonporous gel-type ion-exchange resins. They explained the hysteresis for the macroporous specimens as being related to the freezing of water in pores, but the much lesser hysteresis of the gel-type resins as being at least partly due to structural changes occurring within the resin polymer. This is supported by the fact that the hysteresis extended to temperatures above 273K for resins with low cross-linking densities. Examination of their published data suggests that the hysteresis for the macroporous specimens is mainly due to supercooling of noninteracting water. The hysteresis in the transverse relaxation of the gel-type resins is similar to that found for wool–water specimens (Lynch and Webster, 1979a) with water contents less than, but close to, the saturation water content. Calorimetry confirms the existence of a diffuse fusion/freezing effect for these wool–water specimens, whereby the hysteresis can be explained by incremental supercooling of the diffusely freezing water (see above).

Katayama and Fujiwara (1979) measured the temperature dependence of both the ^{1}H and ^{17}O NMR of water in polyacrylamide gels and observed all the phenomena listed previously. They interpreted their results in terms of a complex freezing/thawing mechanism involving four equilibrium states containing combinations of bulk water, interacting water, unfrozen water, and ice. The depressed freezing point was explained as the effect of increased pressure of bulk water in micropores.

3.7. ^{1}H NMR Discrimination of Physiological States of Tissues

The development of ^{1}H NMR techniques for body scanning (e.g., Mallard *et al.*, 1980; Smith *et al.*, 1981; Edelstein *et al.*, 1981; Doyle *et al.*, 1981) and for the diagnosis of diseased tissues (e.g., Beall *et al.*, 1980; Hollis *et al.*, 1973; Block and Maxwell, 1974) has renewed interest in the nature of biological water. The scanning techniques are based on images formed by the intensity of the ^{1}H NMR signals either measured directly and therefore proportional to the mobile-proton density (mainly water protons, but also in some instances contributed to by protons of lipids, etc.) or after some relaxation of the signals has occurred. The latter technique yields an image resolved partly by variations in the mobile-proton density and partly by variations in the intrinsic magnetic resonance relaxation properties. These properties in turn are determined by the molecular and proton spin interactions involving the tissue water. The different imaging techniques are variously found to perform better for different tissue types. The methods of tissue diagnosis are based on differences observed in the ^{1}H NMR relaxation behavior of diseased and normal tissue. There has been much speculation on how this relaxation behavior is related to the physiological state of the tissue and its water content, and the factors that contribute to the changes.

There are many data in the literature that demonstrate the empirical value of relaxation-time measurements for the diagnosis of diseased tissue (e.g., Damadian, 1971; Hazelwood *et al.*, 1972; Beall *et al.*, 1980; Kiricuta and Simplaceanu, 1975; Peemoeller *et al.*, 1979). Both spin-lattice (R_1) and spin-spin (R_2) relaxation rates are found useful, although most studies have concentrated on the R_1 relaxation. The rates R_1 and R_2 are invariably found to be less for diseased tissues and more often than not there is also an increase in the total water content. This change in water content complicates the understanding of the reasons for the change in the relaxation behavior. A definite implication of the reduction in R_2 and the fact that $R_2 < R_1$ is that the average molecular mobility of the diseased-tissue water is greater. Considering the limited success of extensive NMR studies of the nature of water in heterogeneous systems, the possibility of finding the precise physical reasons for these differences in the ^{1}H NMR relaxation behaviors of diseased and normal tissues is not great. Discussions in the literature are greatly influenced by concepts of bound and free water (e.g., Block and Maxwell, 1974; Peemoeller *et al.*, 1979; Zipp *et al.*, 1976). In some instances efforts have been made to independently assess the "bound" fraction by equating it with the "nonfreezable" water (Section 3.6) by freezing experiments (Rustgi *et al.*, 1978; Morariu and Chiricuta, 1976; Chiricuta and Morariu, 1979; Fung, 1974). Fung found that the fraction of "nonfreezable" water in rat skeletal muscle with a growing tumor was less than in normal tissue and concluded that the reduced R_1 value of the tumor muscle was due to the smaller amount of water in the hydration layer, but that the R_1 relaxation rates of the hydration water were the same for both tissues. Peemoeller *et al.* (1979) arrived at similar conclusions in their study of a range of normal and diseased mice tissues.

Morariu and Chiricuta (1976) found a significant increase in the absolute nonfreezing-water content of a range of diseased lyphilized muscle tissues compared with the normal tissue and attributed this to greater porosity of the diseased tissue. From R_2 measurements of the "nonfreezable" water of these same tissues they (Chiricuta and Morariu, 1979) concluded that there is no difference in the diseased and normal tissues at the "level of the macromolecular–water interface."

Attempts have also been made to estimate the effects of tissue denaturation on the water content and ^{1}H NMR relaxation behavior. Heat-induced conformational changes in tissues have been found to alter R_1 without altering the total water content (Hazelwood *et al.*, 1969; Neville *et al.*, 1974). The same result was obtained for cells denatured with spermine (Beall *et al.*,

1976). Bracacescu *et al.* (1978) found they could increase or decrease the R_1 value for collagen by adjusting levels of γ radiation. They related this to ordering of the structure by low doses and disruption of structure by high doses. Sharimanov *et al.* (1980) experimentally demonstrated transitions in R_1 and R_2 values of collagen associated with thermal transitions in the calorimetrically determined "bound"-water contents.

Many investigators (Kiricuta and Simplaceanu, 1975; Inch *et al.*, 1974; Hollis *et al.*, 1973; Block and Maxwell, 1974), noting the correspondence between a decrease in R_1 and increase in water content, have contended that the values of R_1 and R_2 are simply determined by water content. Hazelwood *et al.* (1972), on the other hand, interpreted the decreased R_1 in malignant tissues as representing a change in the water–macromolecular interaction of the cell. This interpretation is supported by their findings that R_1 increases with the age of rat gastrocnemius muscle, and this increase accompanies a decrease in the water content and an increase in the molecular complexity (Hazelwood *et al.*, 1971); it is also supported by later work in which they demonstrated a relationship between R_1, water content, and the cyclic pattern of cell growth (Beall *et al.*, 1976). They have pointed out that whereas there appears to be a simple relationship between R_1 and water content within narrow physiological limits, the relationship breaks down with gross morphological change.

It is clear from the above that there is no consensus as to the physiological basis of changes in the ^{1}H NMR relaxation of tissue except that there is considerable evidence for reduced water binding by diseased tissues. Further resolution of this question is complicated by the fact that the ^{1}H NMR signals of whole tissues are not from particular cell types and also are contributed to by connective tissues and intracellular water. The measured ^{1}H NMR relaxation rates of tissues are the average from a heterogeneous distribution of water relaxation states. Changes in the tissue structure such as alterations in the macromolecular conformations, ionic concentrations, or water:solid ratio are likely to modify the water–macromolecular interactions, thereby affecting the water mobility and possibly the cross-relaxation effects between the water and macromolecular proton populations (Section 3.4; Fung, 1977) and thus the overall relaxation behavior.

In order to focus on the basic question, it is convenient here to ignore substances and factors extraneous to the cellular water.

Most researchers (e.g., Kiricuta, 1978; Fung, 1974; Peemoeller *et al.*, 1979; Block and Maxwell, 1974) have modeled the cellular water as consisting of a fraction b of "bound" or "hydration" water of relaxation rates R_{ib} undergoing rapid exchange (Section 3.3) with the "unbound" or free cell water that has relaxation rates R_{if}. The basic assumption is that the measured relaxation rates R_i represent the average behavior of all the water. Equation (15) can be used to represent this model in the form

$$R_i = (1 - b)(R_{if} - R_{ib}) + R_{ib} \tag{48}$$

The properties and hence R_{if} of the free or noninteracting water are by definition unaffected by changes in the cell. The variables of the model are therefore the fraction b and the properties of the "bound" or interacting water (i.e., R_{ib}). In absolute terms this means that either a differential change in the quantities of the two water types or a change in the nature of the interacting water affects the values of R_i. The NMR-detected difference ΔR_i in the cellular water can be represented as

$$\Delta R_i = (R_{i1} - R_{i2}) = \Delta b(R_{ib_1} - R_{if}) + b_2 \Delta R_{ib} \tag{49}$$

where $\Delta b = (b_1 - b_2)$ and $\Delta R_{ib} = (R_{ib_1} - R_{ib_2})$ and the indices 1 and 2 refer to the two physiological states of the cell. Indications are that the free-water relaxation rates R_{if} (both R_{1f} and R_{2f}) are an order of magnitude less than the R_{ib}, so this equation can be reduced to

$$\Delta R_i \approx \Delta b R_{ib_1} + b_2 \Delta R_{ib} \tag{50}$$

Biological macromolecules contain between 30 and 60% of their dry weight of nonfreezable water (Kuntz and Kauzmann, 1974; Golton *et al.*, 1981). Spin–lattice relaxation rates of water at this level of hydration are of the order 10 s^{-1} (Lynch and Marsden, 1969; Rustgi *et al.*, 1978; Fung and McGaughy, 1974; Fung *et al.*, 1975). The fraction of nonfreezable water in tissue has been estimated at about 0.1 (Peemoeller *et al.*, 1979; Morariu *et al.*, 1978; Belton *et al.*, 1972; Fung and McGaughy, 1974; Fung, 1977). Thus to account for a change in R_1 of 0.1 s^{-1}, either $\Delta b \sim 0.01$ or $\Delta R_{ib} \approx 1$ s^{-1}. These shifts represent, in percentage terms, a change of 10% in the fraction, or a 10% change in the relaxation rate of the "bound" water.

Estimates of Δb and ΔR_{ib} can be obtained from studies of "nonfreezable" water. The R_1 rates of "nonfreezable" water are found, in agreement with those of sorbed water, to have maxima below 273K—at between 230 and 260K (Fung *et al.*, 1975; Packer and Selwood, 1978; Lynch and Marsden, 1969; Rustgi *et al.*, 1978). However, it is possible, because of different temperature dependences of R_{1b_1} and R_{1b_2}, that ΔR_{1b} determined from the "nonfreezable" water at low temperature is different from that determined at room temperature.

It is of interest to consider the physical nature of changes that could occur in the cellular water. The fact that both interacting and noninteracting water are contained within the cellular structure suggests that the total water content is regulated by factors other than direct molecular interactions or the intrinsic water-binding capacity of the macromolecules. Presumably the water is contained by the membranes or by other nonmolecular structural features of the cell. The saturation water content or maximum water-binding capacity of biological macromolecules has been variously estimated at from 30 to 70% of the dry weight (Kuntz and Kauzmann, 1974; Golton *et al.*, 1981). This quantity depends on the macromolecule and its conformation. The tendency for macromolecules to interact with water and perturb bulk-water structure in their vicinity is determined not only by the nature of the

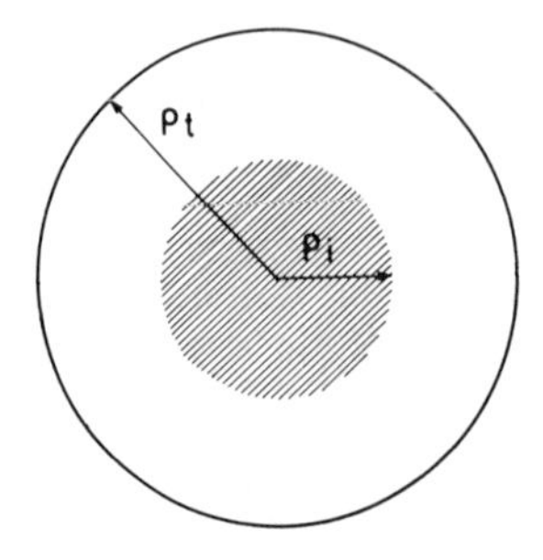

Figure 16. Simple spherical representation of tissue water. The volume of water interacting with the tissue macromolecules is represented in a volume of radius ρ_i, and the free water in the remainder of the total volume of radius ρ_t.

surface but also by the water accessibility, which can be affected by the ability of the macromolecule to change its conformation. For example, cross-linkages can restrict swelling of a structure resulting from water uptake so that there is a smaller volume of interacting water than if the structure is freer in its tendency to swell.

A simple spherical representation delineating the volume of water interacting with the macromolecules and the free-water volume is useful to describe changes in tissue water (Figure 16). Thus a change in the amount of interacting water caused by a change in the quantity of macromolecular substance or of its volume capacity to interact with water can be indicated by a change in ρ_i, and a change in the total water content by a change in ρ_t. The "bound" fraction b is therefore $(\rho_i/\rho_t)^3$. The interacting fraction has some distribution of microdynamic states $P(\tau_c)$ (see Section 3.6 and Figure 15) that determines its ^{1}H NMR relaxation (R_{ib}). A change in ρ_i other than by a simple increase in the quantity of macromolecular substance would likely alter $P(\tau_c)$. Such a change and any change in the ratio ρ_i/ρ_t would result in a change in the average NMR properties of the tissue water.

4. CONCLUDING REMARKS

Although NMR is clearly a powerful tool for the study of heterogeneous water systems, it has not lived up to the expectations that have encouraged its use by many researchers. The reason for this is the complexity of surface water. In its turn, NMR has contributed much to the present awareness of the extent of the dynamic and structural complexity of this water. Despite this, NMR studies not only have continued, but have become more widespread because of the versatility of the technique. Different and more-extensive measurements have been introduced progressively, and more-detailed modeling and refined methods for analyses have been developed.

Thermodynamic, structural, and dynamic aspects are distinguished as necessary to describe the physical nature of hydration water (Berendsen, 1975). However, most NMR observables are directly dependent on the lifetimes of nuclear spin states and are not direct measurements of thermodynamic or molecular properties. Rather, NMR measurements enable a limited description of the system in terms of its molecular dynamics. Some thermodynamic transitions can be detected, but structural information can only be extracted from NMR data by a modeling analysis that is critically dependent on the molecular dynamics. Limitation in the dynamic information obtainable from NMR data results both from resolution and bandwidth restrictions and from the difficulty of fully specifying the factors that affect the nuclear spin system. The theory used for relating NMR observations to the properties of these complex systems is tenuous because of the above and also because of the *ad hoc* physical modeling of the molecular system that practical analysis requires. Concepts of distinct phase states of water and their distribution within the system are ultimately based on operational definitions that only relate to idealized systems and the NMR measurement method.

The general approach in these studies has been to collect a set of data in the hope that some conclusions will result that classify the system by its structural and dynamic properties. There is thus a tendency to describe the water in terms of extractable NMR parameters that cannot confidently be related to basic physical properties of the water. Less often are specific questions posed or particular hypotheses tested. Even with the latter approach, the NMR answers to such apparently simple questions as "How much 'bound' water is in a system?" are not straightforward. In this regard NMR has been rather successful in qualitative and comparative assessments but of limited use in providing quantitative answers. Thus, for example, the occurrence of molecular anisotropy, but not its degree, can be readily assessed.

The most clear-cut conclusions from NMR studies of hydrated systems result from temperature scanning. This is because phenomena related to thermodynamic states and transformations are observed from which deductions can be made independent of a great mass of theory and the assumption of models (see Section 3.6). Temperature scanning has been useful for the assessment of nonfreezable water and by strong implication, the distinction of interacting from noninteracting water components.

Biological tissues containing large amounts of water, and in particular muscle tissues, have been the subjects of intensive NMR studies. The full range of NMR measurement techniques has been used, but still there is no general agreement on the nature of such biological water. A factor contributing to this is the large proportion of the water that is either noninteracting or at best only slightly modified with respect to bulk water. The NMR relaxation processes are insensitive to the microdynamics of such water, and its presence can only complicate NMR studies of that component of the water which is directly a part of the structure and function of the tissue rather than a reservoir or buffer for it. In the author's opinion it has been correctly asserted (Resing *et al.*, 1977) that in such systems NMR reveals very little about the microdynamic nature of the bulk of the water but detects average effects that emanate from the small proportion of interacting water. To the extent that interacting water is of prime interest, methods of isolating it from the noninteracting or bulk water for NMR study are desirable. Zipp *et al.* (1976) suggested studies of frozen systems for this purpose. It has been claimed that models pertaining to unfrozen tissue do not relate to the frozen tissue (Resing *et al.*, 1977), a claim that implies the occurrence of unpredictable transitions or transformations in the process of deep-freezing. However, there is evidence from relevant studies that indicates a close relationship between "nonfreezable" and "bound" water (Section 3.6). Thus to elucidate the nature of vicinal water, efforts should be made to work on systems in which the interacting water component is isolated. One approach to this is the study of water sorbed to biomolecular substances at levels below the saturation water content.

Progress in understanding the complexity of surface water continues as factors that influence the spin and molecular systems are better defined. For the spin system, the effects of spin interaction and transfer of magnetization across the water–substrate interface have only recently been fully appreciated and attempts made to account for them in relaxation analysis (Section 3.4). Further advances can be anticipated in this area. Efforts to interpret NMR data have led to a recognition that the origin of anisotropy in surface water can be intrinsic to the existence of water surfaces as distinct from the influence of specific water–substrate interactions (Walmsley and Shporer, 1978; Woessner and Snowden, 1973; Glasel and Lee, 1974). This realization has introduced new dimensions both to the understanding of vicinal water and to NMR relaxation analysis (Halle *et al.*, 1981).

LIST OF SYMBOLS

A	Attenuation of spin-echo amplitude in presence of an applied field gradient $G(x)$
a	Order of reaction of water for proton exchange between water and labile surface groups
B	Doublet splitting constant
b'	Order of reaction of surface groups for proton exchange between water and labile surface groups
b	Fraction of "bound" cellular water
C	Static nuclear-spin-interaction strength factor
c_a	Proton exchange rate $(= \tau_a^{-1})$
C_i^+, C_i^-	Preexponential parameters for coupled magnetization decay functions resulting from cross-relaxation between two proton populations
D	Self-diffusion coefficient

$(dI/dt)_B$, $(ds/dt)_B$	Differential relaxation functions for bulk zones of two interfacial proton populations affected by cross-relaxation
$(dI/dt)_x$, $(ds/dt)_x$	Differential relaxation functions at the interface zone of two interfacial proton populations affected by cross-relaxation
$\mathbf{F}$	Population ratio of two coupled proton populations $(= n_S/n_I)$
F	Anisotropy scaling factor
$F_a(t)$	$a = 0, 1, 2$. Components of the fluctuating interaction function
f_I, f_S	Fractions of spin populations I and S in the interface zone of two proton populations
$g(H_l')$	Static local-field distribution function
$G(x)$	Linear magnetic-field gradient
$\hbar$	Planck's constant divided by 2π
$H_{\text{eff}}(X)$	Effective magnetic field at position X
H_l'	Proton local magnetic field
$H_l(X)$	Local intrinsic magnetic field of spin system at position X
H_0	Applied steady magnetic field
H_1	Magnetic field strength of spin-locking pulse
$H_{1\rho}$	Resonant spin-locking magnetic field or rotating frame-applied field
$[H_2O]$	Chemical activity of sorbed-water phase
I	Nuclear spin quantum number
$\imath$	Reduced magnetization of spin population I at the interface with spin population S
$\mathbf{I}$	Intensity of ^{1}H NMR signal of mobile water
I_z	The z component of the macroscopic magnetic moment of spin population I
I_0	Equilibrium z component of the magnetic moment of spin population I
$J_a(\omega)$	$a = 0, 1, 2$. Spectral-density or Fourier-intensity functions of the fluctuating spin-interaction components
$k_a(\tau)$	Correlation function of random reorientation process
k_e	Effective rate of magnetization transfer across interface between spin populations
k_f	Chemical rate constant for proton exchange
K_i	Weighting constant for the relaxation component of the ith mode of anisotropic reorientation
k_I, k_S	Magnetization transfer rates for cross-relaxing populations I and S

k'_I, k'_S	Cross-relaxation rates for coupled spin populations I and S, according to Andree's three-zone model
L, M, N	Direction cosines of spin interaction vector
M_I, M_S	Reduced magnetizations of spin systems I and S, respectively
M_{sub}	Reduced magnetization of substrate protons
M_x	Transverse magnetization
M_w	Reduced magnetization of water protons
N	Constant relating the cross-relaxation rate with the population and magnetization transfer rates of cross-relaxing populations
n	Number of substrate protons with which each water proton interacts
N_{H_2O}	Number density of water molecules
n_I, n_S	Number of spins in populations I and S, respectively
p_a, p_b	Fractions of water in the free and bound states, respectively
p_s	Fraction of water in anisotropic surface phase
$P(\tau_c)$	Continuous distribution function of water states
r	Internuclear distance
R_i	$i = 1, 2$. Spin-lattice and spin-spin relaxation rates or reciprocal relaxation times (T_1^{-1}, T_2^{-1}), respectively
R'_i	$i = 1, 2$. Exchange-averaged spin-lattice and spin-spin relaxation rates, respectively
R_{ia}, R_{ib}	$i = 1, 2$. Relaxation rates of free- and bound-water protons, respectively
R_{i1}, R_{i2}	$i = 1, 2$. Relaxation rates of water in biological tissue of physiological states designated 1 and 2 respectively
R_{Ix}, R_{Sx}	Direct spin-lattice relaxation rates for spin populations I and S of Andree's three-zone cross-relaxation model
r_{ij}	Distance between spins i and j
R_S	Spin diffusion rate in solid
R_T	Cross-relaxation rate
R_{1I}, R_{1S}	Direct spin-lattice relaxation coefficients for spins I and S respectively in the interface zone of three-zone model
R_{1s}	Spin-lattice relaxation rate of protons in anisotropic surface water
R^B_{1I}, R^B_{1S}	Intrinsic bulk spin-lattice relaxation rates of populations I and S respectively
$R_{1\rho}$	Rotating-frame spin-lattice relaxation rate or reciprocal of rotating-frame spin-lattice relaxation time ($T_{1\rho}^{-1}$)

R_{2i}	Averaged intermolecular contribution to spin-spin relaxation rate of irrotationally bound water protons affected by cross-relaxation
R'_{2a}, R'_{2b}	Observed relaxation rates of components of spin-spin relaxation when intermediate-to-fast exchange conditions hold between phases a and b
R^{u}_{2b}	Contribution to spin-spin relaxation of dipolar interaction between water and substrate protons
r_{20}, r_{22}	Second-order spherical expansion or anisotropy coefficients
R_3	Rate of transverse relaxation after excitation by $\pi/2 - \pi/2$ radiofrequency pulse sequence
s	Reduced magnetization of spin population S at the interface with spin population I
S_z	The z component of the macroscopic magnetic moment of spin population S
[SuH*]	Chemical activity of labile proton surface groups
S_0	Equilibrium z component of the magnetic moment of spin population S
t_d	Defined time interval for self-diffusion measurement
t_{180}	Duration of 180° RF preparation pulse
w_i, w'_1	$i = 0, 1, 2$. Transition probabilities for spin exchange between dipolar-perturbed eigenstates of a spin pair
w_{1e}, w_{2e}	Transition probabilities for spin exchange between eigenstates of a spin pair due to external effects
X	Location of spin in space for two-dimension model of self-diffusion
x	Variable parameter for isotopic ratio $^1H:^2H$ or enhancement factor with respect to ^1H relaxation for ^2H relaxation
$\gamma_{^1H}$, $\gamma_{^2H}$, $\gamma_{^{17}O}$	Magnetogyric ratios of the proton (^1H), the deuterium nucleus (^2H), and the ^{17}O nucleus, respectively
κ	Cross-relaxation coefficient for a two-spin system
λ^+, λ^-	Exponential parameters for coupled magnetization decay functions resulting from cross-relaxation between two proton populations
ρ_i, ρ_t	Radii of spheres representing volume of interacting water and total water in a tissue
σ_0^2	Rigid-lattice second moment for dipolar interaction within a population of protons
τ	Elapsed-time parameter of correlation function $k_a(\tau)$
τ_a	Lifetime of proton in the free state or reciprocal probability of exchange between a free and bound state

τ_b	Lifetime of water molecule in bound state
τ_c	Nuclear or molecular correlation time
τ_{ca}, τ_{cb}	Molecular correlation times in the free and bound states, respectively
τ_{ci}	Correlation time of *i*th mode of reorientation in anisotropic water phase
τ_{cs}	Correlation time of slow mode of anisotropic reorientation
τ_{ei}	Lifetime of molecule in phase state *i*
τ_F	Correlation time of most mobile water state of a continuous-distribution-of-states model $P(\tau_c)$
τ_R	Correlation time separating rigid-lattice and motionally averaged water states of a continuous-distribution-of-states model $P(\tau_c)$
ω	Frequency of molecular fluctuations
ω_0, ω_1	Resonance frequency in static (H_0) and rotating (H_1) fields respectively

REFERENCES

Abetsedarskaya, L. A., F. G. Miftakhutdinova, and V. D. Fedotov (1968), *Biophysics* **13,** 750.

Ananyan, A. A. (1979), *Colloid J. USSR* **40,** 980.

Andree, P. J. (1978), *J. Magn. Reson.* **29,** 419.

Andrewartha, K. A., R. T. C. Brownlee, and D. R. Phillips (1978), *Arch. Biochem. Biophys.* **185,** 423.

Ashpole, D. K. (1952), *Proc. Royal Soc. London* **A212,** 112.

Avogadro, A., and M. Villa (1977), *J. Chem. Phys.* **66,** 2359.

Barnes, G. T. (1962), *Z. Angew Math. Phys.* **13,** 533.

Basler, W. D., and W. Maiwald (1979), *J. Phys. Chem.* **83,** 2148.

Beall, P. T., C. F. Hazelwood, and P. N. Rao (1976), *Science* **192,** 904.

Beall, P. T., B. B. Aach, D. C. Chang, D. Medina, and C. F. Hazelwood (1980), *J. Natl. Canc. Inst.* **64,** 335.

Belton, P. S., R. R. Jackson, and K. J. Packer (1972), *Biochim. Biophys. Acta* **286,** 16.

Berendsen, H. J. C. (1975), in *Water—A Comprehensive Treatise,* Vol. 5, F. Franks, Ed., Plenum Press, New York, pp. 293–357.

Block, R. E., and G. P. Maxwell (1974), *J. Magn. Reson.* **14,** 329.

Bloembergen, N., E. M. Purcell, and R. N. Pound (1948), *Phys. Rev.* **73,** 679.

Bracacescu, A., D. Bracacescu, D. Demco, V. Simplaceanu, and N. Vilcu (1978), *Trav. Mus. Hist. Natl. 'Grigore Antipa',* **19,** 59.

Brunauer, S., P. H. Emmett, and E. Teller (1938), *J. Am. Chem. Soc.* **60,** 309.

Bryant, R. G., and W. M. Shirley (1980), *Biophys. J.* **32,** 3.

Brynjas-Kraljevic, J., and S. Maricic (1978), *Biochem. Biophys. Res. Commun.* **83,** 1048.

Bystrov, G. S., M. I. Nikolayev, and G. I. Romanenko (1973), *Biophysics* **18,** 508.

Carles, J. E., and A. M. Scallan (1973), *J. Appl. Polym. Sci.* **17,** 1855.

Chang, D. C., and D. E. Woessner (1977), *Science* **198,** 1180.

Chiarotti, G., G. Christiani, and L. Guilotto (1955), *Nuovo Cimento* **1,** 863.

Chiricuta, I. C., and V. V. Morariu (1979), *Oncologia* **18,** 29.

Civan, M. M., and M. Shporer (1975), *Biophys. J.* **15,** 299.

Civan, M. M., A. M. Achlama, and M. Shporer (1978), *Biophys. J.* **21,** 127.

Clifford, J., and B. Sheard (1966), *Biopolymers* **4,** 1057.

Cyr, T. J., W. Derbyshire, J. L. Parsons, J. M. Blanshard, and R. A. Lowrie (1971), *Trans. Faraday Soc.* **67,** 1887.

Damadian, R. (1971), *Science* **171,** 1151.

Dehl, R. E. (1970), *Science* **170,** 738.

Derbyshire, W., and I. D. Duff (1974), *Discuss. Faraday Soc.* **57,** 243.

Derbyshire, W., and J. L. Parsons (1972), *J. Magn. Reson.* **6,** 344.

Deroyane, E. G. (1969), *Bull. Soc. Chim. Belges* **78,** 111.

Doyle, F. H., J. M. Pennock, J. S. Orr, J. C. Gore, G. M. Bydder, R. E. Steiner, I. R. Young, H. Clow, D. R. Bailes, M. Burl, D. J. Gilderdale, and P. E. Walters (1981), *Lancet* **2**(8237), 53.

Edelstein, W. A., J. M.S. Hutchison, F. W. Smith, J. Mallard, G. Johnson, and T. W. Redpath (1981), *Br. J. Radiol.* **54,** 149.

Edzes, H. T., and E. T. Samulski (1977), *Nature* **265,** 521.

Edzes, H. T., and E. T. Samulski (1978), *J. Magn. Reson.* **31,** 207,

Eisenstadt, M., and M. E. Fabry (1978), *J. Magn. Reson.* **29,** 591.

Fennena, O. (1976), in *Food Proteins,* J. R. Whitaker and S. R. Tannerbaum, Eds., Avi Publishing, Westport, CT, pp. 50–90.

Foster, K. R., H. A. Resing, and A. N. Garroway (1976), *Science* **194,** 324.

Freude, D., U. Lohse, D. Michel, H. Pfeifer, and H. J. Zahr (1978), *Z. Phys. Chem.* (Leipzig) **259,** 225.

Fung, B. M. (1974), *Biochim. Biophys. Acta* **362,** 209.

Fung, B. M. (1977), *Biophysical J.* **18,** 235.

Fung, B. M., and T. W. McGaughy (1974), *Biochim. Biophys. Acta* **343,** 663.

Fung, B. M., and T. W. McGaughy (1980), *J. Magn. Reson.* **39,** 413.

Fung, B. M., D. L. Durham, and D. A. Wassil (1975), *Biochim. Biophys. Acta* **399,** 191.

Glasel, J. A., and K. H. Lee (1974), *J. Am. Chem. Soc.* **96,** 970.

Golton, I. C., B. J. Gellatly, and J. L. Finney (1981), *Studia Biophysica* **84,** 5.

Halle, B., T. Anderson, S. Forsen, and B. Lindman (1981), *J. Am. Chem. Soc.* **103,** 500.

Hallenga, K., and S. H. Koenig (1976), *Biochemistry* **15,** 4255.

Haly, A. R., and J. W. Snaith (1969), *Biopolymers* **7,** 459.

Haly, A. R., and J. W. Snaith (1971), *Biopolymers* **10,** 1681.

Haschemeyer, A. E. V., W. Guschlbauer, and A. L. De Vries (1977), *Nature* **269,** 87.

Hazelwood, C. F., B. L. Nichols, and N. R. Chamberlain (1969), *Nature* **222,** 747.

Hazelwood, C. F., B. L. Nichols, D. C. Chang, and B. Brown (1971), *Johns Hopkins Med. J.* **128,** 117.

Hazelwood, C. F., D. C. Chang, D. Medina, D. Clevland, and G. Nichols (1972), *Proc. Natl. Acad. Sci. USA* **69,** 1478.

Hilton, B. D., E. Hsi, and R. G. Bryant (1977), *J. Am. Chem. Soc.* **99,** 8483.

Hollis, D. P., J. S. Economou, L. C. Parks, J. C. Eggleston, L. A. Saryan, and J. L. Czeisler (1973), *Cancer Res.* **33,** 2156.

Hsi, E., and R. G. Bryant (1975), *J. Am. Chem. Soc.* **97,** 3220.

Hsi, E., R. Hossfeld, and R. G. Bryant (1977), *J. Coll. Interface Sci.* **62,** 389.

Hsi, E., G. J. Vogt, and R. G. Bryant (1979), *J. Coll. Interface Sci.* **70,** 338.

Inch, W. R., J. A. McCredie, R. R. Knispel, R. T. Thompson, and M. M. Pintar (1974), *J. Natl. Canc. Inst.* **52,** 353.

Jones, G. P. (1966), *Phys. Rev.* **148,** 332.

Kalk, A., and H. J. C. Berendsen (1976), *J. Magn. Reson.* **24,** 343.

Katayama, S., and S. Fujiwara (1979), *J. Am. Chem. Soc.* **101,** 4485.

Katayama, S., and S. Fujiwara (1980), *J. Phys. Chem.* **84,** 2320.

Kimmich, R. (1980), *Bull. Magn. Reson.* **1,** 195.

Kimmich, R., and Noack, F. (1971), *Ber. Bunsenges. Phys. Chem.* **75,** 269.

Kiricuta, I. C. (1978), *Indian J. Cancer* **15,** 19.

Kiricuta, I. C., and V. Simplaceanu (1975), *Cancer Res.* **35,** 1164.

Koenig, S. H., and W. E. Schillinger, (1969), *J. Biophys. Chem.* **244,** 3283.

Koenig, S. H., K. Hallenga, and M. Shporer (1975), *Proc. Natl. Acad. Sci. USA* **72,** 2667.

Koenig, S. H., R. G. Bryant, K. Hallenga, and G. S. Jacob (1978), *Biochemistry* **17,** 4348.

Kruger, G. J., and G. A. Helcke (1967), in *Magnetic Resonance and Relaxation,* Ed., R. Blinc, North Holland, Amsterdam, pp. 1136–1143.

Krynicki, K. (1966), *Physica* **32,** 167.

Kubo, R., and K. Tomita (1954), *J. Phys. Soc. Japan* **9,** 888.

Kuntz, I. D. (1971), *J. Am. Chem. Soc.* **93,** 514; **93,** 516.

Kuntz, I. D., and T. S. Brassfield (1971), *Arch. Biochem. Biophys.* **142,** 660.

Kuntz, I. D., and W. Kauzmann (1974), in *Advances in Protein Chemistry,* Vol. 28, C. B. Anfinsen, J. T. Edsall, and F. M. Richards, Eds., Academic Press, London and New York, pp. 239–345.

Kuntz, I. D., T. S. Brassfield, G. D. Law, and G. V. Purcell (1969), *Science* **163,** 1329.

Kuprianoff, J. (1958), in *Fundamental Aspects of the Dehydration of Foodstuffs,* Aberdeen, 25–27 March 1958, Society of Chemical Industry, London, pp. 14–23.

Kvlividze, V. I., V. F. Kiselev, A. V. Krasnushkin, A. B. Kurzaev, and L. A. Ushakova (1978), in *3rd International Conference on Permafrost,* NRCC, Ottawa, pp. 113–118.

Leung, H. K., and M. P. Steinberg (1979), *J. Food Sci.* **44,** 1212.

Lindstrom, T. R., and S. H. Koenig (1974), *J. Magn. Reson.* **15,** 344.

Luz, Z., and S. Meiboom (1963), *J. Chem. Phys.* **40,** 2686.

Lynch, L. J. (1981), *Studia Biophysica* **84,** 75.

Lynch, L. J., and K. H. Marsden (1969), *J. Chem. Phys.* **51,** 5681.

Lynch, L. J., and K. H. Marsden (1973), *J. Coll. Interface Sci.* **42,** 209.

Lynch, L. J., and D. S. Webster (1979a), *J. Coll. Interface Sci.* **69,** 238.

Lynch, L. J., and D. S. Webster (1979b), *Fuel* **58,** 429.

Lynch, L. J., and D. S. Webster (1980), *J. Magn. Reson.* **40,** 259.

Lynch, L. J., and D. S. Webster (1982), *Fuel* **61,** 271.

Lynch, L. J., K. H. Marsden, and E. P. George (1969), *J. Chem. Phys.* **51,** 5673.

Mallard, J., J. M.S. Hutchison, W. A. Edelstein, C. R. Ling, M. A. Foster, and G. Johnson (1980), *Phil. Trans. R. Soc. London* **B289,** 519.

Miyake, A. (1957), *J. Chem. Phys.* **27,** 1425.

Morariu, V. V., and I. C. Chiricuta (1976), *Oncologia* **15,** 207.

Morariu, V. V., and I. C. Chiricuta (1977), *Oncologia* **16,** 291.

Morariu, V. V., and R. Mills (1972), *J. Coll. Interface Sci.* **39,** 406.

Morariu, V. V., I. C. Kiricuta, and C. F. Hazelwood (1978), *Physiol. Chem. Phys.* **10,** 517.

Mousseri, J., M. P. Steinberg, A. I. Nelson, and L. S. Wei (1974), *J. Food Sci.* **39,** 114.

Mrevlishvili, G. M., and P. L. Privalov (1969), in *Water in Biological Systems,* L. P. Kayushkin, Ed., Plenum, New York, pp. 63–66.

Mrevlishvili, G. M., and Y. G. Sharimanov (1978), *Biophysics* **23,** 242.

Mrevlishvili, G. M., and Y. G. Sharimanov (1979), *Biophysics* **23,** 733.

Murday, J. S., R. L. Patterson, H. A. Resing, J. K. Thompson, and N. H. Turner (1975), *J. Phys. Chem.* **79,** 2674.

Narebska, A., and W. Streich (1980), *Coll. Polym. Sci.* **258,** 379.

Neville, M. C., C. A. Paterson, J. L. Rae, and D. E. Woessner (1974), *Science* **184,** 1072.

Oakes, J., (1976a), *J. Chem. Soc. Farad. Soc. Trans. I* **72,** 216.

Oakes, J., (1976b), *J. Chem. Soc. Farad. Soc. Trans. I* **72,** 228.

Odajima, A. (1959), *Prog. Theor. Phys. (Kyoto),* **10** (Suppl.), 142.

Packer, K. J., and T. C. Selwood (1978), *J. Chem. Soc. Farad. Soc. Trans. I.* **74,** 1579.

Pakhovchishin, S. V. (1979), *Doklady Phys. Chem.* **245,** 140.

Pearson, R. T., and W. Derbyshire (1974), *J. Coll. Interface Sci.* **46,** 232.

Peemoeller, H., and M. M. Pintar (1979), *Biophys. J.* **28,** 339.

Peemoeller, H., L. J. Schreiner, M. M. Pintar, W. R. Inch, and J. A. McCredie (1979), *Biophysical J.* **25,** 203.

Pfeifer, H. (1972), in *NMR Basic Principles and Progress,* Vol. 7, P. Diehl, E. Fluck, and R. Kosfeld, Eds., Springer-Verlag, Berlin, pp. 53–153.

Pfeifer, H. (1975), *Surface Sci.* **52,** 434.

Ramirez, J. E., J. R. Cavanaugh, and J. M. Purcell (1974), *J. Phys. Chem.* **78,** 807.

Rennie, G. K., and J. Clifford (1976), *J. Chem. Soc. Faraday Soc. Trans. I.* **73,** 680.

Resing, H. A. (1965), *J. Chem. Phys.* **43,** 669.

Resing, H. A. (1967), *Adv. Mol. Relax. Process.* **1,** 109.

Resing, H. A. (1972), *Adv. Mol. Relax. Process.* **3,** 199.

Resing, H. A. (1974), *J. Phys. Chem.* **78,** 1279.

Resing, H. A. (1976), *J. Phys. Chem.* **80,** 186.

Resing, H. A., A. N. Garroway, and K. R. Foster (1976), in *Magnetic Resonance in Colloid and Interface Science,* A.C.S. Symposium Series Vol. 34, H. A. Resing and C. G. Wade, Eds., American Chemical Society, Washington, D.C., pp. 516–529.

Resing, H. A., K. R. Foster, and A. N. Garroway (1977), *Science* **198,** 1181.

Riekel, C., A. Heidemann, B. E. F. Fender, and G. C. Stirling (1979), *J. Chem. Phys.* **71,** 530.

Rubenstein, M., A. Baram, and S. Luz (1971), *Mol. Phys.* **20,** 67.

Rustgi, S. N., H. Peemoeller, R. T. Thompson, D. W. Kydon, and M. M. Pintar (1978), *Biophysical J.* **22,** 439.

Shanbhag, S., M. P. Steinberg, and A. I. Nelson (1970), *J. Food Sci.* **25,** 612.

Sharma, P. K., and S. K. Joshi (1963), *Phys. Rev.* **132,** 1431.

Sharimanov, Y. G., L. L. Buishvili, and G. M. Mrevlishvili (1980), *Biophysics* **24,** 622.

Silbernagel, B. G., and F. R. Gamble (1974), *Phys. Rev. Lett.* **32,** 1436.

Silva Crawford, M., B. C. Gerstein, A.-L. Kuo, and C. G. Wade (1980), *J. Am. Chem. Soc.* **102,** 3728.

Simpson, J. N., and H. Y. Carr (1958), *Phys. Rev.* **111,** 1201.

Smith, F. W., J. R. Mallard, A. Reid, and J. M. S. Hutchison (1981), *Lancet* **1**(8227), 963.

Solomon, I. (1955), *Phys. Rev.* **99,** 559.

Stejskal, E. O., and J. E. Tanner (1965), *J. Chem. Phys.* **42,** 288.

Stoesz, J. D., A. G. Redfield, and D. Malinowski (1978), *FEBS Lett.* **91,** 320.

Sussman, M. V., and L. Chin (1966), *Science* **151,** 324.

Sykes, B. D., W. E. Hull, and G. H. Snyder (1978), *Biophysical J.* **21,** 137.

Tabony, J. (1980), *Prog. NMR Spectrosc.* **14,** 1.

Tanner, J. E. (1978), *J. Chem. Phys.* **69,** 1748.

Tanner, J. E., and E. O. Stejskal (1968), *J. Chem. Phys.* **49,** 1768.

Toledo, R., M. P. Steinberg, and A. I. Watson (1968), *J. Food Sci.* **33,** 315.

Walmsley, R. H., and M. Shporer (1978), *J. Chem. Phys.* **68,** 2585.

Walter, J. A., and A. B. Hope (1971), *Aust. J. Biol. Sci.* **24,** 497.

Wayne, R. C., and R. M. Cotts (1965), *Phys. Rev.* **151,** 264.

Webster, D. S. (1971), Ph.D. thesis, University of New South Wales.

Webster, D. S., and K. H. Marsden (1974), *Rev. Sci. Instr.* **45,** 1232.

Woessner, D. E. (1961), *J. Chem. Phys.* **35,** 41.

Woessner, D. E. (1962), *J. Chem. Phys.* **37,** 647.

Woessner, D. E. (1974a), *J. Magn. Reson.* **16,** 483.

Woessner, D. E. (1974b), in *Mass Spectrometry and NMR Spectroscopy in Pesticide Chemistry,* R. Haque, and F. J. Biros, Eds., Plenum, New York, pp. 279–304.

Woessner, D. E. (1977), *Mol. Phys.* **34,** 899.

Woessner, D. E., and B. S. Snowden (1969a), *J. Chem. Phys.* **50,** 1516.

Woessner, D. E. and B. S. Snowden (1969b), *J. Coll. Interface Sci.* **30,** 54.

Woessner, D. E., B. S. Snowden (1973), *Ann. N.Y. Acad. Sci.* **204,** 113.

Woessner, D. E., and J. R. Zimmerman (1963), *J. Phys. Chem.* **67,** 1590.

Woodhouse, D. R., W. Derbyshire, and P. J. Lillford (1975), *J. Magn. Reson.* **19,** 267.

Zimmerman, J. R., and W. E. Britten (1957), *J. Phys. Chem.* **61,** 1328.

Zimmerman, J. R., and J. Lasater (1958), *J. Phys. Chem.* **62,** 1157.

Zipp, A., I. D. Kuntz, and T. L. James (1976), *J. Magn. Reson.* **24,** 411.

Index

Acetyl CoA synthetase, 28–29
S-Adenosylmethionine synthetase, 48–49
Adenylate cyclase, 24–26
Adenylosuccinate synthetase, 47–48
Aliphatic side chains, 230–234
  NMR studies, 231
Amino acid side chains in proteins:
  aliphatic side chains, 230–234
    NMR studies, 231
  aromatic side chains, 192–205
    NMR studies, 198–199
    tryptophan, 203–205
    tyrosine, 196–203
  basic side chains, 218–224
    arginine, 218–219
    histones, 223–224
    lysine, 219–223
    NMR studies, 220–221
  carboxyl side chains, 205–214
    NMR studies, 212
  conclusions and prognosis, 235–236
  effects on parameters of incorporation, 152–167
    protein microenvironment and chemical shift, 152–156
    protein mobility and relaxation times, 156–167
  histidine, 173–192
    assignment of resonances of RNase, 175–182
    NMR studies, 176–178
    in other proteins, 189–192
    properties of RNase, 182–189
    thermodynamic parameters from NMR titrations, 192
  hydroxyl side chains, 218–224
    NMR studies, 215
  introduction, 131–133
  proline, 234–235
  proteins and, 133–152
    collagen fibrils, 147–149
    membrane-bound proteins, 150–151
    oriented hene-protein microcrystals, 151–152
    proton NMR of aqueous protein solutions, 133–135
    resolution enhancement and spectral simplification, 135–139
    solid-state NMR, 144–147
    two-dimensional FT NMR, 139–144
    viral protein coats, 151
  sulfur-containing, 224–230
    cysteine and cystine, 230
    methionine, 224–230
    NMR studies, 225
  titration curves, 167–173
Anisotropic effects, NMR monitoring of water systems, 258–259
Aqueous protein solutions, Proton NMR of, 133–135
Arginine, 218–219
Aromatic side chains, 192–205
  NMR studies, 198–199
  tryptophan, 203–205
  tyrosine, 196–203
ATPases, 15–18, 30
ATP sulfate adenylytransferase, 22

"Bound" or "nonfreezable" water, 280–290

Carbamyl phosphate synthetase, 33–37
Carboxyl side chains, 205–214
  NMR studies, 212
Carboxypeptidase A, reaction intermediate of, 81–91
Chemical shift, protein microenvironment and, 152–156
Chemical-shift tensor, 97

Chiral phosphoryl transfer, 9–26
  additional group-transfer reactions, 22–24
  adenylate cyclase, 24–26
  ATPases, 15–18
  5′ nucleotidase, 15
  phosphodiesterases and kinases, 19–22
α-Chymotrypsin, 68–73
Collagen fibrils, amino acid side chains and, 147–149
Creatine kinase, 30
Cross-relaxation effects, water-substrate interface, 267–278
Cryoenzymologic approach, enzyme mechanisms, 80–81
Crystals, proteins and enzymes (spin-labeled), 61–80
  chymotrypsin, 68–73
  hemoglobin, 61–68
  lysozyme, 73–79
Cysteine and cystine, 230

Deoxyribonucleic acid (DNA), 95–129
  introduction, 95–96
  nuclear magnetic resonance (NMR) methods, 96–99
    chemical-shift tensor, 97–98
    relaxation parameters, 98–99
  phosphodiester-backbone conformation, 99–113
    DNA in solution, 99–103
    phosphodiester orientation in fibers, 103–113
Digital resolution, 9
Dipalmitoylphosphatidylethanolamine (DPPE), 23–24
Dispersion, NMR monitoring of water systems, 257–258
DNA dynamics in solution, 113–126
  C NMR studies, 120–125
  internal motion and flexibility, 125–126
  parameters observed for poly(dAdT) · poly(dAdt), 123
  P NMR studies, 114–117
  proton NMR studies, 117–120

Glutamine synthetase, 37–40
Guanosine 5′-phosphate synthetase, 40–42

Hemoglobin, 61–69
Histidine residues in proteins, 173–192
  assignment of resonances of RNase, 175–182
  NMR studies, 176–178
  properties of RNase, 182–189
  thermodynamic parameters from NMR titrations, 192
Histones, 223–224
Human carbonic anhydrase (HCA), 189
Hydroxyl side chains, 214–218
  NMR studies, 215

Isotope effects, phosphorus chemical shifts and, 1–52
  chiral phosphoryl transfer, 9–26
    additional group transfer reactions, 22–24
    adenylate cyclase, 24–26
    ATPases, 15–18
    5′nucleotidase, 15
    phosphodiesterase and kinases, 19–22
  exchange studies, 29–30
    ATPase, 30
    creatine kinase, 30
    pyruvate kinase, 30
  experimental requirements, 8–9
  introduction, 2
  kinetic considerations, 30–33
  kinetic studies, 33–40
    carbamyl phosphate synthetase, 33–37
    glutamine synthetase, 37–40
  kinetic studies proposed, 40–49
    S-adenosylmethionine synthetase, 48–49
    adenylosuccinate synthetase, 47–48
    guanosine 5′-phosphate synthetase, 40–42
    phosphoenolpyruvate carboxykinase, 47
    phosphoenolpyruvate carboxylase, 44–47
    pyruvate-phosphate dikinase, 42–44
  orientation studies in P—Q bond-forming, 26–29
    acetyl CoA synthetase, 28–29
    phosphatases, 26–27
    phosphorylases and phosphohydrolases, 27–28
  theoretical considerations, 2–7
Isotopomers, 11

Lysine, 219–223
Lysozyme, 73–79

Membrane-bound proteins, amino acid side chains and, 150–151
Methionine, 224–230

Nitroxide spin-labels, molecular structure of, 55–61
"Nonfreezable" or "bound" water, 280–290

Nuclear magnetic resonance (NMR), 96–99
amino acid side chains in proteins:
aliphatic side chains, 230–234
aromatic side chains, 192–205
basic side chains, 218–224
carboxyl side chains, 205–214
conclusions and prognosis, 235–236
effects on parameters of incorporation, 152–167
histidine, 173–192
hydroxyl side chains, 214–218
introduction, 131–133
proline, 234–235
and proteins, 133–152
sulfur-containing, 224–230
titration curves, 167–173
chemical-shift tensor, 97–98
relaxation parameters, 98–99
water in heterogeneous systems and, 248–304
"bound" or "nonfreezable" water, 280–290
cross-relaxation effects, 267–278
H NMR techniques, 290–294
introduction, 248–251
list of symbols, 296–301
models used, 261–267
relaxation theory, 251–255
scope of measurements, 257–261
surface water proton exchange kinetics, 278–280
5′ Nucleotidase, 15

Orientation studies, P—O bond-forming, 26–29
acetyl CoA synthetase, 28–29
phosphatases, 26–27
phosphorylases and phosphohydrolases, 27–28
Oriented heme-protein microcrystals, amino acid side chains and, 151–152

Phosphatases, 26–27
Phosphodiesterase and kinases, 19–22
Phosphodiester-backbone conformation of DNA, 99–113
DNA in solution, 99–103
phosphodiester orientation in fibers, 103–113
A form of DNA, 108–113
B form of DNA, 105–108
Phosphoenolpyruvate carboxykinase, 47
Phosphoenolpyruvate carboxylase, 44–47
Phosphohydrolases, 27–28
Phosphorus chemical shifts, isotope effects and, 1–52
chiral phosphoryl transfer, 9–26
exchange studies, 29–30
introduction, 2
kinetic considerations, 30–33
kinetic studies, 33–40
kinetic studies proposed, 40–49
orientation studies in P—O bond-forming, 26–29
theoretical considerations, 2–7
Phosphorylases, 27–28
P nuclear magnetic resonance (NMR) spectroscopy, 2, 3
P—O bond-forming, 26–29
acetyl CoA synthetase, 28–29
phosphatases, 26–27
phosphorylases and phosphohydrolases, 27–28
Positional isotope exchange (or PIX), 29–30
application of, 33–40
Proline, 234–235
Protein microenvironment, chemical shift and, 152–156
Protein mobility, relaxation times and, 156–167
Proteins, NMR methods and, 133–152
collagen fibrils, 147–149
membrane-bound proteins, 150–151
oriented heme-protein microcrystals, 151–152
proton NMR of aqueous protein solutions, 133–135
resolution enhancement and spectral simplification, 135–139
solid-state NMR, 144–147
two-dimensional FT NMR, 139–144
viral protein coats, 151
Proteins and enzymes in crystals, spin-labels and, 61–80
chymotryspin, 68–73
hemoglobin, 61–68
lysozyme, 73–79
Proton NMR:
of aqueous protein solutions, 133–135
spectra of DNA, 117–120
Pyruvate kinase, 30
Pyruvate-phosphate dikinase, 42–44

Reaction intermediate of carboxypeptidase A, 81–91
Relaxation parameters, NMR, 98–99

Relaxation times, protein mobility and, 156–167
Resolution of the "enantiomers," 15
Resolution enhancement, NMR methods applied to proteins, 135–139
RNase, properties of histidine residues, 182–189

Solid-state NMR, 144–147
Spectral simplification, NMR methods applied to proteins, 135–139
Spin-label probes, enzyme action and, 53–94
  cryoenzymologic approach, 80–81
  introduction, 54–55
  nitroxide spin-labels, 5–61
  proteins and enzymes in crystals, 61–80
    chymotryspin, 68–73
    hemoglobin, 61–81
    lysozyme, 73–79
Sulfur-containing amino acids, 224–230
  cysteine and cystine, 230
  methionine, 224–230
  NMR studies, 225
Surface water proton exchange kinetics, 278–280

Temperature scanning, NMR monitoring of water systems, 257
Thermodynamic parameters, histidine residues (from NMR titrations), 192
Titration curves, NMR, 167–173
Tryptophan, 203–205
Two-dimensional FT NMR, 139–144
Tyrosine, 196–203

Water relaxation, heterogeneous and biological systems, 248–304
  introduction, 248–251
  list of symbols, 296–300
  NMR in heterogeneous systems, 255–294
    anisotropy tests, 258–259
    "bound" or "nonfreezable" water, 280–290
    cross-relaxation effects, water-substrate interface, 267–278
    dispersion, 257–258
    general considerations, 257
    H NMR techniques, 290–294
    models used, 261–267
    multinuclei studies, 258
    proton local field-multiwindow analysis, 259–260
    scope of measurements, 257–261
    self-diffusion measurements, 260–261
    surface water proton exchange kinetics, 278–280
    temperature scanning, 257
  relaxation theory, 251–255